T0190122

Springer Series in Chemical Physics

Volume 121

Series Editors

Jan Peter Toennies, Max Planck Institut für Dynamic und Selbstorganisation, Göttingen, Germany

Kaoru Yamanouchi, Department of Chemistry, University of Tokyo, Tokyo, Japan

Wolfgang Zinth, Institute für Medizinische Optik, Universität München, München, Germany

The Springer Series in Chemical Physics consists of research monographs in basic and applied chemical physics and related analytical methods. The volumes of this series are written by leading researchers of their fields and communicate in a comprehensive way both the basics and cutting-edge new developments. This series aims to serve all research scientists, engineers and graduate students who seek up-to-date reference books.

More information about this series at http://www.springer.com/series/676

Chang Q Sun

Solvation Dynamics

A Notion of Charge Injection

 Springer

Prof. Dr. Chang Q Sun
School of Electrical and Electronic
Engineering
Nanyang Technological University
Singapore, Singapore

Yangtze Normal University
Chongqing, China

ISSN 0172-6218
Springer Series in Chemical Physics
ISBN 978-981-13-8443-1   ISBN 978-981-13-8441-7   (eBook)
https://doi.org/10.1007/978-981-13-8441-7

This Springer imprint is published by the registered company Springer Nature Singapore Pte Ltd.
The registered company address is: 152 Beach Road, #21-01/04 Gateway East, Singapore 189721, Singapore

Hydration of anions, cations, electrons, lone pairs, protons, and dipoles mediates the O:H–O bonding network and properties of a solution through O:H formation, H↔H fragilization, O:⇔:O compression, screened polarization, solute-solute interaction, and intra-solute bond contraction.

—Chang Q Sun
Solvation Dynamics—A Notion of Charge Injection
Springer Series of Chem Phys 121, 2019

O:H–O bond segmental disparity and O–O repulsivity form the soul dictating the extraordinary adaptivity, cooperativity, recoverability, and sensitivity of water and ice.

—Chang Q Sun and Yi Sun
The Attribute of Water—Single Notion, Multiple Myths
Springer Series of Chem Phys 113, 2016

Bond and nonbond relaxation and the associated energetics, localization, entrapment, and polarization of electrons mediate the performance of substance accordingly.

—Chang Q Sun
Relaxation of the Chemical Bond
Springer Series of Chem Phys 108, 2014

Spectrometrics of electron emission and diffraction, liquid and solid multifield phonon relaxation refines quantitative, ever-unexpected information of the bonding, nonbonding, electronic and lattice dynamics pertaining to chemisorption, solvation, irregular atomic-coordination, mechanical and thermal excitation.

—Chang Q Sun
Electron and Phonon Spectrometrics
Springer, 2019

In memory of and dedicated to my parents

Preface

The high molecular polarity endows water as an excellent solvent capable of impinging into and then dissolving a soluble crystal into charge carriers to disperse in the aqueous solvent matrix. Charge injection in forms of electrons, electron lone pairs, ions, protons, and molecular dipoles by solvation mediates the hydration network and properties of a solution, which is ubiquitously important to our health and life. Pursuing means of fine-resolution detection and consistent insight into solvation dynamics, molecular interactions, and solute functionalities on the hydrogen bonding network and solution properties has become an increasingly active subject area. Numerous issues are to be ascertained, despite intensive investigations since the 1880s, when Franz Hofmeister discovered the effect known as the Hofmeister series order and Svante Arrhenius defined the acid, base, and their combination of adduct.

Interactions between solute charge carriers and water molecules have profound impacts on many chemical, biological, and environmental topics such as DNA engineering, cell culturing, finding cures for diseases and healthcare. The solvation of salts changes the solutions' viscoelasticity, thermal stability, critical temperatures and pressures for solution phase transition, and even lowers the frictional coefficient, when salt solutions are used as wet lubricants. Ionic hydration forms a stiffer volume, which expands the graphene-oxide interlayer spacing from 0.34 up to 1.5 nm and the modulated layer separation varies with the type of cations.

Salt solutions demonstrate the Hofmeister effect on regulating the solution surface stress and its solubility of DNA and proteins. The ability of anions to precipitate certain proteins not only follows the Hofmeister series order, but the extent of solvation modulation also varies with solute concentration. Debating mechanisms for salt solvation include water structure maker and breaker, ionic specification, quantum dispersion, skin induction, and interaction length scale of hydration.

From the perspective of proton and lone pair acceptance and donation, Svante Arrhenius (1884), Brønsted–Lowry (1923) and Gilbert Lewis (1923) defined the acid, base, and adduct. Regarding the transport manner and mobility of protons and lone airs in Lewis solutions, Theodor Grotthuss proposed in 1903 the "structural diffusion" or the "concerted random hopping" from one H_2O motif to another. The proton mobility was thought high compared to the mobility of the H_2O molecules. Later development refined this concept by invoking proton thermal hopping, structural fluctuating, and quantum tunneling.

In 1964, Eigen proposed an $H_9O_4^+$ (H_3O^+:$3H_2O$) complex, in which an H_3O^+ core is strongly bounded to three H_2O molecules and leaves the lone pair of the tetrahedrally-structured H_3O^+ hydronium free. The lone pairs of the neighboring three H_2O molecules were orientated simultaneously toward the H^+ protons, forming the hydrogen bonds (O:H–O or HB with ":" as the electron lone pairs of oxygen). Zundel favors in 1967, however, a $[H_5O_2]^+$ (H^+:$2H_2O$ nonbonded or H^+–$2H_2O$ covalently bonded) complex in which the proton shuttles randomly between two H_2O molecules. Protons frustrate in two positions off midway between adjacent two oxygen anions, which is the same as the transitional quantum tunneling of Bernal–Fowler–Pauling proposed in the 1930s. Protons and lone pairs were thought to follow the same motion dynamics albeit their polarity inversion.

The known mechanisms for proton and lone pair transportation are in contradiction to the fact that an H–O bond possesses ~ 4.0 eV energy, which requires at least 121.6 nm wavelength laser irradiation for the H–O bond to dissociate in the gaseous phase. On the other hand, the conservation of the 2N number of protons and lone pairs and the O:H–O configuration and orientation endow liquid water as a highly ordered fluctuating crystal. One has to constrain the motion of H_2O molecules and protons in water solvent.

Organic molecular solvation makes the situation even more complicated. For instance, carboxylic acids (such as alcohols) modify the HBs of vodka and whiskey and because these distilled spirits to taste differently one from another; aldehydes and carboxylic acids could damage DNA, causing cancer; glycine and its N-methylated derivatives denature proteins; salt (NaCl) and sodium glutamate ($NaC_5H_8NO_4$) could cause hypertension, but ascorbic acids (vitamin C) can lower blood pressure. The addition of sugar could lower the freezing temperature, which enables low-temperature storage of bio-species without crystallization. The understanding of organic molecular solvation is still in its infancy.

From the perspectives of molecular spatial and temporal performance, proton transportation, nuclear quantum interaction, and O:H–O bond and electronic cooperativity, the following multiscale approaches have made important contributions to understanding water and solutions:

1. Classical continuum thermodynamics scheme takes water ice as a collection of neutral particles and correlates its properties directly to the applied stimulus with focus on the Gibbs free energy. This approach embraces the dielectrics, diffusivity, surface stress, viscosity, latent heat, entropy, liquid/vapor phase transition, though this scheme has faced difficulties in dealing with solvation dynamics and the anomalous properties of water and ice.

2. Molecular dynamics (MD) premises treat water ice as a collection of flexible or rigid, polarizable or non-polarizable molecular dipoles. The combination of MD computations and the ultrafast phonon spectroscopies explores the spatial and temporal performance of water and solute molecules as well as the proton and lone pair motion and transportation behavior. Information includes the phonon relaxation or the molecular residing time at sites nearby the solute or under different coordination conditions or under perturbations.

3. Nuclear quantum interactions approach is focused on visualizing the concerted quantum tunneling of the light protons within the water clusters and quantify the impact of zero-point motion on the strength of a single hydrogen bond at a water/solid interface. An interlay of the scanning tunneling microscopy/ spectroscopy (STM/S) has verified the sp^3-orbital hybridization at 5 K and the magic number of $Na·3H_2O$ for interface fast transport. The proton quantum interaction elongates the longer part and shortens the shorter part of the O:H–O bond, being the result of electrostatic polarization.

4. O:H–O bond cooperativity and electron polarization notion is focused on the cooperativity of the intra- and intermolecular interactions and nonbonding electron polarization under perturbation. The cooperativity of the segmented O:H–O bond and its segmental specific heat disparity dictate the phase structures and thermodynamics of water and ice. A combination of the Lagrangian oscillation mechanics, MD and DFT computations with the static phonon spectrometrics seeking for the O:H–O bond cooperativity and its transition from the mode of ordinary water to the conditioned states and the associated electron polarization. Obtained information includes the fraction, stiffness, and fluctuation order transition upon perturbation and their consequences on the solution viscosity, surface stress, phase boundary dispersivity, and the critical pressure and temperature for phase transition.

In fact, solvation is a process of aqueous charge injection. Charged species react in their respective ways but cooperatively with the solvent molecules, which mediates the hydrogen bonding network and properties of the solutions. The key challenge is to gain consistent insight into the discriminative roles of various solutes interacting with solvent and other solutes and to develop efficient means for fine detection.

This book is focused on the solvation dynamics, molecular nonbond interactions, and the extraordinary functionalities of various solutes on the solution bond network and properties from the perspectives of ionic and dipolar electrostatic polarization, O:H nonbond attraction, H↔H anti-HB and O:⇔:O super-HB repulsion. A combination of the O:H–O bond cooperativity notion and the differential phonon spectrometrics (DPS) has enabled quantitative information on the following issues: (i) the number fraction and phonon stiffness of HBs transiting from the mode of ordinary water to hydration; (ii) solute–solvent and solute–solute molecular nonbond interactions; (iii) discriminative functionalities of individual solutes; and, (iv) interdependence of skin stress, solution viscosity, molecular diffusivity, solvation thermodynamics, and critical pressures and temperatures for phase transitions.

Systematic examination of solvation dynamics has clarified the following:

(1) Charge dispersive injection by solvation mediates the hydrogen bond network and properties of a solution through electrostatic polarization, H_2O dipolar shielding, O:H van der Waals bond formation, H$\leftrightarrow$H and O:$\Leftrightarrow$:O repulsion, solute–solute interactions, as well as undercoordinated intra-solute bond contraction.

(2) Lone pair or proton injection breaks the 2N number conservation to form the stereo (H_3O^+, HO^-)·$4H_2O$ motifs, turning one O:H–O bond into the H$\leftrightarrow$H anti-HB point breaker, and the O:$\Leftrightarrow$:O super-HB point compressor and polarizer, in Lewis solutions.

(3) The H$\leftrightarrow$H fragilization disrupts the acidic solution network and surface stress; the O:$\Leftrightarrow$:O compression shortens the O:H nonbond but lengthens the H–O bond in basic and H_2O_2 solutions; yet, bond-order-deficiency shortens and stiffens the solute H–O bond due H_2O_2 and OH^- solutes.

(4) Exothermal solvation of base and H_2O_2 arises from energy emission of solvente H–O bond expansion by O:$\Leftrightarrow$:O compression and energy absorption from the solute H–O bond contraction in solvation. The energy gain heats up the respective solution at solvation. Molecular motion and evaporation dissipate energy capped by the O:H bounding of 0.1–0.2 eV being less than 5% of the H–O bond energy. The H$\leftrightarrow$H repulsion governs the exothermal solvation of alcohols.

(5) As an inverse of a monovalent cation, the hydrated electron is entrapped by H_2O dipoles, which serves as a probe to the molecular-site and cluster-size resolved HB polarization and the lifetime of photoelectron energy dissipation.

(6) Ions each serve as a charge center that aligns, clusters, stretches, and polarizes their neighboring HBs to form supersolid hydration volumes. Ions prefer eccntrically the hollow sites of the cubic unit cell containing four $2H_2O$ motifs. The small cation retains its hydration volume because the hydrating H_2O molecules fully screen its electric field; however, the number insufficiency of the ordered hydrating H_2O molecules only partly screen the local electric field of anions and anion-anion repulsion reduces the hydration volume at higher solute concentrations.

(7) Solvation of alcohols, aldehydes, complex salts, carboxylic and formic acids, glycine and sugars distort the solute–solvent interface structures with involvement of the anti-HB or the super-HB.

(8) Pressure, temperature, charge injection, and molecular undercoordination act jointly on the critical pressures and temperature for the phase transition and quasisolid phase boundary dispersion of a solution.

(9) Oxidation-solvation-hydration of alkali metals (Y) and the molten alkali halides (YX) fosters their aquatic unconstrained explosion that can be formulated by $[Y^+\leftrightarrow Y^+ + 2(H_2O:\Leftrightarrow:OH^-) + \text{heat}] + (H_2 \text{ or } HX)\uparrow$. The $Y^+\leftrightarrow Y^+$ Coulomb fission initiates and the O:$\Leftrightarrow$:O super-HB fosters the explosion.

(10) The combination of the X:H–Y tension and X:$\Leftrightarrow$:Y or H$\leftrightarrow$H repulsion dictates the sensitivity, energy density, and detonation velocity of energetic substance. The X:H–Y tension determines the sensitivity of the constrained explosion.

As key elements to solvation, O:H–O bond polarization and segmental cooperativity, O:H–O segmental specific heats derived five phases over the full temperature range, and the charge injection derived nonbond interactions and the perturbation dispersed phase boundaries, as well as the bond order deficiency resolved intra-solute bond contraction continue to challenge the limitations of available theories and computational methods. A combination of the spatially-resolved phonon spectrometrics for the phonon abundance-stiffness-fluctuation transition with the X-ray absorption, sum frequency generation, and temporally-resolved pump-probe phonon and photoelectron spectroscopies would be even more revealing for solvation bonding energetic-spatial-temporal evolution dynamics.

It is my great pleasure and obligation to share these personal learnings from the perspective of aqueous charge injection and hydrogen bond transition. Further refinement and improvement are welcome, and critiques from readers are much appreciated.

I hope that this volume, underscoring the essentiality of quasisolidity, supersolidity, anti-HB, super-HB, and O:H–O bond cooperativity to solvation could inspire ways of thinking about and dealing with solvation dynamics, which also stimulate more interest and activities toward correlating the bond-electron-phonon cooperativity to the performance of aqueous and living substances. Directing effort to the areas of extraordinary coordination bond engineering, nonbonding electrons, molecular crystals, and life science could be even more challenging, fascinating, promising, and rewarding.

I would like to express my sincere thanks to colleagues, friends, and peers for their encouragement, invaluable input, and support, to my students and collaborators for their contributions, and to my family, my wife Meng Chen and daughter Yi, for their assistance, patience, support, and understanding throughout this fascinating and fruitful journey.

Singapore   Chang Q Sun
March 2019

Contents

About the Author

Chang Q Sun received his BSc in 1982 from Wuhan University of Science and Technology and an MSc in 1987 from Tianjin University and served on its faculty until 1992. He earned his PhD in Physics at Murdoch University in 1997 and then joined Nanyang Technological University.

Dr. Sun has been working on the theme of *Extraordinary Coordination Bonding and Nonbonding Electronics.* He has pioneered theories of Bond Relaxation and Hydrogen Bond (O:H–O) Oscillator Pair Cooperativity and holds multiple patents on the Coordination-Resolved Electron and Multifield Phonon Spectrometrics. His contribution has been documented in *Electron and Phonon Spectrometrics, Springer 2019; Solvation Dynamics: A Notion of Charge Injection, Springer 2019; The Attribute of Water: Single Notion, Multiple Myths, Springer 2016; Relaxation of the Chemical Bond, Springer 2014* and their Chinese versions. He has published over 20 treatises in *Chemical Reviews, Progress in Materials Science*, etc. He was bestowed with the inaugural Nanyang Award in 2005 and the first rank of the 25th Khwarizmi International Award in 2012.

Nomenclature

θ	Contact angle between the solution droplet and a certain substrate features the solution surface stress
$\eta(C)$	Solute concentration C resolved solution viscosity
$\Theta_{DX}(\omega_X)$	Debye temperature (characteristic phonon frequency) of the x segment of the O:H–O bond
ω_x	O:H–O bond x-segment vibration frequency
$\eta_x(T)$	Segmental specific heat featured by the Θ_{DX} and its integral that is proportional to the bond energy E_x
AFM	Atomic force microscopy
BOLS	Bond-order-length-strength correlation for undercoordinated atoms and molecules
CN	Coordination number
DFT	Density functional theory
DPS	Differential phonon spectrometrics
d_x/E_x	O:H–O bond x segmental length and bond energy (x = L for the O:H and H for the H–O bond)
$f(C)$	Solute concentration C resolved volume or number fraction of the hydrogen bonds transiting from the vibration mode of ordinary water to hydration states
FTIR	Fourier transformation infrared absorption
FWHM	Full width at half maximum, Γ
H↔H	Anti-HB due to excessive H^+ injection serves as a point beaker fragilizing the HB network
LBA	Local bond average approach features the Fourier transformation
MD	Molecular dynamics
NEXAFS	Near edge X-ray absorption fine structure spectroscopy, XAS
$n_H(C)$	Number of the first vicinal hydrating H_2O dipoles
O:⇔:O	Super-HB due to excessive lone pair injection serves as a point compressor and polarizer
O:H–O	Hydrogen bond (HB) coupling inter- and intramolecular interactions

P_C/T_C Critical pressure/temperature for phase transition

PIMD   Path-integral molecular dynamics

QS   Quasisolid phase bounded at $(-15, 4)$ °C due to O:H–O segmental
            specific heat superposition, which disperses under external
            perturbation

S/CIP   Separated/contacted ion pair

SFG   Sum frequency generation spectroscopy

STM/S   Scanning tunneling microscopy/spectroscopy

t-2DIR   Time-resolved two-dimensional infrared absorption

T_m/T_N Melting/freezing temperature closing to the intersections of segmental
            specific-heat curves and dominated by the E_H/E_L

VLEED   Very-low energy electron diffraction

Chapter 1
Introduction

Contents

Abstract Overwhelming contributions have been made since two-century-long ago to understanding the solvation dynamics, solute-solute and solute-solvent molecular interactions, and solution properties from various perspectives. Limited knowledge about the solvent water structure and hydrogen bond cooperativity (O:H–O or HB with ":" being the nonbonding electron lone pairs pertained to oxygen upon sp^3-orbital hybridization) hindered the progress. Amplification of the phonon spectroscopy to spectrometrics and of the perspective of molecular motion to hydration bonding dynamics would be necessary towards the solute capabilities of transiting the ordinary O:H–O bond to the hydrating states and their impact to the performance of solutions.

Highlight

- Aqueous charge injection by solvation is ubiquitously important to many subject areas of research.
- Challenge stays on solvation dynamics, solute capabilities, and molecular interactions.
- The hydrogen bond cooperativity and the DPS strategy could propel the progress.
- Amplifying the solvation dynamics from molecular regime to bonding dynamics is promising.

© Springer Nature Singapore Pte Ltd. 2019
C. Q. Sun, *Solvation Dynamics*, Springer Series in Chemical
Physics 121, https://doi.org/10.1007/978-981-13-8441-7_1

1.1 Wonders of Solvation

Solvation or hydration of the dissolved solutes that serve the same as adsorbates or impurities in solid phases is ubiquitously important to our daily life and living conditions. Solvation of acids, bases, salts and complex or organic molecules in water forms functional groups that are crucial to subject areas such as health care and medication [1]. One can imagine what will happen to our health without acid-base balance in our body fluids and blood. For instance, drinking or eating too much salty water leads to hypertension while lacking salt results in heavy perspiring and one feels exhausted when is doing exercises, which is why an athlete drinks a special salty water of nutrition. Solvation is the key to modulating cell functioning such as regulating, channeling, messaging, sensing, signaling and solution-protein interface stressing [2–8].

A well-known example is the Hofmeister series [9, 10] which indicate that salt solutions are capable of dissolving proteins, called solubility, and modulating the solution surface stress in a certain order of ionic size and electronegativity dependence. However, fine-resolution and clarification of the solute-solvent and solute-solute molecular interactions and the functionality of the solutions stay as great challenge. Little has been known or few measures of quantification have been available yet about the solute capabilities of transiting the fraction and stiffness of the solvent hydrogen bond (O:H–O or HB with ":" denoting the electron lone pairs on an oxygen) from the mode of ordinary water to the hydrating states. Correlation between the solvation dynamics and the solution HB network and properties such as the surface stress (called surface tension), solution temperature, solution viscosity, critical pressures and temperatures for phase transitions stay yet poorly understood.

The understanding of solvation can be traced back to 1900s. Arrhenius [11] won the 1903 Nobel prize for his definition of acid-base dissolution in terms of proton H^+ and OH^- donation. After some 20 years, Brønsted–Lowry [12, 13] and Lewis [14] defined subsequently the compound dissolution in terms of H^+ or electron lone pair ":" donation. Tremendous work has been done since then from various perspectives with derivatives of numerous theories debating mainly on the modes of solute drift motion, proton and lone pair transportation, hydration shell size, interfacial dielectrics, phonon relaxation time, etc., – given birth of the modern solvation molecular dynamics, or liquid-state science [15], which will be introduced in the subsequently respective sections.

Experimentally, neutron diffraction, Raman reflection, and infrared (IR) absorption probe the O:H–O bond stretching and bending vibration modes (called phonon bands), ultrafast electron/phonon/photon dynamics in the liquid water and in the aqueous solutions [16–18].

Figure 1.1 exemplifies the full-frequency Raman spectra collected from NaI [19], HI [20], NaOH [21], and H_2O_2 [22] solutions of different concentrations under the ambient conditions [23]. The Raman spectrum covers the phonon bands of O:H stretching vibration at <200 cm^{-1}, the $\angle$O:H–O bending band centered at 400 cm^{-1}, the $\angle$H–O–H bending band at 1600 cm^{-1} that is very insensitive to external excita-

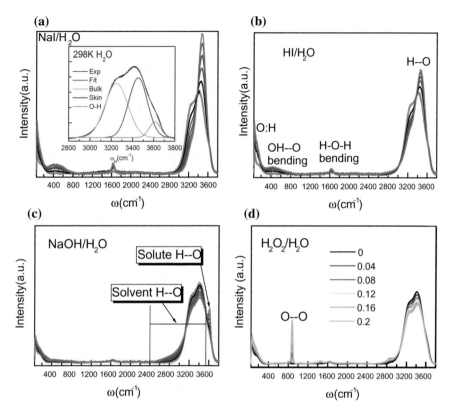

Fig. 1.1 Full-frequency Raman spectroscopy for the concentrated **a** NaI/H_2O [19], **b** HI/H_2O [20], **c** NaOH/H_2O [21] and **d** H_2O_2/H_2O [22] solutions. Inset **a** shows the decomposed H–O phonon band into the bulk (3200 cm^{-1}), skin (3450 cm^{-1}), and the H–O free radical (3610 cm^{-1}) components [24] arising from the site-resolved bond-order-deficiency. Except for the enhanced sharp feature at 3610 cm^{-1} due OH$^-$ solute in the basic solution and the feature at 880 cm^{-1} due the O–O in the H_2O_2/H_2O solution, solvation only shifts the known spectral features of pure water without creating any solute-solvent bond. Using molar fraction C = N_{solute}/$N_{solvent}$ will be more convenient than using the molarity mol/L convention and C = (mol/L)/(1000/18 + mol/L) (Reprinted with copyright permission from [19, 20, 22, 24, 26])

tions, and the H–O stretching band centered at 3200 cm^{-1}, as noted in Fig. 1.1b. The molecular rotational and torsional vibrations and the $\angle$H:O:H bending are within the THz regime, overlapping the mode of O:H stretching.

Figure 1.1 inset (a) shows decomposition of the H–O stretching phonon band into the bulk (3200 cm^{-1}), the skin or the surface having a certain thickness (3450 cm^{-1}), and the surface dangling H–O bond or called free radical (3610 cm^{-1}) directing outwardly of the surface [24]. Likewise, the O:H stretching vibration phonon centered at 75 cm^{-1} features the undercoordination-induced skin O:H elongation and polarization, the ~ 200 cm^{-1} peak features the O:H vibration for the four-coordinated molecules in the ordinary bulk water. Because of the nonbonding lone pair interaction, one may call the O:H nonbond with respect to the intramolecular H–O polar-covalent bond. The O:H nonbond (or vdW bond) is only a part of the O:H–O bond.

Integrating the inter- and intra-molecular interactions of water and ice, the O:H–O bond is segmented into the longer and softer intermolecular O:H nonbond (d_L = 1.7 Å, ω_L = 200 cm^{-1}, $E_L \approx 0.095$ eV) and the shorter and stiffer intramolecular H–O polar-covalent bond (d_H = 1.0 Å, ω_H = 3200 cm^{-1}, $E_H \geq 4.0$ eV), for the 4 °C water [25]. The segmental lengths and their stretching vibration frequencies relax cooperatively while the bending vibration bands are independent of the segmental length or energy. The bending and torsional vibrations have their own independent frequencies and contribute insignificantly to the O:H–O stretching vibrations, which is the advantage of the phonon spectroscopy based on the principle of Fourier transformation that sorts the oscillators of the same force constant into a spectral peak.

The O:H stretching vibration in the THz range overlaps the frequency ($\omega = 2\pi c/\lambda$ or in terms of wavenumber $1/\lambda$ for convenience, 1 THz = 33 cm^{-1}) of molecular rotational or the HB torsional vibrations. The skin molecular undercoordination not only stretches and polarizes the O:H nonbond but also slows down the molecular dynamics in terms of translational, rotational, and torsional motion. Therefore, the O:H phonon redshift from 200 to 75 cm^{-1} is composed of all of them and it is of unrealistic and meaningless for one to decompose the O:H vibration mode at 75–200 cm^{-1} into the uncertain number of components of the Raman spectroscopy.

The Raman spectra in Fig. 1.1 showed also that an impurity solvation only relaxes the known O:H and H–O phonons in terms of their abundance (A, peak area integral), fluctuation (Γ, line-width), and stiffness ($\Delta\omega$, frequency shift) without, however, any excessive spectral features emerging compared with the referential spectrum of pure water. The only exception is the enhanced feature at 3610 cm^{-1} for the H–O dangling bond upon base solvation, arising from the bond-order-deficiency induced H–O contraction of the solute OH^{-} shown in Fig. 1.1c [21]. Base solvation flattens the main peak and broadens the peak towards the lower end of the H–O phonon band. Figure 1.1d shows that H_2O_2 solvation creates a sharp peak at 880 cm^{-1} due to the paring O–O vibration and a hump at 3550 cm^{-1} due to the H–O bond of the undercoordinated H_2O_2 solute.

Salt and acid solvation stiffen the H–O phonon from its bulk component at 3200 cm^{-1} to ~ 3500 cm^{-1} by different abundance. The vibration frequencies of the H–O bond in the hydration shells overlap those of water skin centered at 3450 cm^{-1}. Therefore, the H–O bond in the hydration shell is identical in nature to itself in the undercoordinated molecular supersolid or the semi-rigid skin of pure water, which is shorter and stiffer than those in the ordinary bulk water.

Furthermore, the solute-induced spectral peak width changes in the order of: Γ_{NaI} < Γ_{HI} < Γ_{NaOH} < Γ_{H2O2}. The narrower Γ indicates higher degree of polarization that raises the molecular structure order, lowers the degree of fluctuation, slows down molecular dynamics and prolongs phonon lifetime.

These observations suggest that solvation not only agitates molecular dynamics or water molecular reorientation but also transits the segmental stiffness of a fraction of O:H–O bonds from the mode of ordinary water to hydration without new bonds being formed between the solute and solvent molecules. Coupling of inter- and intramolecular interactions described by the O:H–O bond cooperativity is necessary.

It would suffice for one to focus merely on the cooperativity of the O:H nonbond and the H–O bond stretching vibrations that feature the stiffness of the respective O:H and H–O segments. These segments are highly sensitive to perturbations. Perturbations include the molecular undercoordination (call confinement in occasion) from the standard four in bulk water, though it is sometimes argued more or fewer than four [27], polarization by the electric field of a capacitor or by a point charge of impurity or solute, thermally heating or cooling, mechanical compression or extension, and even electromagnetic and bioelectronic radiation [18, 24]. The phonon frequency shifts only upwardly or downwardly along the frequency axis (in wavenumber for convenient) but the physical origins stemming the frequency shifts are quite different.

One may call the vibration modes phonon bands, but they are the same as the spectroscopic peaks feature the Fourier transformation of bonds vibrating in the same frequency disregarding their locations, orientations or structure phases. The ~ 200 cm^{-1} phonons feature the intermolecular stretching vibration, but one often pays more attention to the intramolecular H–O stretching vibration than to the ~ 200 cm^{-1} phonons because the O:H phonon signal is weak. It is quite often to correlate the H–O frequency relaxation to the motion of solute and solvent molecules. The H–O phonon blue shift in salt solutions is often related to O:H vdW bond weakening, called water structure breaking. The slight H–O phonon redshift is attributed to water structure making.

Figure 1.2 compares the H–O band FTIR spectra of (a, b) NaOH/H$_2$O [28], (c) NaOD/H$_2$O [29], and (d) HCl/H$_2$O [30] solutions of different concentrations in mol/L. Peak intensity normalization derived the similar peak broadening at lower frequencies of NaOD, NaOH, and HCl solutions and an enhanced fine spectral feature at 3610 cm^{-1} for basic solutions. The addition of D$_2$O into the H$_2$O solvent aims to raise the sensitivity of the spectral signals. The D–O bond stretching vibration frequency is featured at ~ 2500 cm^{-1} [31] because of its larger reduced mass of the D–O oscillating dimer [24].

Molecular dynamics (MD) simulations and the pump-probe spectroscopies such as the ultrafast two-dimensional infrared absorption (t-2DIR) probes the solute or water molecular diffusion dynamics in terms of phonon lifetime as a function of the viscosity of the solutions [32, 33]. These conventions treat the solvation as the molecular process and took a molecule as the basic structure unit, and thus often called the O:H vdW interaction the hydrogen bond, rather than the O:H–O that features the cooperativity of inter- and intramolecular interactions [24].

Approaches from the molecular perspective is focused on the solute motion mode and dynamics, the hydration shell size, interfacial dielectrics, molecular orientation, and phonon relaxation time, etc. These conventions lead to detailed and dynamic information on water structure relaxation, water molecular rotational and translational dynamics, and proton translational dynamics. They can also derive the local

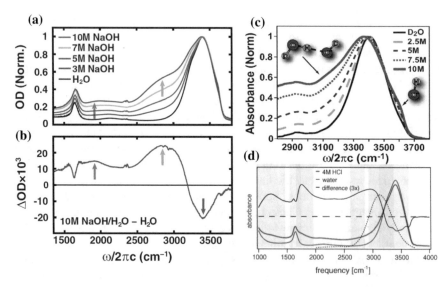

Fig. 1.2 FTIR spectra of the concentrated **a**, **b** NaOH/H$_2$O [28], **c** NaOD/H$_2$O [29], and **d** HCl/H$_2$O [30, 41] solutions. Normalization to the H–O peak intensity at ~3400 cm^{-1} derives the gain and loss of the phonon abundance **a**, **b** upon solvation. Likewise, peak intensity normalization of the 1% HOD contained NaOD/D$_2$O solution derives a feature at 3600 cm^{-1} being attributed to the H–O stretching, and the strong broadening at lower frequency to the D–O bond stretching vibration. **d** HCl solvation evolves the H–O feature with its intensity loss of the main peak at 3500 cm^{-1} (reprinted with copyright permission from [28–30])

electric fields inside the solution and at the hydrated bio-molecular interfaces with derivatives of vibrational relaxation and energy dissipation [34]. For instances, the sum frequency generation (SFG) spectroscopy probes information on the sublayer-resolved dipole orientation, or the skin dielectrics, of a molecularly thin layer at the air-solution interface [35–40].

A systematic ultrafast IR spectroscopy investigation the bulk and the interfacial water using t-2DIF and SFG revealed a strong dependence of the relaxation time on the frequency of the H–O stretching vibration of liquid water in the bulk and at the air/water interface [33]. For the bulk water, the relaxation time increases continuously from 250 to 550 fs when the frequency is increased from 3100 to 3700 cm^{-1}. For water at the air/water interface, the frequency dependence is even stronger. These observations demonstrate that liquid water possesses substantial structural heterogeneity, in the bulk and at the skin, resolved by the bond-order-deficiency induced H–O contraction and stiffness gain, which leads to supersolidity of strong polarization [24]. These spectroscopic observations also correlate the hydrogen bond relaxation to the molecular dynamics. The higher H–O vibration frequency gives rise to the slower of the molecular dynamics and the longer phonon lifetime because of the molecular undercoordination raised polarization and H–O stiffness gain [42].

The vibration energy dissipation lifetime, or population decay, is proportional to the vibration energy or oscillators stiffness.

For aqueous chloride, bromide, and iodide, higher solute concentration stiffens the H–O phonon band towards higher frequency or wavenumber [19, 43], while for the basic solutions, hydroxide OH^- softens the solvent H–O mode to lower frequencies, instead. These spectral changes are usually attributed to the Cl^-, Br^-, and I^- ions that weaken the surrounding HBs (refer to O:H), called structure breakers, or to the OH^- hydroxide that strengthens the surrounding HBs, called structure makers, of the solvent water.

Furthermore, interactions between water and biomolecules can significantly change the structural, dynamic, and thermodynamic properties of the hydrating water compared to its ordinary bulk. Experimental, theoretical, and computational studies showed that water properties change at distances of more than 10 Å from a biomolecule [44]. The effects of biopolymers on hydrating water molecules can be attributed to the chemical nature of the amino acid residues involved, the spatial arrangement of the biomolecule, and its conformational flexibility. A MD investigation [44] suggested that restricting the peptide dynamics can slow down the translational motions of water molecules to a distance of at least 12–13 Å with a slower molecular dynamics (residence time ≥ 100 ps).

However, from the full-frequency IR and Raman and the pump-probe spectroscopies, one is hardly able to gain direct and quantitative information on the solute-solvent and solute-solute molecular interactions, intra-molecular bond relaxation, or the solute capabilities of mediating the solution properties. It is even harder to assume the solvent matrix as in an amorphous state or highly-fluctuated without considering the statistically mean of crystal-like structure [25]. Questions stay concerning the solvation molecular dynamics and its consequences on the behavior of the solutions [28–30, 45–51]:

(1)   What is the role played by the excessive protons or the electron lone pairs?
(2)   What are the exact interacting potentials between the solute and solvent molecules?
(3)   How does the solute change the critical temperature for solution phase transition and why is the solvation some situations exothermic and some other situations endothermic?
(4)   How to discriminate the anion's capability from a cation of mediating the HB network?
(5)   In what way does the water O:H–O bonding dynamics influence the transfer of the excessive proton and the drift motion of the anions in an acidic solution?
(6)   Most importantly, how many bonds and how their stiffness change from the mode of ordinary water to the solute hydration shells?

One can hardly answer these questions without proper knowledge about the structure of liquid water and the O:H–O bond segmental cooperativity when responding to aqueous charge injections or other external stimulations.

1.2   Motivation and Objectives

The current pursuit for solvation is on the spatial and temporal performance of solute molecules, protons, and lone pairs such as the motion and transport dynamics, drift velocity, hydration shell size, and the phonon relaxation time—the population decay of a specific phonon peak. We will look the solvation from an alternative angle of the solvation mediated solute hydrogen-bonding network and properties. Focus will be given on the solute-solvent and solute-solute molecular nonbonding interactions and the solute capabilities of transiting the fraction and stiffness of the O:H–O bonds from the mode of ordinary water to hydrating. One needs to consider the solvation from the perspective of charge injection in forms of anions, cations, electrons, lone pairs, protons, and dipoles. To understand the performance of the H^+, OH^-, electron lone pairs ":", Y^+ and X^- ions, and complex molecular solutes at solvation would be essential. We aim to finding out and eventually grasping with factors dictating the solution properties such as surface solubility and stress, solution viscosity and the critical pressures and temperature for phase transitions.

For the solute molecules, the positive H^+ proton and the negative lone pair ":" form the bases interacting with solvent molecules. Besides the number harmonicas, the spatial-distribution symmetry of H^+ and ":" governs the network performance. Besides the solute H^+ and the ":" that interact with their alike or unlike of the solvent to stabilize molecular interactions, the solute molecules form each a dipole with an anisotropic, local, and short-range potential on the surrounding water molecules and the solute molecules as well in the solution matrix.

From the perspectives of molecular temporal and spatial performance, proton transportation, nuclear quantum interactions, and hydrogen bond and electronic cooperativity, the following multiscale approaches have made substantial progress in understanding the solvation dynamics:

(1) Classical continuum thermodynamics [52–57]: This approach took water ice as a collection of neutral molecules and related the properties of water and ice directly to the stimulus acting on the Gibbs free energy. This approach embraced the dielectrics, diffusivity, surface stress, viscosity, and latent heat, entropy, liquid/vapor phase transition in terms of free energy, though this approach has faced difficulties in dealing with solvation dynamics and the properties of water and ice because of lack of inter- and intramolecular coupling interactions.

(2) Molecular dynamics [30, 58–61]: This approach took water ice as a collection of individually flexible or rigid, polarizable or non-polarizable molecular dipoles. MD computations and the ultrafast phonon spectroscopies are focused on the spatial and temporal performance of water and solute molecules as well as the

proton and lone pair motion and transportation behavior. Information includes the phonon relaxation or the molecular residing time at sites near by the solute, under different coordination conditions or perturbations. Debating on the manner and mobility of solute and molecular motion is less relevant to the bonding network deformation and property change of an aqueous solution. Discrimination of solute functionalities on the solution performance stays a challenge.

(3) Nuclear quantum interactions [62–64]: An interlay of STM/S and the ab initio path-integral molecular dynamics (PIMD) simulations has enabled visualize the concerted quantum tunneling of protons within the water clusters and quantify the impact of zero-point motion on the strength of a single hydrogen bond at a water/solid interface. STM/S observations have verified the sp^3-orbital hybridization at 5 K and the magic number of Na·$3H_2O$ for interface fast transport. The proton quantum interaction elongates the longer part and shortens the shorter part of the O:H–O bond.

(4) O:H–O bond cooperativity and polarization [24, 65–67]: This approach focuses on the cooperativity of the intra- and intermolecular interactions and nonbonding electron polarization under perturbation. A combination of the Lagrangian mechanics, MD and DFT computations with the static phonon spectrometrics seeking for the O:H–O bond cooperativity and its transition from the mode of ordinary water to the conditioned states and the associated electron polarization. Obtained information includes the fraction, stiffness, and fluctuation order transition upon perturbation and their consequence on the solution viscosity, surface stress, phase boundary dispersivity, and the critical pressure and temperature for phase transition.

The aim of this article is to embrace the above approaches with focus on the hydrogen bond transition, solvation bonding and nonbonding dynamics, solute-solvent and solute-solute interface molecular nonbonding interactions, and solute capabilities of transforming the solvent O:H–O bonds into the solute hydration shells as well as physical origin of the solution property variation.

A combination of the known water structure and the O:H–O bond cooperativity [24, 25], the differential phonon spectrometrology (DPS) [68, 69], ultrafast electron/phonon spectroscopy, and surface stress (contact angle) detection strategies has enabled the presently reported progress. Experience suggested that it is essential and efficient to extend the following conventions in dealing with aqueous solvation dynamics:

Convention	Extension	Purposes
Static phonon spectroscopy	DPS	Resolution of the solute capability of bond transition and polarization and their consequence on the solution properties
Dynamic electron/phonon spectroscopy	Electron/phonon energy dissipation	Electron binding energy; phonon energy dissipation by polarization and edge reflection
Molecular regime	Hydration bonding dynamics	Inclusion inter- and intra-molecular interactions and their correlative bond relaxation
Highly-dynamic fluctuating liquid solution	Statistic crystal-like ordered structure	Analogy to chemisorption for quantifying molecular nonbond interaction and intramolecular bond relaxation in terms of stiffness, abundance and fluctuation
Quantum calculation, classical thermal dynamics	Bond relaxation dynamics	Discriminative process of energy absorption emission by bond relaxation and energy dissipation by molecular motion and structure fluctuation upon solvation

1.3 Scope

This work starts with a brief overview in this chapter on the significance, status and challenge in understanding the solvation dynamics of different chemicals. Emphasis is given on the essentiality of concepts and strategies amplification for fine-resolution and comprehension of the bonding and electronic dynamics involved in the solvation and hydration. Chapter 2 introduces the principle and the capability of the patent DPS strategy [69, 70]. The combination of the contact angle detection, the static DPS, and ultrafast electron/phonon spectroscopy offers a powerful set of means to verify theoretical predictions on the site-resolved hydration bonding relaxation and electron polarization. A combination of the experimental and theoretical strategies has enabled quantification on how many bonds transform from the mode of ordinary water to the hydration shells and how the O:H–O bond segmental length and energy change cooperatively upon solvation. Chapter 3 introduces the conservation and registration rules for water structure and O:H–O bond. The 2N number of protons and lone-pairs and the O:H–O conservation and H_2O molecular rotation and proton tunneling restriction defines water as a highly ordered, strongly correlated, and fluctuating crystals covered with a supersolid skin. Charge injection by solvation into such a solute metrics mediates the solution HB network and properties through O:H nonbond formation, H↔H anti–HB fragilization, O:⇔:O super–HB compression, and electrostatic polarization and screen shielding.

Chapter 4 explores the hydration-bonding dynamics in the Lewis H(Cl, Br, I) acidic solutions [20]. A combination of the DPS, surface stress detection, and DFT optimization confirms the prediction of both halogenic anionic polarization and H↔H anti–HB fragilization. The excessive H^+ forms the H_3O^+ firmly without independent hopping or freely tunneling between two oxygen anions. The H–O bond pertained to the H↔H between the H_3O^+ solute and its neighboring H_2O becomes shorter due to the H↔H repulsion and local network termination. The hydration shell consists the central X·H–O and the subsequent O:H–O subshells of greater X·H and O:H intermolecular distances and shorter intra-molecular H–O bond length. Polarization shortens the H–O bond and stiffens its phonon from the water mode at 3200 cm^{-1} to 3450 cm^{-1} for the hydration shells, and meanwhile relaxes the O:H nonbond oppositely. The H↔H anti–HB formation disrupts the solution network and surface stress. Acid solvation has the same effect of thermal fluctuation that relaxes the HBs of the entire network and disrupts the surface stress in the same manner.

Chapter 5 deals with the (Li, Na, K)OH and H_2O_2 solvation that turns out excessive election lone pairs to form the O:⇔:O super–HB point compressor [21]. The O:⇔:O repulsion has the same effect of mechanical compression that elongates the H–O bond and softens its phonon from above 3100 cm^{-1} to its below. Meanwhile, the solute H–O undergoes the bond-order-loss resolved contraction [68, 71] featured at 3550 cm^{-1} for the H_2O_2 and 3610 cm^{-1} for the OH$^-$ solute. The H–O bond due OH$^-$ shares the same stiffness of the dangling H–O bond of water surface. The effect of strong O:⇔:O repulsion annihilates the effect of alkali cation polarization. The exothermic solvation elongates the solvent H–O bond, which emits at least 0.15 eV energy heating up the ":" dominated solutions. The H_2O_2 is less than the OH$^-$ capable of relaxing the O:H–O bond, raising the solution surface stress, and heating up the solutions because of the bond-order-deficiency discrimination of the H_2O_2 from the OH$^-$ solute.

Chapter 6 examines the effect of ionic polarization on the O:H–O relaxation and surface stress for the Na(F, Cl, Br, I) [25], (Na, K, Rb, Cs)I [26], and Na(ClO$_4$, NO$_3$, HSO$_4$, SCN) [72] complex salt solutions. O:H–O bond relaxation is less sensitive than the surface stress to the type of the alkali cations in the order of Na$^+$ > K$^+$ > Rb$^+$ > Cs$^+$. However, O:H–O bond relaxation and surface stress are both sensitive to the type of the halogenic anions, in the I$^-$ > Br$^-$ > Cl$^-$ > F$^-$ ≈ 0 Hofmeister series order but here we refer to the capability of phonon/bond transition. The sp^3-orbital hybridization with production of electron lone pairs is necessary to those P, S, and Cl atoms surrounding N, O, and F elements in the periodic table as well upon reaction, which specifies the solute molecular geometries for the complex salts. Lone pairs surrounding the monovalent complex anions make the complex anions perform oppositely to the halogenic ions in the hydration shell size stability and HB relaxation at solvation. The number of bonds in the hydration shells of the larger complex anions keep almost constant but these HBs are sensitive to the local electric field when the solute concentration is increased.

Chapter 6 also summarizes observations on the solute capabilities of bond transition and polarization in terms of the number fraction (abundance) and stiffness (frequency shift) of O:H–O bonds transiting from the water mode to the hydrating states, and the extent of polarization characterized by contact angle or surface stress. The slope of the fraction coefficient features the number of O:H–O bonds associated to the solute in the hydration shell. The solute capability of bond transition follows: $f_H(C) = 0$, $f_Y(C) \propto C$ and $f_X(C) \propto 1 - \exp(-C/C_0)$ towards saturation. The concentration decay constant C_0 features the solute-solute interaction that changes the local electric field of the hydration shells. The $f_H(C) = 0$ shows that the excessive H^+ neither hop freely nor polarize its surroundings but form the $H \leftrightarrow H$ anti–HB. The $f_Y(C)$ for the small Y^+ cations, OH^- hydroxide, and molecular dipole approaches linear dependence, which suggest their constant hydration shell size. The water dipoles in the hydration shells screen the field of other solutes. For the halogenic ions, the shorter C_0 means the involvement of the strong anion-anion interactions; the H–O vibration frequency in the hydration shell changes insignificantly with solute concentration. Only anion-anion interaction exists in the alkali halide solutions without involvement of cation-anion or cation-cation interactions.

Chapter 7 probes the capabilities of organic molecules of alcohols (methanol ethanol, glycerol) [73], sugars (sucrose, trehalose, maltose) [74], organic acids (methanolic, acetic, methylacetic) and aldehydes (form-, ace-, propionic-) [75], and on bond transition and polarization. Apart from the O:H–O bond and the irregular $H \leftrightarrow H$ anti–HB and the O: $\Leftrightarrow$:O super–HB formation, the anisotropic and short-range molecular dipolar fields play significant roles in the solution HB network relaxation and surface stress disruption. Like the complex anions in salts, atoms of the electronegative elements also subject to the sp^3-orbital hybridization and lone pair production. For the organic molecules, the H^+ and ":" distributed asymmetrically so they form dipoles whose polarizability varies with the CH_2 chain length. The smaller fraction coefficient and high-order of surface stress disruption may hint the manner and origin of DNA fragmentation by aldehyde solvation.

Chapter 8 features the joint effect of salting, heating, compressing and under-coordinating on the O:H–O bond relaxation, surface stress variation, and critical pressures and temperatures for the solution phase transition [19, 43, 76]. Heating and salting have the same effect on bond relaxation but opposite on surface stress construction. Salt solvation polarization raises the surface stress, but thermal fluctuation disrupts it. Salting and heating lengthen the O:H and shortens the H–O but have different origins—salting polarization and O:H thermal expansion. Compression has opposite effect to salt solvation on the O:H–O bond segmental length and stiffness relaxation. Salt solvation raises the critical pressures for the Liquid-VI and the subsequent VI-VII phase transition as the transition requires excessive pressure to restore the polarization-deformed O:H–O bond. Solute of diffrent types performs differently from solute concentration because of the involvement of anion-anion interaction at higher solute concentrations. Correlation between the phonon lifetime and the fraction coefficients for the NaBr solutions and the sized water droplets confirmed that the supersolidity arises from polarization that slows the dynamics of water molecules in the hydration shell and in the skins of the droplets. The $f_{\text{droplet}}(D) \propto D^{-1}$ clarifies the core-shell configuration of water droplet of D size. For salt solutions, the phonon relaxation time $\tau_{\text{solution}}(C) \propto f(C)$ but for nanodroplet $\tau_{\text{droplet}}(D)$ results from skin supersolidity and geometric confinement due to standing wave formation.

Meanwhile, observations revealed the correlation between the solution temperature T(C) and the fraction f(C) of exothermic H–O bond elongation from the state of water to the hydrated by O:$\Leftrightarrow$:O compression in the YOH and H_2O_2 solutions and the H$\leftrightarrow$H repulsion upon alcohol solvation [77, 78]. It is also clarified that sugar solvation lowers the freezing temperature of the solution because of O:H phonon red shifting. The solution viscosity is in the same solute concentration dependence of the fraction coefficient for bonds transiting into the solute hydration shells.

This work ends with Chap. 9 summarizing the gained knowledge based on the provision of O:H–O bond cooperativity, DPS strategy, and contact angle detection with recommendation of the essentiality of concepts and strategies transition from solvation molecular dynamics to the hydration bonding dynamics. Observations prove the essentiality of the O:H nonbond formation, H$\leftrightarrow$H anti–HB fragilization, O:$\Leftrightarrow$:O super–HB compression, bond-order-resolved H–O contraction, and the molecular dipolar interaction between molecules in the aqueous solutions. The surface stress, viscoelasticity, phonon relaxation time, and molecular dynamics are sensitive to the electron polarization. Practice also demonstrates the power of the DPS capable of resolving the solvation dynamics, solute-solute molecular interactions, and solute capabilities in terms of polarization and the quantity and stiffness of hydrogen bonds transition. Extending the experimental and theoretical strategies to water-protein, drug-cell, and molecular interactions in molecular crystals and liquids would be even more challenging, fascinating and rewarding. For reader's convenience, we highlight the key points in front of each section.

References

1. C.Q. Sun, Aqueous charge injection: solvation bonding dynamics, molecular nonbond interactions, and extraordinary solute capabilities. Int. Rev. Phys. Chem. **37**(3–4), 363–558 (2018)
2. P. Jungwirth, P.S. Cremer, Beyond Hofmeister. Nat. Chem. **6**(4), 261–263 (2014)
3. C.M. Johnson, S. Baldelli, Vibrational sum frequency spectroscopy studies of the influence of solutes and phospholipids at vapor/water interfaces relevant to biological and environmental systems. Chem. Rev. **114**(17), 8416–8446 (2014)
4. P. Lo Nostro, B.W. Ninham, Hofmeister phenomena: an update on ion specificity in biology. Chem. Rev. **112**(4), 2286–2322 (2012)
5. J. Ostmeyer, S. Chakrapani, A.C. Pan, E. Perozo, B. Roux, Recovery from slow inactivation in K channels is controlled by water molecules. Nature **501**(7465), 121–124 (2013)
6. J. Kim, D. Won, B. Sung, W. Jhe, Observation of universal solidification in the elongated water nanomeniscus. J. Phys. Chem. Lett., 737–742 (2014)
7. M. van der Linden, B.O. Conchúir, E. Spigone, A. Niranjan, A. Zaccone, P. Cicuta, *Microscopic origin of the Hofmeister effect in gelation kinetics of colloidal silica*. J. Phys. Chem. Lett., 2881–2887 (2015)
8. W.J. Xie, Y.Q. Gao, A simple theory for the Hofmeister series. J. Phys. Chem. Lett., 4247–4252 (2013)
9. F. Hofmeister, Zur Lehre von der Wirkung der Salze. Archiv f experiment Pathol u Pharmakol **25**(1), 1–30 (1888)

10. F. Hofmeister, Concerning regularities in the protein-precipitating effects of salts and the relationship of these effects to the physiological behaviour of salts. Arch. Exp. Pathol. Pharmacol. **24**, 247–260 (1888)

11. S. Arrhenius, *Development of the Theory of Electrolytic Dissociation*. Nobel Lecture (1903)

12. J. Brönsted, Part III. Neutral salt and activity effects. The theory of acid and basic catalysis. Trans. Faraday Soc. **24**, 630–640 (1928)

13. T.M. Lowry, I.J. Faulkner, CCCXCIX.—Studies of dynamic isomerism. Part XX. Amphoteric solvents as catalysts for the mutarotation of the sugars. J. Chem. Soc. Trans. **127**, 2883–2887 (1925)

14. G.N. Lewis, Acids and bases. J. Franklin Inst. **226**(3), 293–313 (1938)

15. D. Chandler, From 50 years ago, the birth of modern liquid-state science. Annu. Rev. Phys. Chem. **68**, 19–38 (2017)

16. J. Li, Inelastic neutron scattering studies of hydrogen bonding in ices. J. Chem. Phys. **105**(16), 6733–6755 (1996)

17. I. Michalarias, I. Beta, R. Ford, S. Ruffle, J.C. Li, Inelastic neutron scattering studies of water in DNA. Appl. Phys. A Mater. Sci. Process. **74**, s1242–s1244 (2002)

18. J.C. Li, A.I. Kolesnikov, Neutron spectroscopic investigation of dynamics of water ice. J. Mol. Liq. **100**(1), 1–39 (2002)

19. Q. Zeng, T. Yan, K. Wang, Y. Gong, Y. Zhou, Y. Huang, C.Q. Sun, B. Zou, Compression icing of room-temperature NaX solutions (X = F, Cl, Br, I). Phys. Chem. Chem. Phys. **18**(20), 14046–14054 (2016)

20. X. Zhang, Y. Zhou, Y. Gong, Y. Huang, C. Sun, Resolving H(Cl, Br, I) capabilities of transforming solution hydrogen-bond and surface-stress. Chem. Phys. Lett. **678**, 233–240 (2017)

21. Y. Zhou, D. Wu, Y. Gong, Z. Ma, Y. Huang, X. Zhang, C.Q. Sun, Base-hydration-resolved hydrogen-bond networking dynamics: quantum point compression. J. Mol. Liq. **223**, 1277–1283 (2016)

22. J. Chen, C. Yao, X. Liu, X. Zhang, C.Q. Sun, Y. Huang, H_2O_2 and HO- solvation dynamics: solute capabilities and solute-solvent molecular interactions. Chem. Select **2**(27), 8517–8523 (2017)

23. Y. Zhou, Y. Gong, Y. Huang, Z. Ma, X. Zhang, C.Q. Sun, Fraction and stiffness transition from the H-O vibrational mode of ordinary water to the HI, NaI, and NaOH hydration states. J. Mol. Liq. **244**, 415–421 (2017)

24. Y.L. Huang, X. Zhang, Z.S. Ma, Y.C. Zhou, W.T. Zheng, J. Zhou, C.Q. Sun, Hydrogen-bond relaxation dynamics: resolving mysteries of water ice. Coord. Chem. Rev. **285**, 109–165 (2015)

25. Y. Zhou, Y. Huang, Z. Ma, Y. Gong, X. Zhang, Y. Sun, C.Q. Sun, Water molecular structure-order in the NaX hydration shells (X = F, Cl, Br, I). J. Mol. Liq. **221**, 788–797 (2016)

26. Y. Gong, Y. Zhou, H. Wu, D. Wu, Y. Huang, C.Q. Sun, Raman spectroscopy of alkali halide hydration: hydrogen bond relaxation and polarization. J. Raman Spectrosc. **47**(11), 1351–1359 (2016)

27. S. Dixit, J. Crain, W. Poon, J. Finney, A. Soper, Molecular segregation observed in a concentrated alcohol–water solution. Nature **416**(6883), 829 (2002)

28. A. Mandal, K. Ramasesha, L. De Marco, A. Tokmakoff, Collective vibrations of water-solvated hydroxide ions investigated with broadband 2DIR spectroscopy. J. Chem. Phys. **140**(20), 204508 (2014)

29. S.T. Roberts, P.B. Petersen, K. Ramasesha, A. Tokmakoff, I.S. Ufimtsev, T.J. Martinez, Observation of a Zundel-like transition state during proton transfer in aqueous hydroxide solutions. Proc. Natl. Acad. Sci. **106**(36), 15154–15159 (2009)

30. M. Thämer, L. De Marco, K. Ramasesha, A. Mandal, A. Tokmakoff, Ultrafast 2D IR spectroscopy of the excess proton in liquid water. Science **350**(6256), 78–82 (2015)

31. S. Park, M.D. Fayer, Hydrogen bond dynamics in aqueous NaBr solutions. Proc. Natl. Acad. Sci. U.S.A. **104**(43), 16731–16738 (2007)

32. M.E. Tuckerman, D. Marx, M. Parrinello, The nature and transport mechanism of hydrated hydroxide ions in aqueous solution. Nature **417**(6892), 925–929 (2002)

33. S.T. van der Post, C.S. Hsieh, M. Okuno, Y. Nagata, H.J. Bakker, M. Bonn, J. Hunger, Strong frequency dependence of vibrational relaxation in bulk and surface water reveals sub-picosecond structural heterogeneity. Nat. Commun. **6**, 8384 (2015)
34. D. Laage, T. Elsaesser, J.T. Hynes, Water dynamics in the hydration shells of biomolecules. Chem. Rev. **117**(16), 10694–10725 (2017)
35. Y.R. Shen, V. Ostroverkhov, Sum-frequency vibrational spectroscopy on water interfaces: polar orientation of water molecules at interfaces. Chem. Rev. **106**(4), 1140–1154 (2006)
36. H. Chen, W. Gan, B.-H. Wu, D. Wu, Y. Guo, and H.-F. Wang, Determination of structure and energetics for Gibbs surface adsorption layers of binary liquid mixture 1. Acetone + water. J. Phys. Chem. B **109**(16), 8053–8063 (2005)
37. S. Nihonyanagi, S. Yamaguchi, T. Tahara, Ultrafast dynamics at water interfaces studied by vibrational sum frequency generation spectroscopy. Chem. Rev. **117**(16), 10665–10693 (2017)
38. N. Ji, V. Ostroverkhov, C. Tian, Y. Shen, Characterization of vibrational resonances of water-vapor interfaces by phase-sensitive sum-frequency spectroscopy. Phys. Rev. Lett. **100**(9), 096102 (2008)
39. S. Nihonyanagi, S. Yamaguchi, T. Tahara, Direct evidence for orientational flip-flop of water molecules at charged interfaces: A heterodyne-detected vibrational sum frequency generation study. J. Chem. Phys. **130**(20), 204704 (2009)
40. Y.R. Shen, Basic theory of surface sum-frequency generation. J. Phys. Chem. C **116**, 15505–15509 (2012)
41. J.A. Fournier, W.B. Carpenter, N.H.C. Lewis, A. Tokmakoff, Broadband 2D IR spectroscopy reveals dominant asymmetric H5O2 + proton hydration structures in acid solutions. Nat. Chem. **10**, 932–937 (2018)
42. C.Q. Sun, X. Zhang, J. Zhou, Y. Huang, Y. Zhou, W. Zheng, Density, elasticity, and stability anomalies of water molecules with fewer than four neighbors. J. Phys. Chem. Lett. **4**, 2565–2570 (2013)
43. X. Zhang, T. Yan, Y. Huang, Z. Ma, X. Liu, B. Zou, C.Q. Sun, Mediating relaxation and polarization of hydrogen-bonds in water by NaCl salting and heating. Phys. Chem. Chem. Phys. **16**(45), 24666–24671 (2014)
44. Y. Gavrilov, J.D. Leuchter, Y. Levy, On the coupling between the dynamics of protein and water. Phys. Chem. Chem. Phys. **19**(12), 8243–8257 (2017)
45. C.Q. Sun, X. Zhang, X. Fu, W. Zheng, J.-L. Kuo, Y. Zhou, Z. Shen, J. Zhou, Density and phonon-stiffness anomalies of water and ice in the full temperature range. J. Phys. Chem. Lett. **4**, 3238–3244 (2013)
46. F. Li, Z. Li, S. Wang, S. Li, Z. Men, S. Ouyang, C. Sun, Structure of water molecules from Raman measurements of cooling different concentrations of NaOH solutions. Spectrochim. Acta A **183**, 425–430 (2017)
47. C.D. Cappa, J.D. Smith, K.R. Wilson, B.M. Messer, M.K. Gilles, R.C. Cohen, R.J. Saykally, Effects of alkali metal halide salts on the hydrogen bond network of liquid water. J. Phys. Chem. B **109**(15), 7046–7052 (2005)
48. W.J. Glover, B.J. Schwartz, Short-range electron correlation stabilizes noncavity solvation of the hydrated electron. J. Chem. Theor. Comput. **12**(10), 5117–5131 (2016)
49. T. Iitaka, T. Ebisuzaki, Methane hydrate under high pressure. Phys. Rev. B **68**(17), 172105 (2003)
50. Y. Marcus, Effect of ions on the structure of water: structure making and breaking. Chem. Rev. **109**(3), 1346–1370 (2009)
51. J.D. Smith, R.J. Saykally, P.L. Geissler, The effects of dissolved halide anions on hydrogen bonding in liquid water. J. Am. Chem. Soc. **129**(45), 13847–13856 (2007)
52. K. Wark, Generalized Thermodynamic Relationships in the Thermodynamics, 5th edn. (McGraw-Hill, Inc., New York., 1988
53. O. Alduchov, R. Eskridge, *Improved Magnus' Form Approximation of Saturation Vapor Pressure, in Department of Commerce* (Asheville, NC (United States), 1997)

54. G. Jones, M. Dole, The viscosity of aqueous solutions of strong electrolytes with special reference to barium chloride. J. Am. Chem. Soc. **51**(10), 2950–2964 (1929)
55. K. Wynne, The Mayonnaise effect. J. Phys. Chem. Lett. **8**(24), 6189–6192 (2017)
56. J.C. Araque, S.K. Yadav, M. Shadeck, M. Maroncelli, C.J. Margulis, How is diffusion of neutral and charged tracers related to the structure and dynamics of a room-temperature ionic liquid? Large deviations from Stokes-Einstein behavior explained. J. Phys. Chem. B **119**(23), 7015–7029 (2015)
57. K. Amann-Winkel, R. Böhmer, F. Fujara, C. Gainaru, B. Geil, T. Loerting, Colloquiu: water's controversial glass transitions. Rev. Modern Phys. **88**(1), 011002 (2016)
58. C. Branca, S. Magazu, G. Maisano, P. Migliardo, E. Tettamanti, Anomalous translational diffusive processes in hydrogen-bonded systems investigated by ultrasonic technique, Raman scattering and NMR. Phys. B **291**(1), 180–189 (2000)
59. J.A. Sellberg, C. Huang, T.A. McQueen, N.D. Loh, H. Laksmono, D. Schlesinger, R.G. Sierra, D. Nordlund, C.Y. Hampton, D. Starodub, D.P. DePonte, M. Beye, C. Chen, A.V. Martin, A. Barty, K.T. Wikfeldt, T.M. Weiss, C. Caronna, J. Feldkamp, L.B. Skinner, M.M. Seibert, M. Messerschmidt, G.J. Williams, S. Boutet, L.G. Pettersson, M.J. Bogan, A. Nilsson, Ultrafast X-ray probing of water structure below the homogeneous ice nucleation temperature. Nature **510**(7505), 381–384 (2014)
60. Z. Ren, A.S. Ivanova, D. Couchot-Vore, S. Garrett-Roe, Ultrafast structure and dynamics in ionic liquids: 2D-IR spectroscopy probes the molecular origin of viscosity. J. Phys. Chem. Lett. **5**(9), 1541–1546 (2014)
61. S. Park, M. Odelius, K.J. Gaffney, Ultrafast dynamics of hydrogen bond exchange in aqueous ionic solutions. J. Phys. Chem. B **113**(22), 7825–7835 (2009)
62. J. Guo, X.-Z. Li, J. Peng, E.-G. Wang, Y. Jiang, Atomic-scale investigation of nuclear quantum effects of surface water: experiments and theory. Prog. Surf. Sci. **92**(4), 203–239 (2017)
63. J. Peng, J. Guo, R. Ma, X. Meng, Y. Jiang, Atomic-scale imaging of the dissolution of NaCl islands by water at low temperature. J. Phys.: Condens. Matter **29**(10), 104001 (2017)
64. J. Peng, J. Guo, P. Hapala, D. Cao, R. Ma, B. Cheng, L. Xu, M. Ondráček, P. Jelínek, E. Wang, Y. Jiang, Weakly perturbative imaging of interfacial water with submolecular resolution by atomic force microscopy. Nat. Commun. **9**(1), 122 (2018)
65. C.Q. Sun, Y. Sun, The attribute of water: single notion, multiple myths. Springer Ser. Chem. Phys., vol. 113 (Springer, Heidelberg, 2016), 494pp
66. Y.L. Huang, X. Zhang, Z.S. Ma, G.H. Zhou, Y.Y. Gong, C.Q. Sun, Potential paths for the hydrogen-bond relaxing with $(H_2O)(N)$ cluster size. J. Phys. Chem. C **119**(29), 16962–16971 (2015)
67. Y. Huang, X. Zhang, Z. Ma, Y. Zhou, G. Zhou, C.Q. Sun, Hydrogen-bond asymmetric local potentials in compressed ice. J. Phys. Chem. B **117**(43), 13639–13645 (2013)
68. X.J. Liu, M.L. Bo, X. Zhang, L. Li, Y.G. Nie, H. TIan, Y. Sun, S. Xu, Y. Wang, W. Zheng, C.Q. Sun, Coordination-resolved electron spectrometrics. Chem. Rev. **115**(14), 6746–6810 (2015)
69. C.Q. Sun, *Atomic Scale Purification of Electron Spectroscopic Information* (US 2017 patent No. 9,625,397B2). United States (2017)
70. Y. Gong, Y. Zhou, C. Sun, *Phonon Spectrometrics of the Hydrogen Bond (O:H–O) Segmental Length and Energy Relaxation Under Excitation*, B.o. intelligence, Editor. China (2018)
71. C.Q. Sun, Size dependence of nanostructures: impact of bond order deficiency. Prog. Solid State Chem. **35**(1), 1–159 (2007)
72. Y. Zhou, Yuan Zhong, X. Liu, Y. Huang, X. Zhang, C.Q. Sun, NaX solvation bonding dynamics: hydrogen bond and surface stress transition (X = HSO_4, NO_3, ClO_4, SCN). J. Mol. Liq. **248**(432–438) (2017)
73. Y. Gong, Y. Xu, Y. Zhou, C. Li, X. Liu, L. Niu, Y. Huang, X. Zhang, C.Q. Sun, Hydrogen bond network relaxation resolved by alcohol hydration (methanol, ethanol, and glycerol). J. Raman Spectrosc. **48**(3), 393–398 (2017)
74. C. Ni, Y. Gong, X. Liu, C.Q. Sun, Z. Zhou, The anti-frozen attribute of sugar solutions. J. Mol. Liq. **247**, 337–344 (2017)

75. J. Chen, C. Yao, X. Zhang, C.Q. Sun, Y. Huang, Hydrogen bond and surface stress relaxation by aldehydic and formic acidic molecular solvation. J. Mol. Liq. **249**, 494–500 (2018)
76. Q. Zeng, C. Yao, K. Wang, C.Q. Sun, B. Zou, Room-temperature NaI/H_2O compression icing: solute–solute interactions. PCCP **19**, 26645–26650 (2017)
77. C.Q. Sun, J. Chen, X. Liu, X. Zhang, Y. Huang, (Li, Na, K)OH hydration bonding thermodynamics: solution self-heating. Chem. Phys. Lett. **696**, 139–143 (2018)
78. C.Q. Sun, J. Chen, Y. Gong, X. Zhang, Y. Huang, (H, Li)Br and LiOH solvation bonding dynamics: molecular nonbond interactions and solute extraordinary capabilities. J. Phys. Chem. B **122**(3), 1228–1238 (2018)

Chapter 2
Differential Phonon Spectrometrics (DPS)

Contents

Abstract An incorporation of the hydrogen bond cooperativity theory to the DPS strategy and surface stress (contact angle) detection could resolve the solvation bonding and nonbonding dynamics and solute capabilities. The enabled information includes bond length and stiffness transition, electron polarization, and the fraction of bonds transformed from the mode of ordinary water to the hydration shells. A combination of the DPS and the ultrafast IR spectroscopy would be more revealing towards solute-solvent and solute-solute molecular interactions, solute capabilities, and solution properties. The DPS is focused on the solvation O:H–O segmental cooperative bonding dynamics and the ultrafast IR on molecular motion dynamics by measuring phonon relaxation time.

Highlight

- DPS distills the fraction and stiffness transition of the O:H–O bonds into the conditioned states.
- Temporally resolved electron/phonon records energy dissipation dynamics and lifetime.
- Spectral peak area normalization minimizes extrinsic artefacts such as scattering cross-section.
- The static and dynamic spectrometrics resolves molecular interactions and solute capabilities.

© Springer Nature Singapore Pte Ltd. 2019   19
C. Q. Sun, *Solvation Dynamics*, Springer Series in Chemical
Physics 121, https://doi.org/10.1007/978-981-13-8441-7_2

2.1 Phonon Abundance and Frequency

2.1.1 Phonons Versus Bond Oscillations

Spectroscopic techniques with electron, phonon, and photon excitation sources have been employed to investigate the solvation dynamics. There are two tracks of spectroscopic approaches for study of solvation. One is the pump-probe, ultrafast dynamic electron/phonon spectroscopies and the other the static photoelectron emission such as X-ray photoelectron spectroscopy (XPS) and near edge soft X-ray absorption/emission fine structure (NEXAFS/XES), and IR phonon absorption and Raman reflection spectroscopies [1–8]. Neutron scattering probes the phonon density-of-states [9, 10].

The ultrafast two-dimensional infrared (2DIR) absorption spectroscopy records the lifetime for molecular motion dynamics as a function of solution viscosity [11]. The 2DIR is focused more on the vibration energy dissipation by molecular motion dynamics, in terms of molecular residing time and the size of hydration volume by probing the relaxation/life time of H–O phonons of a certain frequency. Ultrafast phonon spectroscopy allows one to access more vibrational levels of a given set of anharmonic oscillators, enabling a better characterization of their anharmonic potentials and factors influencing them [12]. The sum frequency generation (SFG) spectroscopy probes information regarding dipolar orientation and the skin dielectrics, at the monolayer air/solution interface [7]. Micro liquid-jet photoelectron spectroscopy (PES) records the molecular site resolved binding energy and energy dissipation lifetime of the hydrated electrons [8].

The principles for the electron/phonon ultrafast spectroscopies are very much the same to the optical fluorescent spectroscopy. The signal lifetime is proportional in a way to the density and distribution of defects and impurities. The impurity or defect states, in terms of quantum entrapment or polarization, prevent the thermalization of the electrons transiting from the excited states to the ground state for an exciton, or electron-hole pair, recombination [13]. Density of defects and impurities and their spatial distribution in a solid inhibits the electron transition from the excited states to the ground for exciton recombination, which determines the photon relaxation time.

The phonon spectrometrics probes the bond relaxation in length d and energy E in the form of $\Delta\omega^2 \propto E/(\mu d^2)$ of an oscillator on average with the reduced mass of $\mu = m_1 m_2/(m_1 + m_2)$.

As an emerging approach, the DPS is focused more on the information distillation on the energetics of bonding and nonbonding cooperativity in terms of phonon fraction-stiffness-order transition of O:H–O bonds from the mode of ordinary water to being conditioned [14–19].

NEXAFS or XAS resolves the effect of H–O bond relaxation and charge polarization on the O-K pre-edge peak shift [20]. In contrast, the photoelectron spectroscopy fingerprints the energy shift of electrons in a νth level or band: $\Delta E_\nu \propto E$, due bond relaxation, with ν varying from the innermost to the outermost energy level/band. The time-resolved XPS probes the energy. The XAS K-edge absorption/emission detects the energy level shift difference between the unoccupied valence band and a core band, which is proportional to the bond energy: $\Delta E_{XAS} \propto \Delta E_{val} - \Delta E_v \propto E$ and $|\Delta E_{val}| > |\Delta E_v|$ because the weak shielding of a high-energy band than a deeper band [16].

The specific O:H–O bond couples the initially independent intermolecular O:H nonbond and the intramolecular H–O polar-covalent interactions [21]. Properties such as bonding thermodynamics, surface stress, solubility, polarizability, viscosity, thermal stability, and critical pressures and temperatures for phase transition of water ice and aqueous solutions depend functionally on the bonding and nonbonding cooperativity and the associated electron energetics. These electron/phonon static and dynamic spectroscopies complement each other for comprehension of solvation dynamics.

In contrast, the phonon relaxation time depends on the viscosity of the solution. Cooling the solution or increasing the salt concentration results in polarization, which raises the viscosity and lowers the Stokes-Einstein drift diffusivity [22]. The lower diffusivity slows down molecular motion dynamics and elongates the H–O phonon lifetime. From the perspective of energy dissipation, defects and impurities modulate the local potential by polarization and quantum entrapment, resulting in the inhomogeneous dielectrics, elastic modulus, and resistivity, to absorb energy dissipation causing the photon/phonon/electron population decay.

The pump-probe time-dependent phonon spectroscopy is widely used for probing the decay time of a known intramolecular H–O bond vibration to derive the molecular motion information. One can switch off/on the pump/probe to record the intramolecular H–O vibrational band intensity change with time and then refer the decay time to intermolecular interactions and molecular dynamics. Alternatively, the population decay corresponds to the energy dissipation depending on the local coordination and energetic environment, phonon wave propagation and boundary reflection.

Complementally, a Raman reflection or an IR absorption spectral peak, called a vibration mode or a phonon band, features the Fourier transformation of all bonds vibrating in the same frequency from the real space, irrespective of their locations or orientations in the liquid, solid, or vapor phase of the substance. As the network of aqueous solutions and liquid water are highly ordered, strongly correlated, and highly fluctuating [21], unlike those disordered, amorphous, high-entropy alloys gained by rapid freezing, the phonon spectroscopies collect well-resolved characteristic spectral peaks. The spectral Fourier transformation sorts the dimer vibration modes according to their stiffness or force constants of the vibrating dimers, or curvatures of the resultant potentials acting on the vibrating dimer. The repulsive interaction shows no direct vibration features but deforms the O:H–O bond on which the repulsive force is acting. The consequence of repulsion can be resolved spectrometrically as it does in the Lewis acid and basic solutions [19, 23].

A Raman ratio-spectrometer was developed to determine the hydration shell size of a solution, as described as follows [24]:

$$I_{hydration}(\omega_H) = I_{solution}(\omega_H) - \alpha I_{H_2O}(\omega_H)$$

Fig. 2.1 The first moment of Raman spectra of the 0.1 M SO_4^{2-}, Cl^-, SCN^-, NO_3^-, and ClO_4^- hydration shells in the aqueous sodium salt solutions. Reprinted with permission from [24]

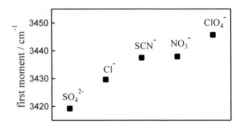

$$\alpha = I_{\text{solution}}(\omega_H < 3100\text{cm}^{-1})/I_{H_2O}(\omega_H < 3100\text{cm}^{-1}) = \text{constant}$$

This method dividing the H–O stretching vibration spectral intensity of a concentrated aqueous solution by the spectrum of water to find the constant, α and then subtract the $I_{\text{solution}}(\omega_H)$ by the $\alpha I_{H2O}(\omega_H)$. Applying this approach to the 0.1 M $NaNO_3$, $NaSCN$, $NaClO_4$, Na_2SO_4, $NaCl$ aqueous solutions, results in the first momentum (statistic average of the frequency) shifts from 3420 to 3445 cm^{-1} with the Hofmeister sequence of $SO_4^{2-} < Cl^- < SCN^- < NO_3^- < ClO_4^-$, as shown in Fig. 2.1.

Integrating the $I_{\text{hydration}}(\omega_H)$ over the spectral peak is the number fraction of the H_2O molecules with the hydration shells and the first ω_H moment of the H–O bond hydrating molecules. Calculations lead to the number of H_2O molecules in the hydration shells of the ion pairs of Na^+ and the Cl^- (8.4 ± 0.1), SO_4^{2-} (12.0 ± 0.2), and NO_3^- (11.4 ± 0.1), agreeing with MD simulations [25, 26].

The combination of the time-resolved electron/phonon/photon spectroscopy [27], the XAS [28], the SFG [29] and the DPS for O:H–O bond transition would be much more revealing. The former probes the molecular spatial and temporal distribution kinetics and the latter resolves the on-site O:H–O bond length and energy dynamics. An interplay of both will give comprehensive information on the performance of the molecules and the O:H–O bonds. A further extension of the developed knowledge and strategies to the solvation bonding and nonbonding dynamics of complicated solutes, solution-protein tractions, drug-cell targeting, biomolecular cell activating and inactivating molecular interactions towards discoveries in molecular crystals and liquids would be certainly even more fascinating, revealing, and rewarding.

What differs an aqueous solution from a regular solid substance is the O:H–O bond that is segmented into the intermolecular O:H nonbond and the intramolecular polar-covalent H–O bond. The repulsion between the lone pairs of oxygen ions couple these two parts relaxing cooperatively. The O:H–O bond segmental disparity and the O–O repulsivity defines the O:H–O bond cooperativity in segmental relaxation, which differentiates the liquid water and aqueous solution from other regular substance [30] in responding to perturbation.

One can mimic an aqueous solution as a highly-ordered crystal-like with impurities evenly dispersed in the ordinary O:H–O bonding network. The addition of solutes distorts the local HBs by polarization or disruption, as defects and impurities do in solid substance [16]. Unlike chemisorption that breaks old bonds and forms new ones, solvation derives no spectral peaks for solute-solvent bond formation in the full-frequency regimes except for the solute intramolecular bonding such as C–H $(2900 \pm 100 \text{ cm}^{-1})$, H–N $(3464–3534 \text{ cm}^{-1})$, and O–O $(877 \text{ cm}^{-1}$ in $H_2O_2)$ interactions. Therefore, taken as impurities, solutes form no regular bonds with solvent water molecules.

2.1.2 Hydrogen-Bond Phonon Characteristics

2.1.2.1 Bond Stiffness Versus Phonon Frequency

Three parameters characterize a specific phonon band. One is the peak frequency that shifts upon perturbation, upward or downward, along the ω axis (in unit of wavenumber cm^{-1}). The $\Delta\omega_x$ shift is proportional to the square root of the curvature of the dimer potential at equilibrium in the harmonic approximation, according to the Taylor series. The dimer potential U(r) sums all the possible contributions including polarization, long-range attraction and repulsion, and nuclear quantum interaction, etc. The high-order nonlinear vibrations in the Taylor series contribute to the transport dynamics, which becomes only significant to a system being away from the equilibrium. The involvement of the high-order approximation only offsets frequencies of the characteristic vibration peaks without producing excessive vibrational features [31].

The segmental stiffness $(Yd)_x$ is the product of the segmental length d_x and its elastic modulus Y_x. The elastic modulus Y_x whose unit in Pascal equals the force crossing its acting area, $[F/S = (F \cdot d)/(S \cdot d) = E/d^3]$, which is simply proportional to the local energy density in dimension. E_x is the cohesive energy of the specific x segment of the O:H–O bond. The x = L and H represent for the intermolecular low-frequency O:H nonbond phonon centered at ~200 cm^{-1} and the intramolecular high-frequency H–O bond vibration at 3200 cm^{-1}, respectively, for the bulk water. It is reasonable to take this harmonic and dimensionality approximation in resolving the frequency-shift dependence on the bond stiffness $(Yd)_x$ of a segment closing to its equilibrium [21]. The Taylor series yield the frequency shift [30],

$$(\Delta\omega_x)^2 \propto \frac{d^2u(r)}{\mu_x dr^2}\bigg|_{r=d_x} ;$$

$$\Delta\omega_x \propto \sqrt{\frac{E_x}{\mu_x d_x^2}} \propto \sqrt{\frac{Y_x d_x}{\mu_x}} \propto \sqrt{\frac{k_x + k_C}{\mu_x}}$$

where $\mu_x = m_1 m_2/(m_1 + m_2)$ is the reduced mass of an oscillating dimer. The μ_L $=18^2/(18+18) = 9$ unit for the $H_2O:H_2O$ molecular dimer is different from the μ_H $= 1 \times 16/(1+16) = 16/17$ unit for the H–O atomic oscillator. For the O:H–O bonded system, the μ_x conserves unless isotope Deuterium is involved, which shifts the mode from 3200 to 2500 cm^{-1}. The k_x and k_C is the force constants, or curvatures of the potentials and the O–O Coulomb potentials, respectively. The sum of $k_x + k_C$

features the O:H–O bond cooperativity within the harmonic approximation, with x_x being the amplitude of vibration of the xth atom [32, 33],

$$2V = k_L x_L^2 + k_H x_H^2 + k_C (x_L - x_L)^2$$

Thus, one may take the O:H–O bond as a primary structural unit that stores energy physically by relaxation under non-conservative force such mechanical compression and thermal excitation. A Lagrangian-Laplace treatment and then a reverse transformation can mathematize the strong correlation of the O:H–O oscillator pair, which yields,

$$\omega_x = (2\pi c)^{-1} \sqrt{\frac{k_x + k_C}{\mu_x}}$$

It is generally true that if a segment becomes shorter and stronger, a blue shift occurs to its specific phonon, and vice versus, as the cohesive energy of a segment is inversely proportional to a certain power m of its length, $E \propto d^{-m}$ [34]. Therefore, the phonon frequency shift offers direct information on bond stiffness gain or loss upon perturbation or charge injection by solvation. The perturbation by solvation may be polarization, attraction, or even repulsion through nonbond interactions, which varies from situation to situation [35].

2.1.2.2 Local Bond Average Versus Gibbs Energy

One can translate the Fourier transformation into the local bond average (LBA) scheme [36] with focus on bonds vibrating in the real space into the characteristic peaks in the reciprocal space according to their frequencies. For a given specimen whether it is a crystal, non-crystal, or a liquid, with or without defects or impurities, the nature and the total number of bonds keep unchanged when stimulated without phase transition. However, the length and strength of the involved bonds will response to the stimulation accordingly. If the functional dependence of a detectable property on the bonding identities is clear, one would be able to know the performance of the entire specimen by focusing on the length and strength response to stimulation of the representative bonds at different sites or their average.

The LBA approach seeks for the relative change of a quantity under stimulation with respect to the known bulk standard. The LBA focuses merely on the performance of the local representative bonds disregarding the exact number of bonds that remains in the given specimen. The presence of broken bonds, defects, impurities, or the non-crystallinity in the referential standard will affect the reference values of concern rather than the nature of observations. Contributions from long-range or the high-order approximations can be simplified by folding them into the bond partition between a specific atom and one of its nearest neighbors. The LBA approach represents and simplifies the true situations of measurements or theoretical computations that collect statistic information from bonds according to their stiffness. Given with the LBA, difficulties encountered by the classical and quantum approaches for the fluctuating object under stimulation could be overcome.

For a defect-free elemental crystal, one can focus on the performance of the representative bond in terms of its length and energy that determines its vibration frequency [36]. For the defected, limited sized crystals, one can consider more than one representative bonds for different coordination conditions. For water and ice, one representative O:H–O bond has two segments. Solutions may need more O:H–O bonds representing for those in the network of ordinary non-hydrated water and those in the hydration shells.

External stimulus such as pressure P, temperature T, coordination number (bond order) z, magnetic B and electric E field relaxes the segmental length and energy cooperatively of the representative O:H–O bonds in their respective manner, so the vibration frequency shifts forward or backward depending functionally on the stimulus [37]:

$$\omega_x(E_x(P, T, z, E, M, \ldots); d_x(P, T, z, E, M, \ldots)). \tag{2.1}$$

These stimuli are also variables of the Gibbs free energy in the classical thermodynamics for a continuum medium [36]. For instance, the detectable quantities are related directly to the external stimulus such as the coupling of temperature and entropy (TS), pressure and volume (PV), surface area and surface energy (Aγ), chemical potential and composition ($\mu_i n_i$), charge and electric field (qE), etc., without needing consideration of atomistic origin:

$$G(T, P, A, n_i, E, B, \ldots) = \Omega(ST, VP, \gamma A, \mu_i n_i, qE, \mu_B B, \ldots)$$

The LBA deals with the bond length and bond energy relaxation by stimulation rather than the entire Gibbs scheme minimizing the total energy. Strikingly, besides the stronger covalent bond and the weaker O:H nonbond interactions, H↔H inter-proton and O:⇔:O inter-lone-pair repulsions play important roles in dictating the performance of water and aqueous solutions despite their positive interaction energies. The sum of all sorts of energies in the classical thermodynamics annihilates the significance of the extraordinary nonbonding weak or repulsive interactions. Therefore, the total binding energy minimization or classical thermodynamics is facing unexpected limitations in dealing with liquid water and aqueous solutions [38]. Hence, the LBA resolves the site and bond nature dependent vibrations in terms of Fourier transformation.

2.1.2.3 Phonon Abundance Versus Number of Bonds

The second characteristics of a phonon spectrum is the integral of the peak area or the phonon abundance that sums over all bonds contributing to the mode of vibration. The peak intensity at a certain frequency only is the probability of distribution without giving information on the number of the contributing bonds. The integration from $I(\omega_{xm}) = 0$ to $I(\omega_{xM}) = 0$ includes all the phonons contributing to the specific ω_x peak of detection,

$$A(\omega_x) = \int_{\omega_m}^{\omega_M} I(\omega_x)d\omega \tag{2.2}$$

The peak central position or the center-of-gravity (COG) is given by [39],

$$\omega_{COG} = \frac{\int_{\omega_m}^{\omega_M} \omega I(\omega)d\omega}{\int_{\omega_m}^{\omega_M} I(\omega)d\omega}$$

There are often two methods for peak normalization. One is done by a maximal peak intensity across and the other by the peak area alone. The peak intensity normalization crossing a set of spectra of all solutions deforms the peak shape and deviates the spectral fine-structure information from the true situation. However, the independent peak area normalization not only overcomes the limitation of intensity normalization but also removes artifacts of detection, which is of physically meaningful. The difference between the area-normalized peak for the solution and the referential peak for the ordinary water distils exactly the fraction-stiffness-fluctuation of bonds transiting from the mode of the ordinary water to the hydration states.

2.1.2.4 Linewidth Versus Molecular Fluctuation Orders

The third characteristics of a spectral peak is the linewidth (Γ) or linewidth that features the order of molecular order fluctuation. A narrower Γ means a higher molecular structure order, or slower motion dynamics [30]. Electric polarization by salt solvation or by molecular undercoordination stiffens the H–O bond, which narrows its Γ and shifts the phonon to higher frequency. The phonon frequency blue shift by polarization also prolongs its lifetime and slows down molecular translational and rotational dynamics. Thermal fluctuation broadens the Γ and lowers the structure order but speeds up molecular motion though the H–O phonon undergoes a blue shift.

However, one can only probe the statistic mean of the vibration peak shift, abundance, and its Γ without being able to discriminate the sources of stimuli. For example, base solvation [18] has the same effect to mechanical compression [40] as both stimuli soften the H–O phonon and stiffen slightly the O:H oscillator; liquid heating [41], molecular undercoordination [42], aciding [19] and salting [43] shorten the H–O bond and soften the O:H phonon in the same manner albeit different mechanisms. Thermal fluctuation by heating and fragilization by H↔H point breaking disrupt the solution network and surface stress though both heating and acid solvation have the same effect on H–O phonon stiffening. From the conventional full-frequency Raman spectroscopy or peak maximum normalization approach, one could hardly be able to gain quantitative information on the stiffness, fraction, and fluctuation of bonds transiting from the referential standard to the functionalized states.

2.2 DPS Strategy and Derivatives

2.2.1 From Spectroscopy to Spectrometrics

Spectrometrics deals with the principle, process, and practical applications of spectrometrology, aiming at the dynamic, molecular-scale and quantitative information on bond and electronic response to perturbation [15]. Conventionally, one needs to correct the spectral background using the standard Tougaard method [44–46] by employing Gaussian-, Lorentz-, or Doniach-Sunjic-type functions before decomposing the spectroscopic peak into multiple components needing unclear specification. The spectrometrics strategy saves, however, such tedious processes of background correction, peak decomposition, component specification, and peak energy error-trial tuning.

The DPS is thus developed to calibrate the solute capability of O:H–O bond transformation by separating the phonon abundance transiting from the water mode to the hydration shells. The DPS applies to any situation with bond vibration under perturbation is involved. Figure 2.2a illustrates the DPS process extended from the coordination-resolved electron spectrometrics [15, 16].

Generally, the phonon population is asymmetrical such as the cases of acid [19] and basic solvation [23] because of the H$\leftrightarrow$H and O:$\Leftrightarrow$:O repulsion on the local hydrogen bonding network deformation. It would be misleading to focus on the asymmetry of the spectral peak in terms of nonlinear effect without exploring the intrinsic nature of bond transition due to charge injection. The nonlinear contribution does not create any new features to the existing bonds but shifts only the spectral peak by a certain wave numbers due to harmonic contribution [31, 47].

The DPS distils the phonon mode gain only in the closest hydration shells as a component above the x-axis, which equals the abundance loss of ordinary water as a valley below the axis in the DPS spectrum. This process removes the spectral features shared by the ordinary water and the high-order hydration shells. The removal simplifies the situation as one is concerned only the number of bonds transiting into the first hydration shells or the topmost layer of water surface or a skin of water droplet. A hydration shell may have one, two or more subshells extending to the bulk, depending on the charge quantity and the size of the solute. The solute size and charge quantity determine its local electric field that is subject to the screening by the local hydrating H$_2$O dipoles and the solute-solute interactions. The $\Delta\Gamma$ between the DPS peak and the DPS valley tells directly the fluctuation order. If $\Delta\Gamma < 0$, molecular structure order is enhanced in the hydration shells than it is in the reference water.

(a)   **(b)**

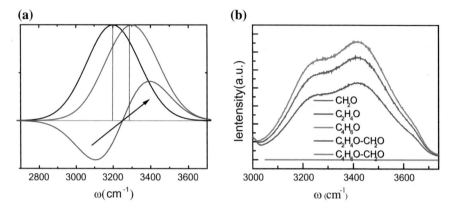

Fig. 2.2 Illustration of the **a** DPS process [15] that **b** removes the artefacts of cross-section of mode reflectivity and the frequency dependent transition polarizability. The DPS profile (blue curve, a) is the phonon abundance difference between the conditioned (red curve) and the referential standard (dark curve) upon both spectral peaks being area normalized. The DPS peak area features the fraction and the stiffness (frequency shift) transition from the ordinary H–O bond vibration mode (valley) to the hydration shells (peak). Solvation of the same concentration aldehydes of different CH_2 chain length resolves the reflectivity but the DPS normalization removes this effect as their difference derives a straight line

As artefacts of phonon spectroscopy, the cross-section of mode reflectivity and the frequency dependence of transition polarizability (intensity increases with frequency) contribute to the spectral peak shape and intensity. However, experimental artefacts never contribute to the intrinsic transition of phonon abundance and stiffness upon solvation. As Fig. 2.2b illustrated, the spectral intensity varies with the type of the aldehyde solutes at the same concentration due to the mode reflectivity. The DPS of the aldehyde solutions shows negligible spectral difference. One can thus remove such artifacts in detection by normalizing the specific peak area. Furthermore, the relative reflectivity changes in the 10^{-3} order when the NaCl solution is heated from 0 and 300 °C under high pressure, which is within the tolerance of most available spectroscopic methods [48].

2.2.2 Phonon Abundance Transition

The DPS purifies merely the transition in the phonon abundance and bond stiffness by conditioning, which monitors the phonon relaxation both statically and dynamically with high accuracy and high sensitivity without needing any approximation or assumption. The fraction coefficient, being the DPS peak integral, represents the fraction of bonds, or number of phonons transiting from water to the hydrating at a solute concentration C,

$$f_x(C) = \int\limits_{\omega_m}^{\omega_M} \left[\frac{I_{solution}(C, \omega)}{\int_{\omega_m}^{\omega_M} I_{solution}(C, \omega)d\omega} - \frac{I_{H_2O}(0, \omega)}{\int_{\omega_m}^{\omega_M} I_{H_2O}(0, \omega)d\omega} \right] d\omega.$$

The slope of the fraction coefficient, $df(C)/dC$, is proportional to the number change of bonds per solute in the hydration shells, which characterizes the hydration shell size and its local electric field. The local electric field is subject to the screening of the surrounding H_2O dipoles aligned oppositely along the electric field. Solute-solute interactions will change the local field by electric fields superposition.

2.2.3 Criteria for Solute Interactions

The $f_x(C)$ for bond transition and polarization is so critical that determines the solution viscosity, skin stress, phonon lifetime, and solute diffusivity consequently [49]. The $f_x(C)$ offers direct information on the local electric field, the hydration shell size, and the manner of solute-solute interaction, as following elaborates:

$f_x(C)$	$\equiv 0$	**No bond transition or polarization occurs.** The x solute creates no field acting on the HBs, which is the case of the excessive H^+ in the HX acid solutions, because of $H \leftrightarrow H$ bond formation [19]
	$\propto C\ [f''_x(C) = 0]$	**The local electric field and the hydration shell size keep constant and respond not to interference of other solutes.** The quasi-constant slope indicates that the number of bonds per solute conserves in the hydration shell. The electric field of a small Y^+ alkali cation is fully screened by the ordered H_2O dipoles in its hydration shells; thus no cation-anion or cation-cation interaction is involved for the alkali solutions [14]. The electric field of a giant complex molecular dipole is anisotropic and short-range. Screened by the hydrating H_2O dipoles, the molecular solute does not interact with other solutes
	$\propto 1 - \exp(-C/C_0)$ $[f''_x(C) < 0]$	**Repulsive solute interaction becomes dominant.** The number of H_2O molecules in the hydration shells is insufficient to fully-screen the solute local electric field because of the geometric limitation to molecules packed in water. The solute can thus interact with their alike—solute-solute repulsion comes into play to weaken the local electric field. Therefore, the $f_x(C)$ increases approaching saturation, the hydration shells size turns to be smaller, which limits the solute capability of bond transition. The fraction coefficient of the X^- halogenic anions [14] and the H_2O_2 hydrogen peroxide solutes [50] show $f_x(C)$ saturation with different C_0 values
	$[f''_x(C) > 0]$	**Attractive solute interaction becomes dominant.** This happens to the high-valent and complex salt solutions with exponential growth of the viscosity, such as $CaCl_2$ [51], $NaClO_4$, and $LiCiO_4$ [52] aqueous solutions. Solutes agglomerate at higher concentrations because of solute-solute attraction

2.3 Polarization Versus Surface Stress

2.3.1 Contact Angle

Gibbs [53] first defined the surface stress as the amount of the reversible work needed
to elastically stretch a pre-existing unit area surface. A similar term called "surface
free energy", which represents the excessive free energy per unit area needed to create
a new surface, is easily confused with "surface stress". Since both terms represent a
force per unit length, they are referred as "surface tension", which contributes further
to the confusion in the literature [54].

Conventionally, the surface stress of a solution is correlated to the angle between
the solution surface and the solid surface of contacting, which follows Young's equation, as illustrated in Fig. 2.3 [55],

$$Cos\theta = (\gamma_{SV} - \gamma_{SL})/\gamma_{LV} \qquad (2.3)$$

where the γ_{SV}, γ_{SL} and γ_{LV} describe stresses of the solid/vapor, solid/liquid, and
liquid/vapor interfaces, respectively. One can also regard the γ as force acting on the
water drop at the triple-phase contact line. Thomas Young derived this equation in
1804 when he firstly observed the constancy of the contact angle between a liquid
surface and a specific solid surface and then he formulated the capillary phenomena
on the principle of force equilibrium. The free surface of the liquid droplet meets
a solid to form the triple-phase contact line that can move along the solid surface,
leading to the "wetting" or "de-wetting" phenomena.

The size of Young's contact angle results from the total energy minimization. If
the liquid-vapor interface stress is smaller than the solid-vapor interface stress (γ_{LV}
< γ_{SV}), the liquid-solid interface stress will increase to minimize the energy. As the
liquid drop wets the surface, the contact angle approaches zero. Other γ_{LV}/γ_{SV} ratios
will lead to different shapes of the drops. A hydrophilic solid surface is defined as θ
< 90°; and a hydrophobic surface is $\theta \geq$ 90°.

Further examination of Young's equation suggests that for a specific solid sub-
strate, γ_{SV} is a constant but the γ_{SL} and the γ_{LV} vary with the solute type and solute
concentration of the liquid solution. Differentiating the ln(cosθ) yields,

Fig. 2.3 Illustration of
Young equation for the
surface energy. The γ_{SV}, γ_{SL}
and γ_{LV} describe stresses of
the solid/vapor, solid/liquid,
and liquid/vapor interfaces,
respectively

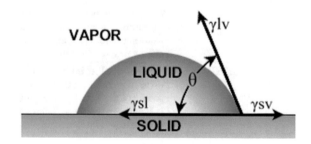

$$\Delta\theta \propto \Delta\gamma_{LV}/\gamma_{LV} + (\Delta\gamma_{SL}/\gamma_{SL})/(\gamma_{SV}/\gamma_{SL} - 1), \quad (\text{with } \Delta\gamma_{SV} = 0)$$

Typically, if $\Delta\theta > 0$, both terms are positive, $\Delta\gamma_{LV} > 0$, $\gamma_{SL} < \gamma_{SV}$ and $\Delta\gamma_{SL} > 0$; or either term becomes positively dominant, and vice versa. However, one can hardly determine the γ_{SL} and the γ_{LV} simultaneously, and particularly when chemical reaction occurs between the solution and solid substrate. One can only approximate that the γ_{SL} change insignificantly and conduct comparative investigation qualitatively.

2.3.2 Polarization Raises Surface Stress

Contact angle between the surface of a solution and the surface of a solid substrate is a direct measure of the surface stress or called surface tension that varies not only with the nature of the solid substance, the interface conditions, and the pH value of the solution. Having a certain thickness, the skin or the interface may be subject to compression or tension depending on the local bond relaxation and charge polarization [30].

What determines the surface stress of the aqueous solution?

First, having the same dimension of elastic modulus, the surface stress is proportional to the energy density of the skin of a certain thickness, or the third order differentiation of the pairing potential at equilibrium [36]. Formulating the temperature dependence of the water surface stress [56], in terms of energy density, has derived the O:H cohesive energy of 0.095 eV and the Debye temperature of 192 K for water, being consistent with 185 ± 10 K as derived using helium scattering from ice in the temperature range of 150–191 K [57].

Second, electron polarization of the contacting skins dictates the superhydrophobicity, superfluidity, superlubricity, and supersolidity of the solid-solid and liquid-solid contacting interfaces [58, 59]. Repulsion between the high-density polarized electrons and the high elasticity of the soft O:H phonons ensuring skin adaptivity of deformation dictate the slipperiness of ice and the toughness of water skin. The skins of 25 °C water and $-(15\text{–}20)$ °C ice share the same supersolidity featured by H–O phonon frequency of 3450 cm^{-1} and strong polarization. The contact angle between the water surface and the nanoscale-patterned Ag substrate is greater than the water droplet on a smooth Ag surface. The roughened solid surface is more hydrophobic because of the curvature enhanced solid and water skin polarization [30].

Therefore, the solution polarization caused by both molecular undercoordination and O–O inter-lone-pair repulsion enhances the viscoelasticity of the skin of a solution. The high elasticity and the high density of surface dipoles are essential to the hydrophobicity of a contacting interface, which is beyond the scope of Young's equation.

What one can do is to compare qualitatively the contact angle variation with the solute type and solute concentration of solutions on the same substrate to minimize the effect of interface reactions. Solvation may raise or depress the solution surface stress by polarization or depolarization. Ionic polarization raises the solution surface stress but the H↔H anti-HB fragilization or thermal fluctuation disrupt the solution network and its surface stress. It is unrealistic for one to seek for the effect of the solution-substrate reaction on the its interface stress momently. However, the concentration trends of the contact angle for different solutions would suffice for one to understand the surface polarization or network disruption.

2.4   Summary

A combination of the spatially resolved DPS, contact angle detection, and the temporarily resolved pump-probe spectroscopy shall enable resolution of the solvation dynamics and the solute capabilities as a function of solute type and solute concentrations of the following:

(1) Transition of the fraction and stiffness of hydrogen bonds from the mode of ordinary water to the hydration shells.
(2) Resolution of the solute-solute interaction and the screening of the solute electric field by the H_2O dipoles in the hydration shells.
(3) Direct information on solution network disruption or polarization.
(4) The division of the $f_x(C)/C$ is proportional to the number of bonds per solute in its hydration shell, whose variation reflects the screening of the hydrated H_2O dipoles and the influence of solute-solute molecular interactions.
(5) The linewidth Γ features the molecular structure order of the solution; polarization narrows the Γ, blueshifts the H–O vibration frequency, slows down molecular dynamics, and prolongs the phonon lifetime.
(6) At the equilibrium, harmonic approximation is sufficient to describe bond transition and any spectral symmetrical population features the intrinsic nature of bonding dynamics.
(7) Phonon stiffness relaxation will disperse the Debye temperature of the segmental specific heat and the quasisolid phase boundaries that determine the critical temperatures for melting and homogeneous ice nucleation.
(8) Vibration and electron energy dissipation is subject to the local environment of charge injection, which fingerprints the charge polarization and bond relaxation.

References

1. C.M. Johnson, S. Baldelli, Vibrational sum frequency spectroscopy studies of the influence of solutes and phospholipids at vapor/water interfaces relevant to biological and environmental systems. Chem. Rev. **114**(17), 8416–8446 (2014)
2. M. Thämer, L. De Marco, K. Ramasesha, A. Mandal, A. Tokmakoff, Ultrafast 2D IR spectroscopy of the excess proton in liquid water. Science **350**(6256), 78–82 (2015)
3. Z.S. Nickolov, J. Miller, Water structure in aqueous solutions of alkali halide salts: FTIR spectroscopy of the OD stretching band. J. Colloid Interface Sci. **287**(2), 572–580 (2005)

4. S. Park, M.B. Ji, K.J. Gaffney, Ligand exchange dynamics in aqueous solution studied with 2DIR spectroscopy. J. Phys. Chem. B **114**(19), 6693–6702 (2010)

5. T. Brinzer, E.J. Berquist, Z. Ren, 任哲, S. Dutta, C.A. Johnson, C.S. Krisher, D.S. Lambrecht, S. Garrett-Roe, Ultrafast vibrational spectroscopy (2D-IR) of CO_2 in ionic liquids: carbon capture from carbon dioxide's point of view. The J. Chem. Phys. **142**(21), 212425 (2015)

6. Y.R. Shen, Basic theory of surface sum-frequency generation. The J. Phys. Chem. C **116**, 15505–15509 (2012)

7. Y.R. Shen, V. Ostroverkhov, Sum-frequency vibrational spectroscopy on water interfaces: polar orientation of water molecules at interfaces. Chem. Rev. **106**(4), 1140–1154 (2006)

8. J. Verlet, A. Bragg, A. Kammrath, O. Cheshnovsky, D. Neumark, Observation of large water-cluster anions with surface-bound excess electrons. Science **307**(5706), 93–96 (2005)

9. I. Michalarias, I. Beta, R. Ford, S. Ruffle, J.C. Li, Inelastic neutron scattering studies of water in DNA. Appl. Phys. A Mater. Sci. Process. **74**, s1242–s1244 (2002)

10. J.C. Li, A.I. Kolesnikov, Neutron spectroscopic investigation of dynamics of water ice. J. Mol. Liq. **100**(1), 1–39 (2002)

11. S.T. van der Post, C.S. Hsieh, M. Okuno, Y. Nagata, H.J. Bakker, M. Bonn, J. Hunger, Strong frequency dependence of vibrational relaxation in bulk and surface water reveals sub-picosecond structural heterogeneity. Nat. Commun. **6**, 8384 (2015)

12. J. Wang, Ultrafast two-dimensional infrared spectroscopy for molecular structures and dynamics with expanding wavelength range and increasing sensitivities: from experimental and computational perspectives. Int. Rev. Phys. Chem. **36**(3), 377–431 (2017)

13. R.A. Street, *Hydrogenated Amorphous Silicon* (Cambridge University Press, 1991)

14. X. Zhang, Y. Xu, Y. Zhou, Y. Gong, Y. Huang, C.Q. Sun, HCl, KCl and KOH solvation resolved solute-solvent interactions and solution surface stress. Appl. Surf. Sci. **422**, 475–481 (2017)

15. C.Q. Sun, Atomic scale purification of electron spectroscopic information (U.S. 2017 Patent No. 9,625,397B2) (United States, 2017)

16. X.J. Liu, M.L. Bo, X. Zhang, L. Li, Y.G. Nie, H. TIan, Y. Sun, S. Xu, Y. Wang, W. Zheng, C.Q. Sun, Coordination-resolved electron spectrometrics. Chem. Rev. **115**(14), 6746–6810 (2015)

17. Q. Zeng, C. Yao, K. Wang, C.Q. Sun, B. Zou, Room-temperature NaI/H_2O compression icing: solute–solute interactions. PCCP **19**, 26645–26650 (2017)

18. Q. Zeng, T. Yan, K. Wang, Y. Gong, Y. Zhou, Y. Huang, C.Q. Sun, B. Zou, Compression icing of room-temperature NaX solutions (X = F, Cl, Br, I). Phys. Chem. Chem. Phys. **18**(20), 14046–14054 (2016)

19. X. Zhang, Y. Zhou, Y. Gong, Y. Huang, C. Sun, Resolving H(Cl, Br, I) capabilities of transforming solution hydrogen-bond and surface-stress. Chem. Phys. Lett. **678**, 233–240 (2017)

20. M. Nagasaka, H. Yuzawa, N. Kosugi, Development and application of in situ/operando soft X-ray transmission cells to aqueous solutions and catalytic and electrochemical reactions. J. Electron Spectrosc. Relat. Phenom. **200**, 293–310 (2015)

21. Y.L. Huang, X. Zhang, Z.S. Ma, Y.C. Zhou, W.T. Zheng, J. Zhou, C.Q. Sun, Hydrogen-bond relaxation dynamics: resolving mysteries of water ice. Coord. Chem. Rev. **285**, 109–165 (2015)

22. J.C. Araque, S.K. Yadav, M. Shadeck, M. Maroncelli, C.J. Margulis, How is diffusion of neutral and charged tracers related to the structure and dynamics of a room-temperature ionic liquid? Large deviations from Stokes-Einstein behavior explained. J. Phys. Chem. B **119**(23), 7015–7029 (2015)

23. C.Q. Sun, J. Chen, X. Liu, X. Zhang, Y. Huang, (Li, Na, K)OH hydration bondin thermodynamics: solution self-heating. Chem. Phys. Lett. **696**, 139–143 (2018)

24. Y. Wang, W. Zhu, K. Lin, L. Yuan, X. Zhou, S. Liu, Ratiometric detection of Raman hydration shell spectra. J. Raman Spectrosc. **47**(10), 1231–1238 (2016)

25. V. Vchirawongkwin, B.M. Rode, I. Persson, Structure and dynamics of sulfate ion in aqueous solution an ab initio QMCF MD simulation and large angle X-ray scattering study. J. Phys. Chem. B **111**(16), 4150–4155 (2007)

26. N. Galamba, Mapping structural perturbations of water in ionic solutions. J. Phys. Chem. B **116**(17), 5242–5250 (2012)

27. A. Bragg, J. Verlet, A. Kammrath, O. Cheshnovsky, D. Neumark, Hydrated electron dynamics: from clusters to bulk. Science **306**(5696), 669–671 (2004)
28. M. Nagasaka, H. Yuzawa, N. Kosugi, Interaction between water and alkali metal ions and its temperature dependence revealed by oxygen K-edge X-ray absorption spectroscopy. J. Phys. Chem. B **121**(48), 10957–10964 (2017)
29. Y. Otsuki, T. Sugimoto, T. Ishiyama, A. Morita, K. Watanabe, Y. Matsumoto, Unveiling subsurface hydrogen-bond structure of hexagonal water ice. Phys. Rev. B **96**(11), 115405 (2017)
30. C.Q. Sun, Y. Sun, The attribute of water: single notion, multiple myths. Springer Ser. Chem. Phys. **113** (2016)
31. Y. Shi, Z. Zhang, W. Jiang, Z. Wang, Theoretical study on electronic and vibrational properties of hydrogen bonds in glycine-water clusters. Chem. Phys. Lett. **684**, 53–59 (2017)
32. Y. Huang, X. Zhang, Z. Ma, Y. Zhou, G. Zhou, C.Q. Sun, Hydrogen-bond asymmetric local potentials in compressed ice. J. Phys. Chem. B **117**(43), 13639–13645 (2013)
33. Y.L. Huang, X. Zhang, Z.S. Ma, G.H. Zhou, Y.Y. Gong, C.Q. Sun, Potential paths for the hydrogen-bond relaxing with (H$_2$O)(N) cluster size. J. Phys. Chem. C **119**(29), 16962–16971 (2015)
34. C.Q. Sun, Size dependence of nanostructures: Impact of bond order deficiency. Prog. Solid State Chem. **35**(1), 1–159 (2007)
35. C.Q. Sun, J. Chen, Y. Gong, X. Zhang, Y. Huang, (H, Li)Br and LiOH solvation bonding dynamics: molecular nonbond interactions and solute extraordinary capabilities. J. Phys. Chem. B **122**(3), 1228–1238 (2018)
36. C.Q. Sun, Thermo-mechanical behavior of low-dimensional systems: the local bond average approach. Prog. Mater Sci. **54**(2), 179–307 (2009)
37. X.X. Yang, J.W. Li, Z.F. Zhou, Y. Wang, L.W. Yang, W.T. Zheng, C.Q. Sun, Raman spectroscopic determination of the length, strength, compressibility, Debye temperature, elasticity, and force constant of the C-C bond in graphene. Nanoscale **4**(2), 502–510 (2012)
38. K. Amann-Winkel, R. Böhmer, F. Fujara, C. Gainaru, B. Geil, T. Loerting, Colloquiu: water's controversial glass transitions. Rev. Mod. Phys. **88**(1), 011002 (2016)
39. A. Wong, L. Shi, R. Auchettl, D. McNaughton, D.R. Appadoo, E.G. Robertson, Heavy snow: IR spectroscopy of isotope mixed crystalline water ice. Phys. Chem. Chem. Phys. **18**(6), 4978–4993 (2016)
40. X. Zhang, P. Sun, Y. Huang, T. Yan, Z. Ma, X. Liu, B. Zou, J. Zhou, W. Zheng, C.Q. Sun, Water's phase diagram: from the notion of thermodynamics to hydrogen-bond cooperativity. Prog. Solid State Chem. **43**, 71–81 (2015)
41. C.Q. Sun, X. Zhang, X. Fu, W. Zheng, J.-L. Kuo, Y. Zhou, Z. Shen, J. Zhou, Density and phonon-stiffness anomalies of water and ice in the full temperature range. J. Phys. Chem. Lett. **4**, 3238–3244 (2013)
42. C.Q. Sun, X. Zhang, J. Zhou, Y. Huang, Y. Zhou, W. Zheng, Density, elasticity, and stability anomalies of water molecules with fewer than four neighbors. J. Phys. Chem. Lett. **4**, 2565–2570 (2013)
43. Y. Zhou, Y. Huang, Z. Ma, Y. Gong, X. Zhang, Y. Sun, C.Q. Sun, Water molecular structure-order in the NaX hydration shells (X= F, Cl, Br, I). J. Mol. Liq. **221**, 788–797 (2016)
44. S. Hajati, S. Coultas, C. Blomfield, S. Tougaard, XPS imaging of depth profiles and amount of substance based on Tougaard's algorithm. Surf. Sci. **600**(15), 3015–3021 (2006)
45. M.P. Seah, I.S. Gilmore, S.J. Spencer, Background subtraction—II. General behaviour of REELS and the Tougaard universal cross section in the removal of backgrounds in AES and XPS. Surf. Sci. **461**(1–3), 1–15 (2000)
46. X.B. Zhou, J.L. Erskine, Surface core-level shifts at vicinal tungsten surfaces. Phys. Rev. B **79**(15), 155422 (2009)
47. Z. Zhang, D. Li, W. Jiang, Z. Wang, The electron density delocalization of hydrogen bond systems. Adv. Phys.: X **3**(1), 1428915 (2018)
48. X. Wu, W. Lu, W. Ou, M.C. Caumon, J. Dubessy, Temperature and salinity effects on the Raman scattering cross section of the water OH-stretching vibration band in NaCl aqueous solutions from 0 to 300 °C. J. Raman Spectrosc. **48**(2), 314–322 (2016)

49. Y. Zhou, Y. Huang, L. Li, Y. Gong, X. Liu, X. Zhang, C.Q. Sun, Hydrogen-bond transition from the vibration mode of ordinary water to the (H, Na)I hydration states: molecular interactions and solution viscosity. Vib. Spectrosc. **94**, 31–36 (2018)
50. J. Chen, C. Yao, X. Liu, X. Zhang, C.Q. Sun, Y. Huang, H_2O_2 and HO- solvation dynamics: solute capabilities and solute-solvent molecular interactions. Chem. Sel. **2**(27), 8517–8523 (2017)
51. D.R. Lide, *CRC Handbook of Chemistry and Physics*, 80th edn. (CRC Press, Boca Raton, 1999)
52. Q. Wei, D. Zhou, H. Bian, Negligible cation effect on the vibrational relaxation dynamics of water molecules in $NaClO_4$ and $LiClO_4$ aqueous electrolyte solutions. RSC Advances **7**(82), 52111–52117 (2017)
53. J.W. Gibbs, On the equilibrium of heterogeneous substances. Am. J. Sci. **96**, 441–458 (1878)
54. R.C. Cammarata, Surface and interface stress effects in thin films. Prog. Surf. Sci. **46**(1), 1–38 (1994)
55. T. Young, An essay on the cohesion of fluids. Philos. Trans. R. Soc. Lond. 65–87 (1805)
56. M. Zhao, W.T. Zheng, J.C. Li, Z. Wen, M.X. Gu, C.Q. Sun, Atomistic origin, temperature dependence, and responsibilities of surface energetics: an extended broken-bond rule. Phys. Rev. B **75**(8), 085427 (2007)
57. M.T. Suter, P.U. Andersson, J.B. Pettersson, Surface properties of water ice at 150–191 K studied by elastic helium scattering. J. Chem. Phys. **125**(17), 174704 (2006)
58. C.Q. Sun, Y. Sun, Y.G. Ni, X. Zhang, J.S. Pan, X.H. Wang, J. Zhou, L.T. Li, W.T. Zheng, S.S. Yu, L.K. Pan, Z. Sun, Coulomb repulsion at the nanometer-sized contact: a force driving superhydrophobicity, superfluidity, superlubricity, and supersolidity. J. Phys. Chem. C **113**(46), 20009–20019 (2009)
59. X. Zhang, Y. Huang, Z. Ma, L. Niu, C.Q. Sun, From ice supperlubricity to quantum friction: electronic repulsivity and phononic elasticity. Friction **3**(4), 294–319 (2015)

Chapter 3
Theory: Aqueous Charge Injection by Solvation

Contents

Abstract Solvation is a process of aqueous charge injection in the forms of H^+, electrons, electron lone pairs, cations, anions, or molecular dipoles with long- and short-range interaction. A solute interacts with its neighboring H_2O molecules through the O:H vdW, O:⇔:O super-HB compression, H↔H anti-HB fragilization, ionic or dipolar polarization with screen shielding, and solute-solute interaction and their combinations. The hydration H_2O dipoles tend to be aligned oppositely along the electric field screen in turn the electric fields of the solute. The ionic size, charge quantity, and the numbers and spatial distribution of H^+ and ":" determine the form of solute-solvent interaction. A solute may be sensitive or not to interference of other solutes depending on the solute size and its extent of screening. The intermolecular nonbond and intramolecular bond cooperative relaxation determines the performance of a solution in terms of surface stress, solution viscosity, energy absorption-emission-dissipation at solvation, solvation temperature, thermal stability, critical pressures and pressures for phase transition.

Highlight

- Ionic polarization, O:H vdW, H↔H and O:⇔:O repulsion comprise molecular interactions.
- Molecular nonbonding interactions relax cooperatively the intramolecular covalent bond.
- Ions align, cluster, stretch, and polarize the O:H–O bonds to form supersolid hydration shells.
- Cooperative nonbond-bond relaxation and electron polarization dictate solution properties.

3.1   Solvent Water Structure

3.1.1   Molecular Dipole Premises

Generally, a H_2O molecule in its liquid or in its crystallite performs often as a dipole of fixed moment moving restlessly in the "dipole sea" [1, 2]. The dipole-dipole interaction is subjecting to the Lennard-Jones (L–J) or the van der Waals (vdW) potentials. Neither regular bonds nor electron exchange occurs between molecules with two pairs of ":" and two H^+ as their terminators. For the ordinary water, the rigid non-polarizable [3, 4] and the otherwise [5] approximations are often used to describe their molecular motion dynamics and the properties of the liquid.

Among various models for the structure of water ice, the rigid and non-polarizable TIPnP (n varies from 1 to 5) series [3, 4] and the polarizable models [5] are widely used in calculations. In the TIPnQ models, for instance, the V-shaped H_2O geometry with a bond length of $d_{O-H} = 0.9572$ Å and a bond angle of $\angle H-O-H = 104.52°$ describes a water molecule in the gaseous phase compared with $d_{O-H} = 1.0004$ Å for large volume water [6]. The TIP4Q/2005 model [7] simplifies the H_2O molecule as a dipole (O^+-M^-) with a fixed H^+ point charge. However, the manner and mobility of solute or molecular motion is less relevant to the bonding network relaxation in geometry and energy or the properties of the water and aqueous solutions. As the basic structural unit, H_2O molecular motion dissipates energy capped by the O:H cohesion at 0.1 eV level [8]. Such a dipole-dipole approximation ignores the O–O

Coulomb repulsion that couples the intermolecular and the intramolecular interactions, described as the asymmetrical, short-range, and coupled O:H–O oscillator pair.

The coupling of the initially independent O:H and H–O enables the O:H–O to relax cooperatively when stimulated. The O:H and the H–O segmental disparity and the O–O repulsivity form the soul dictating the extraordinary adaptivity, cooperativity, recoverability, and sensitivity of water and ice subjecting to perturbation [9]. However, from the perspective of molecular dynamics, the H_2O motif, whether rigid or not, serves as the primary structural unit in the liquid, which isolates the intramolecular bond from comprehension the intermolecular nonbond and their cooperativity. Therefore, an amplification of the solvation dynamics from the perspective of molecular dynamics to the hydration bonding dynamics would certainly be proper.

3.1.2 Oxygen Electronic Orbital Hybridization

The sp^3-orbital hybridization occurs to atoms of electronegative elements such as C, N, O and F associated with directional bonding orbitals and electron lone pairs, which follows the basic rules [10–12]:

(1) The sp^3-orbital hybridization of an electronegative atom occurs once its sp orbitals are fully occupied at reaction.
(2) The tetrahedral-oriented electronic orbitals can be occupied by sharing electron pairs with the bonding partners or by the nonbonding lone pairs of the electronegative atom.
(3) The number of lone pairs follows the "$4 - n$" rule where n is the atomic valence. Zero, one, two and three lone pairs are associated with a C, N, O, and F atom at reaction or their neighboring ones in the Periodic Table.
(4) Upon the sp^3 orbital hybridization, the atom can never form more than one bond with a specific neighboring atom because of the orientation specification of the tetrahedral bonding orbitals.
(5) The lone pairs tend to point outwardly to the open end of a molecular cluster, serving as terminators of a molecule but it is subject to the oreintation conservation of water ice.
(6) The solute Y^+ or X^- interacts with the solvent molecules through the $X \cdot H^+$ or Y^+:O Coulomb attraction, the $Y^+ \leftrightarrow H^+$ and $X^- \leftrightarrow$:O repulsion, and the $H \leftrightarrow H$ or the $O:\leftrightarrow:O$ repulsion. Unfavorable charge exchange or orbital overlapping could happen between the solute and the solvent H_2O molecules.

3.1.3 Electron Lone Pairs and Nonbond Interaction

Electron lone pairs are created upon the sp^3-orbital hybridization of the electronegative elements at reaction, which has the localized or delocalized attribute according to coordination environment. Coulomb O:H attraction and delocalization are dominance so the O:H shares some covalent and polarization feature form the perspective of wave function in density functional theory (DFT) calculations [13]. Calculations [14] reveal that the electronic polarization between the targeted solutes and water molecules is the primary many-body effect, whereas the charge-transfer term only makes a small fraction of the total solute–solvent interaction energy.

Charge injection derives the attraction, polarization and repulsion, which is subject to the solute coordinating counterpart. If the H^+ is replaced with a metal M or a M^+ ion core, situation changes. For O:M, the $O^{2-}:M^{dipole}$ forms with repulsion without charge exchange or orbital overlapping [11]. For M^+, the $O^{2-}:M^{+/dipole}$ forms with both attraction and polarization. The binding energy of the electron lone pairs remain as they stay in orbitals of their isolated parent atoms [2], which is a few eVs below the Fermi energy [10, 11]. The number difference and the spatial distribution of the lone pairs and protons discriminate the molecules performance one from another, such as CH_4, NH_3, H_2O, H_2O_2, and HF.

Lone pairs and protons are the primary ingredients for atoms of the electronegative elements or larger molecular solutes interacting with the less electronegative elements. The nonbonding lone pairs, dipoles or the nonbonding single electron in graphite (π bond) do not follow the dispersion of Schrödinger equation either but they add impurity states in the energy gap between the conduction band and the valence band or above the Fermi level of the specimen [15].

The nonbonds also include the O:H vdW bond, inter-proton H↔H and inter-lone-pair O:⇔:O repulsion, and ionic or dipolar polarization, which are different from those regular covalent bond, ionic bond, and metallic bond [10] featured with exchange interactions or valence charge sharing. The characteristics of the nonbonds is their weak interactions or repulsions that contribute significantly neither to the Hamiltonian in the quantum theory nor to the Gibbs free energy in the classical thermodynamics or the total energy minimization in continuum mechanics. The ordinary O:H cohesive energy is around -0.1 eV at the ambient and reaches -0.25 eV under 25 GPa compression, according to the potential path of the compressed O:H–O bond [16]. In comparison, quantum computations [17] suggested that the O:H energy is around 0.2 eV. The H↔H and O:⇔:O repulsion has positive energy. The O:⇔:O repulsion is four times that of the H↔H if they are of the same separation [18, 19].

These nonbonding electrons and interactions govern, however, the performance of the molecular crystals and the edges of the undercoordinated systems such defects, surfaces, and nanostructures of various shapes, such as high-T_C superconductors, atomic catalysts, and topological insulators [10]. These defect states also slow down the transition dynamics from the excited states to the ground to prolong the photon and phonon relaxation lifetime in the ultrafast spectroscopies [20]. Therefore, Gibbs free energy approximation, or the least energy principle is subject to a certain extent of limitation in dealing with water ice, aqueous solutions, and molecular crystals and liquids because of their high fractions of such emerging short-range, weak, or repulsive three-body nonbonding interactions [21, 22].

However, the repulsive nature of the H↔H anti-HB and the O:⇔:O super-HB defines that these two kinds of nonbonds cannot exist alone but must be accompanied with the presence of the O:H–O bond under tension. A combination of these irregular nonbonds not only equilibrates molecular interactions to form a molecular crystal but also deforms the intramolecular covalent bonds cooperatively in their crystals and liquids. Such a combination of the anisotropic, attractive-repulsive, local, and weak interactions makes the nonbond-contained system more complicated in the DFT computations.

3.1.4 Regulations for Solvent H_2O

3.1.4.1 Conservation and Restriction

Water prefers the statistic mean of the tetrahedrally–coordinated, two-phase structure in a core–shell fashion of the same geometry but different O:H–O bond segmental lengths in the covering skin [9, 16]. Figure 3.1a illustrates the $2H_2O$ unit cell of C_{3v} symmetry having four hydrogen bonds bridging oxygen anions. Transiting from the V-shaped H_2O motif of C_{2v} symmetry to the $2H_2O$ unit cell of C_{3v} symmetry aims to resolving the O:H–O cooperativity. As the basic structure and energy storage unit, the O:H–O bond integrates the intermolecular weaker O:H nonbond (or called van der Waals bond with ~0.1 eV energy) and the intramolecular stronger H–O polar-covalent bond (~4.0 eV). The O:H–O bond forms an asymmetrical, short-range, oscillator pair coupled with the Coulomb repulsion between electron pairs on adjacent oxygen ions [16].

O:H–O bond angle and length determines the crystal geometry and mass density of water ice. The segmental stretching vibration frequencies ω_x determine the respective Debye temperatures Θ_{Dx} of their specific heats through the Einstein relation, $\Theta_{Dx} \propto \omega_x$, (X = L and H denotes, respectively, the O:H and the H–O interactions). The segmental binding energy E_x correlated to the thermal integration of the specific heat $\eta_x(T/\Theta_{Dx})$ in Debye approximation. O:H–O bond length and its containing angle relaxation changes the system energy, but fluctuation contributes little to the system energy on average. It is therefore essential to treat water as a crystalline-like structure with well-defined lattice geometry, strong correlation, and high fluctuating network.

Molecules packed in water ice follow the basic regulations:

(1) Number conserves for protons and electron lone pairs. For a specimen containing N oxygen atoms, the 2N numbers of protons H^+ and lone pairs ":" and the O:H–O configuration conserve regardless of structural phase [23] unless excessive H^+ or lone pair ":" is introduced. Excessive H^+ or ":" will transit one O:H–O into the H↔H or the O:⇔:O without altering the rest O:H–O bond configuration or roientation [24]. The O:H–O configuration reserves irrespective of the structure phase or fluctuation, even in the H_3O^+:OH^- super ionic phase [25] or the Xth phase of identical O:H and H–O length [26].

(2) The rotation of a H_2O molecule or the motion of a proton H^+ is subject to restriction. If the molecule in the centre rotates above 60° around its C_{3v} sym-

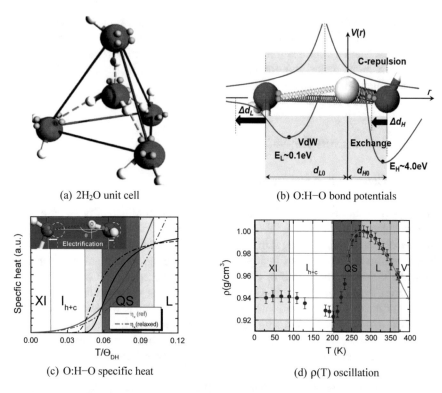

(a) 2H₂O unit cell

(b) O:H−O bond potentials

(c) O:H−O specific heat

(d) ρ(T) oscillation

Fig. 3.1 **a** The 2H₂O primary unit cell contains four oriented O:H−O bonds defines liquid water as a crystal-like with molecular and proton motion restrictions [16]. **b** The asymmetrical, short-range, coupled three-body potentials for the segmented O:H−O bond [21, 22]. **c** Superposition of the segmental specific heat $\eta_x(T)$ defines five phases of water ice. **d** The phases of density oscillate from high temperature downward of the Vapor ($\eta_L = 0$, not shown), liquid and ice I$_{h+c}$ ($\eta_L/\eta_H < 1$), quasisolid (QS) ($\eta_L/\eta_H > 1$), XI ice ($\eta_L \cong \eta_H \cong 0$), and the QS boundaries ($\eta_L/\eta_H = 1$) closing to T_m and T_N, respectively [33]. Electrification (ionic polarization) [24] or molecular undercoordination [30] disperses the QS boundaries outwardly, but compression does contrastingly through the Einstein relation, $\Theta_{Dx} \propto \omega_x$

metrical axis of the 2H₂O, there will be H↔H anti-HB and O:⇔:O super-HB formed between water molecules, which is energetically forbidden. In fact, single molecular orientational defect causes a long-range disorder of the I$_h$ ice [27, 28]. Because of the H−O bond energy of ~4.0 eV, translational tunneling of the H⁺ between adjacent H₂O molecules is also forbidden. 2H₂O →H₃O⁺:OH⁻ superionic transition by one of the four O:H-O bonds directional inversion takes places and holds under extramely high pressure and temperature conditions [29], which evidences against the Boner-Fowler-Pauling proton transitional tunneling under the ambient conditions. Breaking the H−O or the D−O bond in vapor phase requires a 121.6 nm laser radiation [27, 28], estimated 5.1 eV because of the extremely low molecular coordination numbers.

(3) The O:H–O angle, segmental length and energy fluctuate and relax under perturbation. The H–O bond relaxation absorbs or emits energy but the O:H vibrating at THz regime dissipates energy. One can only relax the O:H–O segmental length, O–O repulsion, charge polarization, and the bending angles of the molecules by applying stimulation or solvation, which determines the phase structure and properties of water ice.

(4) Fluctuation does not change the system average energy or molecular orientation but dissipates energy that is caped with the O:H interaction at 0.1–0.2 eV.

3.1.4.2 O:H–O Coupled Oscillator Pair

Figure 3.1b illustrates the asymmetrical, short-range, coupled three-body potentials for the segmented O:H–O bond [21, 22]. The proton serves as the coordination origin. The left-hand side is the O:H vdW interaction and the right-hand side is the H–O polar-covalent bond. The Columbus repulsion between electron pairs on neighboring O^{2-} couple the O:H–O bond to be an oscillator pair.

With the known H–O and O:H segmental length relaxation derived from the V–P profile measured from compressed ice [26, 29] and the framework of tetrahedral-coordination for a water molecule [30], one can correlate the size d_H, separation d_{OO}, bond geometry, and mass density ρ of molecules packed in water and ice in the following manner [6],

$$\begin{cases} d_{OO} = 2.6950\rho^{-1/3} & (Molecular\ separation) \\ \frac{d_L}{d_{L0}} = \frac{2}{1+exp[(d_H-d_{H0})/0.2428]}; & (d_{H0} = 1.0004\ and\ d_{L0} = 1.6946\ at\ 4\,^{\circ}C) \end{cases} \quad (3.1)$$

where the d_{H0} and the d_{L0} are the referential standard at 4 °C. This formulation correlates the d_{OO}, d_L and the d_H to the mass density, ρ. This relation yields the segmental lengths projecting along the O–O line of $d_H = 0.889$ Å, $d_L = 2.076$ Å, and 0.75 g cm^{-3} for the skin of water from the measured d_{OO} of 2.965 Å [31]. The skin mass density for water and ice is much lower than the least value of 0.92 g cm^{-3} for the homogeneous ice formed at 258 K [32]. The value $d_H = 1.0004$ Å and $d_L = 1.6946$ Å is the reference standard of $\rho = 1$ g cm^{-3} for water at 4 °C [6].

3.1.4.3 Specific Heat and Density Oscillation

Figure 3.1c shows the superposition of the specific heat $\eta_x(T/\Theta_{Dx})$ of Debye approximation for the two segments (X = L and H for the O:H and H–O segment, respectively) [33]. The segmental specific heat meets two conditions. One is the Einstein relation $\omega_x \propto \Theta_{Dx}$ and the other is the thermal integration being proportional to the bond energy E_x. The (ω_x, E_x) is (200 cm^{-1}, ~0.1–0.2 eV) for the O:H nonbond and (3200 cm^{-1}, ~4.0 eV) for the H–O bond. The Debye temperatures and the specific heat curves are subject to the ω_x that varies with external perturbation. Thus, the superposition of segmental specific heat η_x defines (c) the phases of density oscillation, for bulk water (T $\geq$ 273 K) and 1.4 nm water droplet (T < 273 K) [32], from high temperature downward of Vapor (not shown), Liquid, Quasisolid (QS), I_{h+c} ice, XI, and the QS boundaries of extreme densities and closing to the T_m and T_N, respectively [33].

The hydrogen bonding thermodynamics is subject to its segmental specific heat difference in terms of the η_L/η_H ratio. The segmental having a lower specific heat follows the regular thermal expansion but the other segment responds to thermal excitation oppositely. As shown in Fig. 3.1d, the thermodynamics and density oscillation of bulk water ice under the ambient pressure proceed in the following phases [33]:

(1) In the Vapor phase ($\geq$373 K), $\eta_L \cong 0$, the O:H interaction is negligible though the H_2O motifs still hold.

(2) In the Liquid phase (277, 373 K), $\eta_L/\eta_H < 1$, O:H cooling contraction and H–O elongation take place, but the O:H contracts more than H–O contracts, cooling contraction of water take place, reaching to the value of $d_{OO} = d_H + d_L = 1.0004 + 1.6946$ Å.

(3) In the QS phase (258, 277 K), $\eta_L/\eta_H > 1$, H–O cooling contracts less than the O:H expands; O:H–O expands, which triggers ice floating.

4) At the QS boundaries (258, 277 K), $\eta_L/\eta_H = 1$, the density drops from its maximum 1.0 to minimum of 0.92 g cm^{-3}. No apparent singularity presents to the specific heat or to the density profile, so the 277 K is recommended be the temperature for phase transition from Liquid to the QS phase. The 258 K corresponds to the T_N for the homogeneous nucleation of ice.

(5) The I_{c+h} ice (258, 100 K), $\eta_L/\eta_H < 1$, repeats the thermodynamics of the Liquid phase at a lower transition rate, transition the density from 0.92 to 0.94 g cm^{-3}.

(6) In the XI phase, $\eta_L \cong \eta_H \cong 0$, neither O:H nor H–O responds sensitively to the temperature change, so the density remains almost constant except for the slight $\angle$O:H–O angle expansion at cooling.

One should note that any relaxation of the O:H–O bond under perturbation will offset the phase boundaries through the Einstein's relation: $\Theta_{DX} \propto \omega_X$; and therefore, the thermodynamics of water ice and aqueous solutions is closely correlated to the O:H–O bond oscillation and relaxation.

3.1.4.4 Supersolidity and Quasisolidity

The concept of supersolidity was initially extended from the ^{4}He fragment at mK temperatures or ultra-cold rare gases [34]. The interface between the rotating ^{4}He fragments is elastic, repulsive, and frictionless [35] because of atomic undercoordination induced local densification of charge (electrostatic repulsion) and energy (elasticity) and the associated polarization [36]. The concepts of supersolidity and quasisolidity were firstly defined for water and ice in 2013 by the current group of practitioners and then intensively verified [30, 33].

The quasisolidity describes phase transition from Liquid density maximum of one g cm^{-3} at 4 °C to Solid density minimum of 0.92 g cm^{-3} at -15 °C. The quasisolid is characterized by the cooling expansion because the specific heat ratio $\eta_L > \eta_H$. The H–O bond contraction drives the O:H expansion and the $\angle$O:H–O angle relaxation from 160° to 165° [33]. The QS boundaries are subject to dispersion by external stimulus, which is why water ice perform abnormally in thermodynamics.

The supersolidity features the behavior of water and ice under polarization by coordination number reduction (also called confinement by the hydrophobic interface) or charge injection. When the nearest CN number is less than four the H–O bond contracts spontaneously associated with O:H elongation and strong polarization. At the surface, the H–O bond contracts from 1.00 to 0.95 Å and the O:H expands from 1.70 to 1.95 Å associated with the O:H vibration frequency transiting from 200 to 75 cm^{-1} and the H–O from 3200 to 3450 cm^{-1} [37]. Salt solvation derives cations and anions dispersed in the solution [24]. Each of the ions serves as a source center of electric field that aligns, stretches and polarizes the O:H–O, resulting the same supersolidity in the hydration shell.

The shortened H–O bond shifts its vibration frequency to a higher value that increases further with the reduction of the molecular CN, which disperses the QS boundaries outwardly, causing the supercooling at freezing and superheating at melting [16]. Perturbation by molecular undercoordination or electrostatic polarization disperses the QS boundaries outwardly, depressing the T_N for homogenous ice nucleation and raising the T_m for melting. The T_C for I_c-XI transition temperature also drops from 100 to 60 K as the droplet size is reduced [38, 39]. However, compression has contrasting effect on the critical T_m and T_N, resulting in the ice Regelation—compression depresses the T_m that reverse when and the pressure is relieved [40]. The high thermal diffusivity of the supersolidity skin governs thermal transportation in the Mpemba paradox—warm water cools faster [41].

3.2 O:H–O Bond Cooperativity

3.2.1 Segmental Length Relaxation

What vary upon perturbation, such as charge injection in the present discussion, are the $\angle$O:H–O containing angle, the O:H–O segmental lengths and energies, and the O–O coupling interaction, which result in various structures of water and ice in the P–T phase diagram [23]. The perturbation includes compression, thermal excitation, molecular undercoordination, electric and magnetic radiation [16].

The O:H nonbond and the H–O bond segmental disparity and the O–O coupling allow the segmented O:H–O bond to relax oppositely–an external stimulus dislocates both O ions in the same direction but by different amounts. Figure 3.2 shows the MD calculations performed using the force field of Sun [42]. The O:H–O segments relax

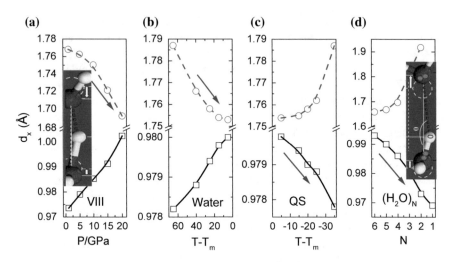

Fig. 3.2 MD-derived O:H–O segmental cooperative relaxation under **a** mechanical compression, **b** Liquid cooling, **c** QS cooling, and **d** $(H_2O)_N$ cluster molecular undercoordination. Arrows denote the master segments that drive the cooperative relaxation under perturbation. The O:H always relaxes more than the H–O in opposite slopes and curvatures, irrespective of the stimulus applied or the structural phase because of the persistence of O–O repulsive coupling [16]

under various stimuli with arrows noting the master segment that drives the other part as a "slave". One may note that the slopes and the curvatures of coupled segments $(dd_L/dq)/(dd_H/dq) < 0$ and $(d^2d_L/dq^2)/(d^2d_H/dq^2) < 0$. The d_L and d_H always relax oppositely in a manner of contrasting curvatures. The softer O:H nonbond always relaxes more than the stiffer H–O bond with respect to the H^+ coordination origin.

The $\angle$O:H–O angle θ relaxation contributes only to the geometry and mass density. The O:H–O bond bending has its specific vibration mode that does not interfere with the H–O and the O:H stretching vibrations [16]. The O:H–O bond cooperativity determines the properties of water and ice under external stimulus such as molecular undercoordination [43–47], mechanical compression [26, 48–51], thermal excitation [33, 52, 53], solvation [54, 55] and determines the molecular behavior such as solute and water molecular thermal fluctuation, solute drift motion dynamics, or phonon relaxation.

Figure 3.3 displays the experimentally derived O:H–O bond segmental length cooperativity. O:H–O bond relaxation (a) under mechanical compression of 80 K ice [26], (b) molecular undercoordination for $(H_2O)_N$ clusters computed using different DFT packages (PW91 [56]) without (red) and with (dark) HB correlation-dispersion correction (OBS-PW [57]) yield two sets of data showing the same trend of O:H–O relaxation, albeit accuracy due to potential functions used in calculations [30]. Panel (c) was derived from thermal excitation [33] derived using (3.2) from the measured mass density $\rho(T)$ of various phases [32]. The d_L–d_H length cooperativity of (3.2) in panel (d) reconciles experimental observations of ice under compression ($d_H > 1.00$ Å) [58], water ice at cooling ($0.96 < d_H < 1.00$ Å) [32, 59], and water skin, molec-

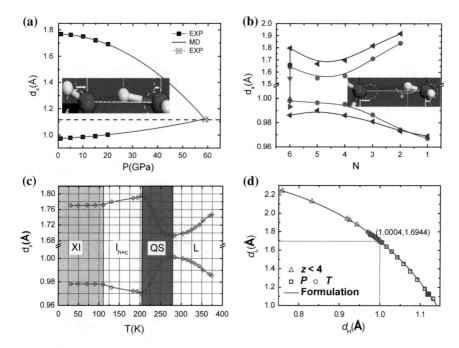

Fig. 3.3 Experimentally derived O:H–O segmental bond length cooperativity. The segmental length changes with **a** mechanical compression of 80 K ice [26], **b** molecular undercoordination of $(H_2O)_N$ clusters [30], and **c** thermal excitation [33] and **d** the d_L–d_H correlation in the liquid (L), quasisolid (QS), ice I and ice XI phases under the ambient pressure. The d_H = 1.0004 Å and d_L = 1.6946 Å correspond to the reference standard at ρ = 1 g cm^{-3}. Insets **a** and **b** illustrate the O:H–O bond relaxation. Solid line of (3.2) reconcile all available measurements of the O:H–O segmental length in (**d**). Reprinted with permission from [6]

ular clusters, and a monomer (d_H < 1.00 Å) [30, 31, 60–68]. Further compression after transiting to the X phase shortens both the d_L and the d_H slightly in the same rate, according to DFT calculations [69]. The size trends in panel b of the O:H–O bond relaxation remain the same despite the calculation code-resolved quantification accuracy. The extraordinary manner of O:H–O segmental relaxation of water and ice in panel c is subject to the O:H and H–O specific heat disparity. Results confirm consistently the expected O:H–O bond cooperativity under external excitations, as illustrated in Fig. 3.3d.

3.2.2 Phonon Frequency Cooperative Relaxation

Figure 3.4 shows the ω_x DPS for deionized water subjected to (a, b) heating [41], (c) compression [50] and (d) molecular undercoordination [37, 70]. Heating stiffens the ω_H from 3200 to 3500 cm^{-1} and meanwhile softens the ω_L from 180 to 75 cm^{-1}.

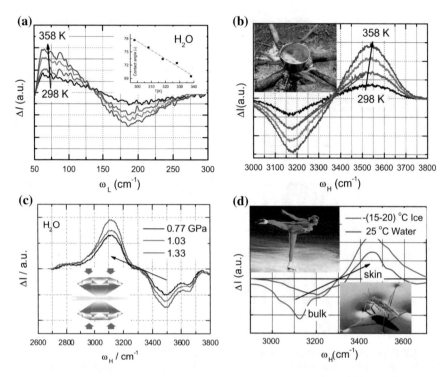

Fig. 3.4 Thermal excitation **a** softens the ω_L from 180 to 75 cm^{-1} and **b** stiffens the ω_H from 3200 to 3500 cm^{-1} for deionized water [41]. **c** Mechanical compression transits the ω_H of liquid water from above 3300 cm^{-1} to its below [50] and, **d** molecular undercoordination transits the ω_H from 3200 for water (at 25 °C) and 3150 cm^{-1} for ice (−20 and −15 °C) to 3450 cm^{-1} for the skins of water and ice [37, 70]. Inset a shows the thermal depression of the contact angle that is the same in trend of surface tension [71]

The O:H nonbond in liquid water follows the regular rule of thermal expansion but the H–O bond is subjecting to thermal contraction because of the O–O repulsive correlation [33]. Mechanical compression softens the ω_H from the skin value of 3450 to 3100 cm^{-1}, which confirms that mechanical compression shortens the O:H nonbond and lengthens the H–O bond. The compression effect retains regardless of the structure phase of water [50] or ice [26]. These observations prove the O–O repulsivity that correlates the inter- and intra-molecular interactions.

The ω_H DPS of water and ice gained by varying the angle between the surface normal and the direction of light reflection distils the monolayer-skin bonding information of (d) 25 °C water and ice (−20 and −15 °C). The phonon transits its frequency from the bulk water (3200 cm^{-1}) and bulk ice (3150 cm^{-1}) to their skins sharing the identical H–O bond stiffness featured at 3450 cm^{-1}, which clarifies that the length and energy of the H–O bond in both skins are identical disregarding temperature or structure phase [37]. An integral of the skin DPS abundance suggests that the skin of ice is 9/4 times thick of liquid water. However, heating reduces the

Table 3.1 O:H–O segmental cooperative relaxation in length, vibration frequency, and surface stress with respect to $d_{L0} = 1.6946$ Å, $d_{H0} = 1.0004$ Å, $\omega_{H0} = 3200$ cm^{-1}, $\omega_{L0} = 200$ cm^{-1}, $\Theta_{DH} = 3200$ K, $\Theta_{DL} = 198$ K upon excitation by heating, compression, molecular undercoordination (skin, cluster, droplet, nanobubble) ($\Delta\Theta_{Dx} \propto \Delta\omega_x$)

	Δd_H	Δd_L	$\Delta\omega_H$	$\Delta\omega_L$	Remark	References
$\Delta T > 0$ (Liquid: 277, 377 K) $I_c + I_h$ (100, 258 K)	<0	>0	>0	<0	Liquid and solid thermal expansion	[33]
$\Delta T < 0$ (QS: 258, 277 K)	>0	<0	<0	>0	QS negative thermal expansion	
XI (0, 100 K)	$\cong 0$				Debye specific heat $\cong 0$	
$\Delta z < 0$; $\Delta E \neq 0$ (charge)	<0	>0	>0	<0	Polarization; supersolidity	[24, 30]
$\Delta P > 0$	>0	<0	<0	>0	d_L and d_H symmetrization	[50]

surface stress and the contact angle between water and glass shown inset **a** as a result of thermal fluctuation [71]. Table 3.1 compares the O:H–O segmental length and stiffness relaxation under perturbation.

Strikingly, measurements in panel (d) confirm the "molecular undercoordination resolved two-phase structure of water and ice" [9]. Covered with a supersolid skin, water is a uniform, strongly correlated, fluctuating crystal-like. This observation is consistent with the wide- and small-angle X-ray scattering observations from amorphous ice [72] and liquid droplet [73]. Nanodroplets and amorphous states are naturally the same as they share undercoordinated molecules in different manners. The only difference between amorphous and a nanostructure is the distribution of the undercoordinated defects—one is ordered at the domain skins and the other is randomly distributed in the bulk. Therefore, the two-phase structure, low-density supersolid skin and the normal bulk, holds for water and ice. This observation may clarify the long-debating two-phase structural model as "coordination-resolved" other than "domain-resolved" nature [72, 73].

Table 3.2 Experimentally-derived skin supersolidity (ω_x, d_x, ρ) of 298 K water and 253 K ice

	Water (298 K)		Ice (253 K)	Ice (80 K)	Vapor
	Bulk	Skin	Bulk	Bulk	Dimer
ω_H(cm^{-1})	3200 [70]	3450 [70]	3125 [70]	3090 [33]	3650 [74]
ω_L(cm^{-1}) [33]	220	~180 [30]	210	235	0
d_{OO}(Å) [6]	2.700 [61]	2.965 [31]	2.771	2.751	2.980 [31]
d_H(Å) [6]	0.9981	0.8890	0.9676	0.9771	0.8030
d_L(Å) [6]	1.6969	2.0760	1.8034	1.7739	$\geq$2.177
ρ(g cm^{-3}) [6]	0.9945	0.7509	0.92 [75]	0.94 [75]	$\leq$0.7396

With the measured d_x and ω_x as input [37]. The skin of 253 K ice and the skin of 298 K water share the same H–O vibration frequency [70]

Numerical reproduction of the Mpemba effect—hot water cools faster, evidences directly the essentiality of the 0.75 unit mass density of the supersolid skin that promotes heat conduction outward the water of heat source [41]. Exothermic reaction proceeds by bond elongation and dissociation while endothermic reaction proceeds by bond contraction and bond formation. The Mpemba effect integrates the O:H–O bond energy "storage-emission-conduction-dissipation" cycling dynamics. The energy storage is proportional to the H–O bond heating contraction and the rate of energy emission at cooling is proportional to its first storage. The skin higher thermal conductivity due to lower mass density benefits heat flow outward the solution, and the source-drain non-adiabatic dissipation ensures heat loss at cooling.

Lagrangian resolution of the O:H–O coupled oscillator pair [21] resulted in the $\omega_x(k_x)$ the dispersion for the coupled oscillators, with k_x and k_C being the segmental force constant and the O–O coupling potential,

$$\omega_x = (2\pi c)^{-1}\sqrt{\frac{k_x + k_C}{m_x}} \tag{3.2}$$

Table 3.2 summarizes the d_{OO}, d_x, ρ, and ω_x for the skin and the bulk water and ice in comparison to those of ice at 80 K and water dimers with the referenced data as input. However, for other regular substance, the surface density increases as only the single-segment interatomic bond undergoes contraction [36]. Geometric configuration may change by the bond angle and segmental length relaxation but the numbers of protons and ":" and the basic O:H–O configuration conserve, irrespective of geometric phase structures of water and ice [23].

3.2.3   Site-Resolved H–O Length and Stiffness

DFT calculations of the site-resolved electronic binding energy and H–O stretching vibration frequency for medium sized $(H_2O)_n$ clusters ($n = 17, 19, 20, 21, 23$ and 25) [13, 76] classified the H–O bonds into five groups: the dangling H–O bonds, the H–O bonds associated with the dangling H_2O molecules, those with undercoordinated H_2O molecules without dangling H–O bonds, those forming the tetrahedral-coordination of the central H_2O and the others in its neighbor four molecules. The neighboring four molecules and the outer undercoordinated molecules form cages covering the central H_2O. Computations resolved that these two regions interact competitively, showing complementary interaction energy with the change of cluster size. Raman spectra in Fig. 3.5 reveal the site-resolved vibration frequencies. The dangling bond frequency features keep constant while others vary with cluster size. Observations confirm the effect of coordination environments on the H–O bond contraction, H–O bond energy gain, and its vibration frequency blue shift [16].

Computational examination [77] of the anharmonic correction contribution to the H–O vibration frequency revealed that the addition of the anharmonic vibrational potential terms (the third and the fourth order differentiation of the Taylor series of the interatomic potentials) only shifts the H–O phonon by about 100 cm^{-1} downward without any new features being derived. Therefore, it is adequate for one to focus on the origin, the trend of the vibrations with proper consideration of the higher-order anharmonic contributions.

SFG measurements, shown in Fig. 3.6 [78], further confirmed that in the outermost subsurface between the first (B1) and the second (B2) bilayer, the O:H nonbond of the O_{B1}–H:O_{B2} is weaker than that of the O_{B1}:H–O_{B2}. The subsurface O–O distance is laterally altered, depending on the direction of the H–O bond along the surface normal: H-up or H-down, which is in contrast with bulk O:H nonbonds. This discovery evidences for the DFT derivatives [77] and the present O:H–O bond cooperativity notion expectations [21, 22]. The least coordinated outermost sublayered H–O_{B1}

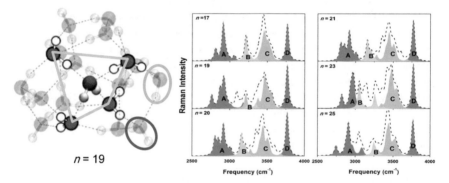

$n = 19$

Fig. 3.5 Computational H–O stretching vibration modes in the $(H_2O)_n$ clusters. The black dashed lines convolute the H–O vibration modes of the entire clusters. The sharp feature D corresponds to the H–O dangling bonds, C to the H–O bonds associated to the undercoordinated molecules. Features A and B to the H–O bonds inside the clusters. Reprinted with copyright permission from [76]

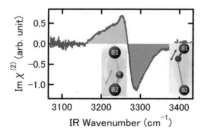

Fig. 3.6 SFG spectrum for the H–O oscillators lined along the c-axis of ice Ih(0001). Insets illustrate the O_{B1}:H–O_{B2} and the O_{B1}–H:O_{B2} segmental bond lengths, orientations, and the H–O stretching vibration frequencies. The positive peak (<3270 cm^{-1}) corresponds to the H–O_{B2} (shaded in green) and the valley (>3270 cm^{-1}) to the O_{B1}–H (shaded in blue). The less coordinated O_{B1}–H is shorter and stiffer and its H:O_{B2} is longer and softer than the inverse. Courtesy of Dr. Toshiki Sugimoto [78]

is shorter and stiffer than the second sublayered $H-O_{B2}$ and hence the $O_{B2}:H$ is longer and softer than the $O_{B1}:H$.

3.2.4 Three-Body Cooperative Potentials

The O:H–O bond forms the primary structural unit for ice, water and aqueous solvent, as the bonding unit combines and correlates the inter- and intra-molecular interactions. Intramolecular covalent bond relaxation absorbs or emits energy the intermolecular nonbonds fluctuation dissipate energy with negligible amount of energy consumption. The $\angle$O:H–O containing angle and the O:H–O segmental length and energy determine the geometry of the H_2O motif and its packing order in a specific structural phase. Figure 3.3b illustrates the O:H–O bond three-body potentials [16].

The $\angle$O:H–O containing angle is valued at $160°$ for the liquid phase with high fluctuation and it increases up to $175°$ when cooled to 0 K, according to quantum optimization [33]. By averaging the surrounding background of long-range interactions from other H_2O molecules and the nuclear quantum fluctuations [79, 80], one can focus on the asymmetrical and short-range interactions in one representative O:H–O bond with the "floating" H^+ as the coordination origin of the O:H–O bond [30]. The short-range interactions include the O:H nonbond van der Waals force [81], the H–O polar-covalent bond exchange interaction [82], and the Coulomb repulsion between the electron pairs attached to the adjacent oxygen ions [21, 22]:

$$
\begin{cases}
V_L(r_L) = V_{L0}\left[\left(\frac{d_{L0}}{r_L}\right)^{12} - 2\left(\frac{d_{L0}}{r_L}\right)^{6}\right] & \text{(L-J potential } (V_{L0}, d_{L0})) \\
V_H(r_H) = V_{H0}\left[e^{-2\alpha(r_H-d_{H0})} - 2e^{-\alpha(r_H-d_{H0})}\right] & \text{(Morse potential } (\alpha, V_{H0}, d_{H0})) \\
V_C(r_C) = \frac{(2\delta)^2}{4\pi\varepsilon_r\varepsilon_0 r_C} & \text{(Coulomb potential } (q_., q_-))
\end{cases}
$$

$$(3.3)$$

where the V_{L0} and the V_{H0}, denoted as the E_{L0} and the E_{H0}, are the potential well depths, or bond energies, of the respective segment at equilibrium. The r_x and d_{x0} (x = L, H, and C) denote the interionic distances (corresponding to the spring's length) at arbitrary position and at equilibrium, respectively. The α parameter determines the width of the potential well. The ε_r and ε_0 are the relative and the vacuum dielectric constant, respectively. The 2δ denotes the partial charge of a pair of electron pairs attached to O^{2-} anions.

Because of the asymmetrical, short-range and cooperative interaction of the O:H–O oscillator pair, one must consider the effective potential for the specific segment—shown in Fig. 3.1b the specific area. One must switch off a potential and on the other at once when one moves to the boundary of the specific region. No spatial decay of any potential is allowed at transition. These asymmetrical and short-range interactions also bring difficulty in quantum computations—how to switch the

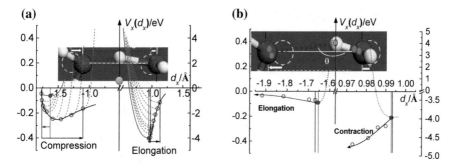

Fig. 3.7 Potential paths (red circles) for the **a** contracted [21] and **b** elongated [22] O:H–O bond. Plot (**a**) is derived from mechanically compressed ice at 80 K (r. to l.: P = 0 to 60 at 5 GPa step) [21] and (**b**) from $(H_2O)_N$ clusters at the ambient (r. to l.: N = 6, 5, 4, 3, 2). Blue dots are states at the $V'_x = 0$ equilibrium without involvement of O–O repulsion. Red circles in the leftmost (**a**) and in the rightmost (**b**) are states at $V'_x + V'_C = 0$ equilibrium with the O–O repulsion being involved. Rest red circles are states subjecting to the $V'_x + V'_C + f_{ex} = 0$. The f_{ex} is the non-conservative force due stimulus, V'_x is the gradient of the inter- and the intra-molecular potential and V'_C is the gradient of O–O Coulomb potential. Both O dislocate in the same direction but by different amounts. Reprinted with permission from [21, 22]

specific potential from one to the other as an atomic site is a calculation challenge. The O:H energy (0.1–0.2 eV) is less than 5% of the H–O energy (4.0 eV).

Figure 3.7 shows the potential paths for the O:H–O bond relaxation under non-conservative stimulations [21, 22]. Lagrangian-Laplace resolution of the O:H–O oscillating dynamics under the non-conservative compression and molecular under-coordination transforms the measured segmental length and phonon frequency (d_x, ω_x) at each point of equilibrium into the segmental force constant and binding energy, which derives the potential paths of O:H–O bond relaxation [26, 30]. The O:H–O relaxation proceeds only in one of the two manners under any stimulation [16]:

(1) O:H–O segmental disparity and O–O Coulomb repulsivity define the segmental potentials and their extraordinary adaptivity, cooperativity, recoverability, and sensitivity when responding to perturbation.

(2) O:H–O bond elongation occurs under electrification (ionic polarization of the O:H–O bond), hydrophobic capillary confinement, molecular undercoordination, liquid and solid heating, quasisolid cooling, or mechanical tension.

(3) O:H–O bond contraction in the opposite manner takes place under mechanical compression, base solvation, liquid and solid cooling, quasisolid heating.

(4) O:H–O bond contraction enlarges the H_2O molecular size (d_H) but reduces their separations (d_L) associated with H–O bond stretching phonon softening and O:H phonon stiffening.

(5) The unprecedented O:H–O bond cooperative relaxation arise from the O–O coupling and the externally applied non-conservative activation.

3.3   Concerns of Solvation

3.3.1   What Is Solvation?

Solvation, or hydration, dissolves a substance into solutes of either charged parti-cles or molecular dipoles injecting to the solvent. Charge injection in the forms of anions, cations, dipoles, electrons, lone pairs, protons mediate the HB network and properties of solutions through electrostatic polarization, O:H formation, H↔H and O:⇔:O repulsion, solute-solute interactions and hydrating H_2O dipole screening. The solutes disperse regularly in the seemingly-disordered solvent with or without skin preferential occupation. For instance, ions serves each as a source of electric field that aligns, clusters, stretches and polarizes their neighboring solvent molecules to form a supersolid [30] or semi-rigid [83, 84] hydration shell. Protons and lone pairs never staying lonely or moving freely but they bond to a H_2O to form the H_3O^+ hydronium and the OH^- the hydroxide, which break the conservation rule of water to create the H↔H anti-HB and the O:⇔:O super-HB, respectively.

A molecular solute is surrounded by numbers of protons H^+ and electron lone pairs ":" symmetrically or not to form a dipole. Each of the H^+ and the ":" interacts with its alike or unlike of the neighboring solvent molecules to form the short-range X:H vdW bond of attraction, H↔H inter-proton repulsion, or X:⇔:Y inter-lone-pair compression. A dipolar solute may distort the O:H–O bonds at the solute-solvent interface because its H^+ and ":" separation may not match the geometric requirement of the ordered solvent molecules. The X and Y refer here to those atoms of electronegative elements like C, N, O, F, and their surrounding elements in the Periodic Table. These atoms hybridize their electronic sp orbitals upon reaction with production of electron lone pairs [11, 12]. Compared with solid surface chemisorption or programmed doping, no regular bonds, i.e., covalent or ionic bonds, form between the solute and the solvent molecules but only form the nonbonds with dominance of induction, repulsion, and polarization [18, 85–87].

It would be efficient to deal with aqueous solvation in the same way of handling with chemisorption and defect formation by equaling the solutes to the adsorbates, dopants, point defects, or impurities in solid phase, disregarding the solvent structural fluctuation and the solute drifting motion. In fact, what determines the properties of a solution is both the intra- and the inter-molecular interactions and charge polariza-tion rather than only the solute or solvent molecular dynamics. Thermal fluctuation or solute drift motion dissipates, rather than absorbs or emits, energy from a ther-modynamic point of view. It is furthermore essential to treat the solvent water as a crystal-like static structure rather than the highly dynamic disorder. A solution, or a water with impurities, conserves its numbers of H^+ and ":" and the O:H–O configuration unless excessive H^+ or ":" is introduced [16].

3.3.2 Extended Hydrogen Bond X:H–Y

Customarily, in the literature, an HB is represented as X–H$\cdots$Y, where a partially electropositive H atom bound covalently with X, is bridged weakly with Y, and the latter is either an electronegative atom or electron-rich segment of the same or of a different molecule. In conventional hydrogen bonds, X represents a highly electronegative atom, viz. O, N and F, so that the X–H bond, termed as hydrogen bond donor, is sufficiently polar, and Y the acceptor is also an electronegative element [88–90]. The energy range of such hydrogen bridged linkage is typically 5–15 kcal/mol (1 kcal/mol = 0.22 eV) [91], which is suitable for association and dissociation of the H-bonded partners in liquids at ambient temperatures. The much less polar X–H groups, e.g. a C–H bond has also been considered as HB donor, and diffused electron densities present in π molecular orbitals have been included as acceptors in the manifold of the molecular entities that could engage in HB.

In the revised IUPAC definition of hydrogen bonding, X could be any atom that is more electronegative than H, namely F, N, O, C, P, S, Cl, Se, Br and I, and Y could be any of these atoms or π-electrons [92, 93]. However, from energetic viewpoints, HBs corresponding to less electronegative X and Y groups belong to the category of weak hydrogen bonds [91, 94, 95], and their bond energies are typically less than 5 kcal/mol. Nevertheless, it has been realized that the collective contributions of such weak hydrogen bonds could be highly significant in determining the structure, conformational and tautomeric preferences of many organic molecules in the liquids and crystals and the functional structures biological macromolecules within living cells [96–99].

The conventional X–H$\cdots$Y bond treats the X–H as a donor and Y an acceptor independently with oversight the X–Y coupling by interaction between the negative charged anions. The presently described O:H–O (X:H–Y) bond integrates the intra- and intermolecular interaction [9, 16, 100]. The X:H–Y applies to any situation where the lone pair ":" can be replaced by any nonbonding interaction without charge sharing or orbital overlapping. The H can be replaced by a cation of an arbitrary element.

Microscopically, a molecule has multiple atoms of different elements, like alcohols, aldehydes, complex salts, drugs, organic acids, sugars, and explosive crystals. The stronger covalent bonds assembled these atoms as a motif whose dangling H^+ and electron lone pairs ":" expose to their alike or unlike partners of its molecular neighbors. $X^{2\delta-}:H^{\delta+}–Y^{2\delta-}$ hydrogen bond, $X^{2\delta-}:\Leftrightarrow:Y^{2\delta-}$ and $H^{\delta+}\leftrightarrow H^{\delta+}$ repulsive nonbonds will form in addition to the ionic and dipolar polarization. One can approximate the partial charge δ (= 0.62–0.66 e for water) as unity, for discussion convenience [9]. One can call them $X:\Leftrightarrow:Y$ super-HB and $H\leftrightarrow H$ anti-HB. It is not necessary for a super-HB to have an H atom. The X:H–Y is naturally the same to the O:H–O bond in water but it has different segmental lengths, energies, force constants, vibration frequencies, and reduced masses of the oscillator pair. Electron lone pair ":" is the primary element of life. The X:H–Y equilibrates its own through the asymmetrical, short-range, and X–Y coupling potentials, as the O:H–O bond does in water ice.

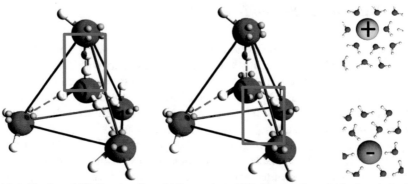

(a) Acid solvated 2H$_2$O unit cell   (b) Base solvated 2H$_2$O unit cell   (c) Ionic polarization

Fig. 3.8 The central H$_2$O replacement of the 2H$_2$O primary unit cells with **a** an H$_3$O$^+$ hydronium tp form an H$_3$O$^+$·4H$_2$O motif for the acid solvation and with **b** an OH$^-$ hydroxide to form an OH$^-$·4H$_2$O motif for the base solvation [50], and **c** the X$^-$ anion and Y$^+$ cation interstatial eccentric occupancy in the acidic, basic and salt solutions [102]. Framed in **a** is the H$^+$↔H$^+$ anti-HB point breaker and **c** the O:↔:O super-HB point compressor. Ions in **c** serve each as a screened point polarizer that aligns, clusters, stretches and polarizes the O:H–O bonds in the hydration shells, which has the same effect of molecular undercoordination on the H–O bond contraction, O:H nonbond elongation, and strong polarization [30]

3.4   Lewis Acidic and Basic Solutions: Conservation Disruption

3.4.1   Excessive Protons: H↔H Anti-HB Fragilization

Upon solvation, an HX acid molecule dissolves into an H$^+$ proton and an X$^-$ anion. The H$^+$ does not stay freely but bonds firmly to a H$_2$O molecule to form a tetrahedral H$_3$O$^+$ hydronium, being NH$_3$–like tetrahedron with one lone pair. The H$_3$O$^+$ replaces a H$_2$O in the center of the 2H$_2$O unit cell or a H$_3$O$^+$·4H$_2$O motif while its four neighbors keep their orientations because they will keep the O:H–O regulation with their rest neighbors [9, 16], as illustrated in Fig. 3.8a. The alteration of the ":" with the H$^+$ in the H$_2$O →H$_3$O$^+$ transition breaks the 2N conservation with derivative of 2N + 1 number of protons and 2N − 1 lone pairs in the solution. The excessive 2N + 1 − (2N − 1) = 2 protons form uniquely an H$^+$↔H$^+$ anti-HB without any other choice. The H$_3$O$^+$ remains the tetrahedron configuration having three H–O bonds and one lone pair which is similar to the situation of H$_{2n+1}$O$_{2n}^+$ cluster formation with n = 2 and 4 but no shuttling between oxygen ions [101] or freely hopping from one site to another.

The strong H–O bond at 4.0 eV energy and over prevents the H$^+$ from random hopping among H$_2$O molecules or from transitional tunneling between two H$_2$O molecules by H$^+$ and ":" alteration [16]. One needs at least a laser radiation of 121.6 nm wavelength to dissociate the H–O bond or the D–O bond [27, 28]. Importantly, the H_3O^+ has four tetrahedrally coordinated H_2O neighbors with two protons and two pairs of ":" directing to it. Most importantly, the $H^+ \leftrightarrow H^+$ disrupts the HB network in a point-by-point manner, making acidic solution corrosive, dilutive, and surface stress destructive, as metal fragilization by H doping [103]. Ab initial-based MD computations [104] confirmed that HCl hydration fragments water clusters into smaller ones could be direct evidence for the $H \leftrightarrow H$ formation and fragilization.

3.4.2 Excessive Lone Pairs: $O: \Leftrightarrow :O$ Super-HB Compression

Likewise, the YOH base is dissolved into a Y^+ cation and a hydroxide (HO^- is HF-like tetrahedron with three lone pairs), see Fig. 3.8b. The HO^- addition transits the 2N conservation into $2N + 3$ number of lone pairs ":" and $2N + 1$ number of protons. The excessively unbalanced two lone pairs can only form the $O: \Leftrightarrow :O$ point compressor between the central OH^- and one of its H_2O neighbors that reserve their molecular orientations. This $O: \Leftrightarrow :O$ compressor pertaining to each hydroxide or to the H_2O_2 hydrogen peroxide has the same effect of but stronger mechanical compression on the network HB relaxation [50].

The Y^+ in the basic solution and the X^- in the acidic solution perform the same as they do in the salt solutions to polarize the surrounding water molecules to form the hydration shells with their directional electric fields. The electronegativity, ionic size and the hydrating H_2O screening discriminate the hydration shell size and the intensity of the local electric field in the hydration shells.

Figure 3.9 illustrates the $H \leftrightarrow H$ anti-HB and the $X: \Leftrightarrow :Y$ super-HB configuration and their repulsive interactions. These unprecedented repulsive nonbonds cannot stay alone. The presence of the $H \leftrightarrow H$ anti-HB and the $X: \Leftrightarrow :Y$ super-HB repulsion must go with X:H–Y stretching or any other type of tension to equilibrate the molecular interactions, such as the H_2O_2 aqueous solution [105]. The $X: \Leftrightarrow :Y$ compressive force is some four times that of the $H \leftrightarrow H$ at the same distance because of the charge quantity of $(2\delta/\delta)^2$ involved in their Coulomb potentials. Being similar to hydrogen-induced metal fragilization [103, 106, 107], the presence of the $H \leftrightarrow H$ serves as a point breaker that disrupts the hydration network and the skin stress of the acidic solutions [18]. The stronger $X: \Leftrightarrow :Y$ repulsion [50] has the same effect of mechanical compression [26] on its neighboring O:H–O bonds associated with strong polarization due the 4e change of the closely contacted lone pairs.

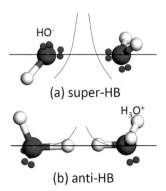

Fig. 3.9 Illustration of the **a** O:⇔:O super-HB and **b** the H↔H anti-HB [18] and their respective repulsive interactions that disallow them to stay alone but must be accompanied with X:H–Y under tension, which not only balance the molecular interaction but also deform the intramolecular bonds. Reprinted with copyright permission from [86]

Table 3.3 Ionic radius R(Å) and electronegativity η of alkali and halide elements [109]

	I^-	Br^-	Cl^-	F^-	Li^+	Na^+	K^+	Rb^+	Cs^+	H^+	O^{2-}
R/Å	2.20	1.96	1.81	1.33	0.78	0.98	1.33	1.49	1.65	0.53	1.32
η	2.5	2.8	3.0	4.0	1.0	0.9	0.8	0.8	0.8	2.2	2.5

Ions of lower η and larger R show stronger Hofmeister effect

3.5 Hofmeister Salt Solution: Polarization and Screen Shielding

3.5.1 Ionic Polarization: Hydration Shell Formation

Water dissolves an YX salt into a monovalent Y^+ alkali cation and an X^- halogenic anion. The X^- and the Y^+ may form the separated or the contacted ion pairs (SIP and CIP). For the SIP, the anion and cation create each a radial electric field pointing outward or inward the center of the ion; for the CIP dipole, the local field should be anisotropic and shorter than that of the SIP. The ion pair separation depends on the salt concentration, and the electronegativity difference $\Delta\eta$ between the X and Y elements. If the $\Delta\eta$ is too high, the Y^+ and X^- tend to stay closer, which creates a CIP dipole, such as NaF ($\Delta\eta = 4.0$–$0.9 = 3.1$). Table 3.3 summarizes the ionic radius and electronegativity of the alkali and halide elements for comparison.

The Y^+ and the X^- serve each as a charge center of different diameters. Anions and cations create each a radial electric field directing from positive to negative. Figure 3.8c illustrates that the radial electric field of an ion aligns, clusters, and stretches the O:H–O bonds in its hydration shell through polarization without any new bond being formed between the ionic solute and its surrounding water molecules [108].

Electronic polarization has the same effect to molecular undercoordination on HB relaxation and polarization, as illustrated in Fig. 3.8c. The electronic polarization stretches the O:H nonbond and shortens the H–O bond. The strong polarization of the nonbonding electrons leads to the local supersolidity. The supersolidity of the hydration shell and the solution skin is responsible for the longer H–O phonon lifetime, higher stress, higher viscoelasticity, higher solubility, higher thermal stability, but lower molecular dynamics and mass density [9]. However, focusing more on the behavior of the O:H–O bonds in the closest hydration shells rather than on those in the higher-order shells would be efficient to understanding the solvation nature and bonding dynamics.

MD computations [110] confirmed that an external field in the 10^9 V/m order slows down water molecules and even crystallize them. The fields generated by the Na^+ ions act locally but can reorient and even hydrolysis their neighboring water molecules. This observation is within the present expectation of solute ionic polarization.

Raman probing over the full frequency of the spectrum justifies that no such bonds could form between the solvent and the Y^+ or the X^- ions as no phonon peaks emerge due to formation of any possible bond [102, 111]. Without any charge sharing, the ionic field induction turns out the hydration shell of $X^- \cdot H–O$ or $Y^+ : O–H$ and a subsequent shells of O:H–O by stretching and polarizing the H_2O dipoles [18]. The solvent H_2O molecules in the closest hydration shells are preferably oriented with the H^+ towards anions and the lone pairs of O^{2-} towards cations due to the directional fields. The thickness of the hydration layer depends on the intensity of the local ionic fields that is subject to screening by its neighboring H_2O dipoles and subject to influence of other solutes.

3.5.2 Solute Electric Field Screening by Water Dipoles

There are two ways of screen weakening the electric field of a hydration shell. One is the H_2O dipole tend to be oriented oppositely to the electric field in the hydration shell and the other is the same kind solute-solute repulsion. The latter becomes more significant at higher solute concentrations. The ionic separation and the ionic electric field intensities vary with not only the solute concentration but also with the solute type [102, 111, 112]. Smaller ions are hardly-influenced by the solutes because the H_2O screening but larger one is a case of another. The unlike solute-solute attraction will enhance the solute local electric field, instead. However, there does not exist the cation-anion attractions and cation-cation repulsion in the YX salt and base solutions but only anion-anion repulsion to YX and HX solutions [24, 86, 87].

The H_2O dipoles aligned oppositely along the solute electric field will screen and spatially weaken the ionic field. The solute electronic field decays thus faster than it is in the vacuum because of the screening effect. The alike ionic repulsion weakens but the unlike ionic attraction enhances the field in the hydration shells. The strength of the local electric field determines the solute capability of transforming the solvent O:H–O bonds in terms of their stiffness and their number fraction from the mode of ordinary water to the hydration shells. The DPS strategy can resolve the effect of the solute local electric field.

The closer anion-cation distance makes the ion pairs more like the close-ion-pair showing the dipole nature. For single ion, the electric field should be long-range but for a mirror pairing ions, the range of the electric field is shorter. Therefore, the thickness of the hydration shells may vary from situation to situation, and argument on the shell thickness would be not very fruitful [113, 114]. One should be focused more on the HB transition from the mode of ordinary water to the mode of the closest hydration shells by omitting the intermediate states between the pure water and the high order hydration shells.

3.6 Complex and Organic Dipolar Solute

3.6.1 Nonbonding Formation

For a complex molecular solute, the number of the positive protons may be identical or not to that of the negative lone pairs, but the spatial distribution of the charged terminators distributed asymmetrically—a real molecular dipole forms. In this case, the positive end performs differently from the negative end in transforming the nearest O:H–O bonds apart from the influence of the entire molecular dipole.

If huge numbers of the dangling H^+ and ":" are involved and they distributed asymmetrically surrounding a molecule, a molecular dipole would form. The dipole induction of its surrounding adds an important factor to determining the performance of the molecular solute dipole. One end is clouded with ":" and the other with H^+, the giant dipole will create its own electric field that is anisotropic and short-range in order. The molecular nonbond interaction and the dipole field induction determine the solute capability of resolving the solution behavior. Therefore, the number harmonicity of the H^+ and the ":" dictate the numbers of the O:H vdW bond, $H \leftrightarrow H$ anti-HB, and $O: \leftrightarrow :O$ super-HB between molecules in their crystals or solutions; the spatial distribution symmetry of the H^+ and the ":" determine the solute local electric field. The mismatch of the solute terminators H^+ and ":" with their neighboring alike or unlike will also distort the local geometric structures at the solute-solvent interfaces.

3.6.2 Intermolecular Nonbond Counting

An organic molecular solute is attached with H^+ and lone pairs ":" to form a dipole. If a molecular motif contains $p(H^+)$ number of dangling protons and $n(:)$ number of lone pairs, and if $p(H^+) < n(:)$, there will be 2p HBs and n−p super-HBs; or else if $p(H^+) > n(:)$, there will be 2n HBs and p−n anti-HBs between molecules in their molecular crystals. H_2O is the simplest case of $p = n$, neither anti-HB nor super-HB could form unless excessive H^+ or ":" are introduced to break the rule of 2N number and the O:H–O configuration conservation.

Upon aqueous solvation, a molecule with unequal number of H^+ and ":" interacts with the solvent water molecules that have each only two H^+ and two pairs of ":". The protons and the lone pairs of the solute have each ½ probability to interact with a H^+ or a ":" to form the afore-mentioned HB, anti-HB or super-HB. Therefore, a molecular solute will form $2p + (n-p)/2$ HBs and $(n-p)/2$ anti-HBs for the case of $p < n$; otherwise, $2n + (p-n)/2$ HBs and $(p-n)/2$ super-HBs will form with its neighboring solvent water molecules.

Although it is a rough estimation, this nonbond counting helps one to understand the intermolecular interaction in molecular crystals and their solutions. The nonbond counting is summarized in Table 3.4 and epitomized in Table 3.5 that counts the numbers of HB, anti-HB and super-HB for typical molecules formed in their crystals and in their aqueous solutions.

According to the nonbonds counting, an ethanol (C_2H_5OH) motif is surrounded by six H^+ and two ":", there will be four H↔H anti-HBs and four O:H–O bonds in its pure liquid but forms two H↔H and six O:H–O bonds when the molecule is solvated in water. For the explosive TNT ($2C_7H_5N_3O_6$) motif, there are ten H^+ and $2 \times (3 + 2 \times 6) = 30$ lone pairs ":", each motif will interact with its surrounding neighbors with 20 HBs and 20 super-HBs in its own crystal but it forms 30 HBs and 10 super-HBs when solvated in water. The super-HBs compression must co-exist with the O:H–O bonds under tension, which not only balances the molecular interactions but also shortens the intramolecular covalent bonds. The O:H–O extension is realized by O:H elongation and H–O contraction, as illustrated in b, and the super-HB compression shortens its direct intramolecular bonds as well to store energy.

Table 3.4 Basic rule for nonbond counting in aqueous solutions and molecular crystals

Molecular solution	X:H–Y	H↔H	X:⇔:Y
$p(H^+) > n$ (:)	$2n + (p-n)/2$	$(p-n)/2$	0
$p(H^+) < n$ (:)	$2p + (n-p)/2$	0	$(n-p)/2$
Molecular crystal			
$p(H^+) > n$ (:)	$2n$	$p-n$	0
$p(H^+) < n$ (:)	$2p$	0	$n-p$

Table 3.5 Intermolecular X:H–Y bond, X:⇔:Y super-HB, and H↔H anti-HB counting in aqueous solutions and typical molecular crystals

	Molecules	$p(H^+)$	$n(:)$	Aqueous solution			Molecular crystal		
				X:H-Y	X:⇔:Y	H↔H	X:H-Y	X:⇔:Y	H↔H
Methane	CH_4	4	0	2	0	2	0	0	4
Ammonia	NH_3	3	1	3	0	1	2	0	2
H-fluoride	HF	1	3	3	1	0	2	2	0
Hydronium	H_3O^+	3	1	3	0	1	–		
Hydroxide	OH^-	1	0	3	1	0			
H-peroxide	H_2O_2	2	4	5	1	0	4	2	0
Explosive crystal	$2C_7H_5N_3O_6$	10	30	30	10	0	20	20	0
	$4CH_2N_2O_2$	8	24	24	8	0	16	16	0
	$3CH_2N_2O_2$	6	18	18	6	0	12	12	0
	$C_7H_5N_5O_8$	5	21	13	13	0	10	16	0

3.7 Solvation Thermodynamics and QS Boundary Dispersion

3.7.1 Exemplary H_xO_y Solvation

Figure 3.10 shows the sp^3-orbital hybridized molecular structures of H_2O, hydroxide (OH^-), hydronium (H_3O^+), and hydrogen peroxide (H_2O_2) with different numbers of H^+ and ":" surrounding the central O^{2-}. Except for the H_2O having two H^+ and two ":", the numbers of the positive H^+ and the negative ":" are not equal for other motifs. The number anharmonicity and the spatial symmetry of the H^+ and the ":" determine the forms of H_xO_y molecular interactions in their crystallites and performance of their aqueous solutions.

Upon solvation, protons and lone pairs have each a half probability to interact with their alike or unlike to form the O:H vdW bond, the H↔H anti-HB, and the O:⇔:O super-HB with its neighboring oxygen anions, without other choice. For pure water example, the 2N number and the O:H–O configuration conservation allows the H_2O to interact with its four neighbors through four O:H–O bonds with neither the super-HB nor the anti-HB being involved. Therefore, H_2O is the simplest primary molecular crystal. It is hard to understand the performance of a molecular crystal without knowing well water and ice.

However, the H_3O^+, the OH^- and the H_2O_2 are associated with different numbers of H^+ and ":". The excessive H^+ creates the anti-HB and the unbalanced ":" forms a super-HB with one of their coordinated H_2O neighbors. In contrast, the numbers of H^+ and ":" for the superionic H_3O^+: OH^- water that is only stable at extremely

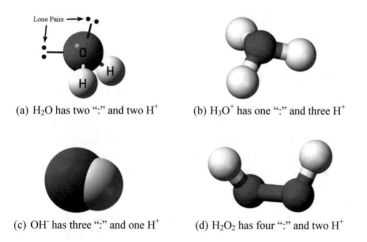

(a) H_2O has two ":" and two H^+ (b) H_3O^+ has one ":" and three H^+

(c) OH^- has three ":" and one H^+ (d) H_2O_2 has four ":" and two H^+

Fig. 3.10 Number harmonicity of the positive H^+ and the negative electron lone pair ":" for the H_xO_y systems. Besides the O:H–O bonds, aqueous solvation of these molecules turns out **b** an H↔H anti-HB, and **c, d** an O:⇔:O super-HB with its neighboring solvent oxygen anions

high pressure (2 TPa) and high temperature (2000 K) [25] also meet the 2N and the O:H–O configuration conservation.

For the H_2O_2 molecule, the two O^{2-} anions are paired up covalently and each of them bonds to an H atom. The sp orbitals of the oxygen hybridize then to form a tetrahedral pair. The H_2O_2 molecule has two protons and four pairs of ":". The excessive two pairs of ":" have half probability to interact with a ":" or a proton, so one O:$\Leftrightarrow$:O super-HB is formed in its aqueous solution. In its own liquid there will be two O:$\Leftrightarrow$:O formed. The strong O:$\Leftrightarrow$:O repulsion not only weakens the O–O bond of the H_2O_2 but also endows its hydrophobicity. The H_2O_2 is readily evaporated. The H_2O_2 solution at higher concentration is easily explosive and less stable because the O:$\Leftrightarrow$:O repulsion stretches the intermolecular O:H–O bond. Both O:$\Leftrightarrow$:O repulsion and O:H–O tension shortens the intramolecular covalent bond that stresses stores energy for the (CNHO) explosive materials—H_2O_2 is the simplest example.

H_3O^+ and OH^- solvation create additionally unexpected H$\leftrightarrow$H anti-HB and O:$\Leftrightarrow$:O super-HB in their respective solutions. The repulsive H$\leftrightarrow$H anti-HB and O:$\Leftrightarrow$:O super-HB formation may also help understanding the corrosive nature of acid and base solutions from the extraordinary nonbond configuration point of view. One can imagine that NH_3 and HF solvation share the respective H_3O^+ and OH^- bond structural configuration. One can also imagine that a tetrahedron of N_4 or a CH_4 molecule is structurally stable, but they can hardly be packed up yo form a crystal because of the molecular H$\leftrightarrow$H and N:$\Leftrightarrow$:N repulsive interactions without involvement of any attraction to balance. Because of its surrounding lone pairs, the CF_4 molecule serves as an efficient anticoagulant for synthetic blood [10].

3.7.2   Exothermic Solvation and QS Phase Boundary Dispersion

One often observes that solvation can heat up or cool down the solution, called exothermic or endothermic solvation. Exothermic and endothermic reactions are of great importance to both basic and engineering sciences [115, 116], as well as drug efficiency at function [117]. Solute-solvent interactions govern the path, ultimate outcome, and the efficiency of chemical reaction upon solvation [118]. Rosenholm [119] reviewed in 2017 the possible mechanisms available to date for the chemical thermodynamics occurred in liquids, solids, and semiconductor materials. Current understandings are mainly within the framework of classical thermodynamics in terms of enthalpy [120, 121] and Gibbs energy [122]. The heat generation at reaction is mainly attributed to the intra- and inter-molecules electron displacements [119], charge transfer [123], water molecular dynamics [124], H–O interaction [125], or the solute-water molecular vdW interactions [126]. However, correlation between the solution temperature and the network bond relaxation and the microscopic mechanism for the exothermic and endothermic dynamics remain open for examination.

According to Pauling [89], energy stores in the chemical bonds and the energy emission or absorption proceeds by bond relaxation—the equilibrium atomic distance and binding energy change under non-conservative fields [10] such as aqueous solvation or mechanical compression. Bond dissociation and bond elongation emit energy, but bond formation and bond contraction absorb energy, leading to the respective exothermic and endothermic reaction.

At solvation, HX, YX and YHO dissociation take place injecting charge into the solvent, which deform the HB network and mediate solution properties though O:H–O relaxation. H–O bond contraction under polarization and H–O bond elongation by H$\leftrightarrow$H and the O:$\Leftrightarrow$:O repulsion occur, which will emit or absorb energy. Under multiple process of the bonding dynamics, the dominant process governs the thermic solvation will heat up or cool down the solution. As showed shortly, solvation of base, H_2O_2 and alcohols heats up the solutions whose temperature is proportional to the fraction of the H–O bonds transition from the water mode to their elongation by H$\leftrightarrow$H and O:$\Leftrightarrow$:O repulsion.

Solvation such as sugars and salts can modulate the critical temperatures for freezing and melting, which is essential to reserving living organisms at very low temperatures and preventing slipperiness of ice on roads under heavy snows. Serving as excellent bio-protectant agents, sugars serve as agent for anti-frozen proteins making the living organisms to survive in cold and snowy climates [127, 128].

The critical temperatures for phase transition are proportional to a certain kind of energy in dimension. One needs to consider how to correlate the freezing and the melting temperature to the cohesive energy of the O: H nonbond or the H–O bond. In fact, the O:H and the H–O segment has independent specific heat that is featured by the respective Debye temperature Θ_{Dx} and its thermal integral over the full temperature range [16]. The Θ_{Dx} determines the rate of the specific heat curve reaching saturation; the integral is proportional to the segmental cohesive energy (~0.1 and ~4.0 eV, respectively) [129]. The correlation between the specific heat integral and the bond energy shall be updated to text books.

The segment with a lower Θ_{Dx} value on the specific heat curve will approach saturation faster than the other one does. According to Einstein's relation: $\hbar\omega_x = k\Theta_{Dx}$, where $\hbar$ and k is the reduced Planck's constant and Boltzmann's constant, respectively. Any relaxation of the phonon frequency ω_x mediates directly the Debye temperature Θ_{Dx}. The superposition of the specific heat curves, results in two intersecting temperatures corresponding to the quasisolid-phase boundaries nearby the T_N and T_m. The quasisolid phase whose boundaries correspond to the extreme mass densities of one unit at 4 °C and 0.92 unit at −15 °C for bulk specimen [16]. Therefore, from the frequency shift of the O:H and the H–O phonons under perturbation, one can infer easily to the Θ_{DL} and Θ_{DH} and the T_N and T_m variation by perturbation.

3.8 Hydration Versus Oxidation

It is known that the solubility of a substance is proportional to its polarity under certain conditions such as temperature and pressure. The neutral metals and diamond can hardly be dissolved by pure water, but they are easily oxidized by oxygen attacking. One of the key concerns in solvation is the dynamics of ionic bond or molecular dissociation when the solute is dispersed into solvent water matrix. One must answer how the solvent water dipoles intrude into the crystal to break the bonds between cation and anion by weakening their interactions or peel off the solute from the parent bulk. Knowledge about the dynamics of metal and diamond oxidation should shed light to the understanding of hydration.

3.8.1 Oxidation: Guest Active

3.8.1.1 Oxygen Bonding Dynamics

Very-low-energy electron diffraction (VLEED) and STM investigations revealed that oxidation occurs in four discrete stages [11]. An oxygen atom forms one bond after the other. Two bonds formation is then followed by the sp^3-orbital hybridization with generation of two pairs of lone pairs. The lone pairs in turn polarize their neighboring host atoms into dipoles. The dipoles are readily eroded away from the surface causing the oxygen attacked corrosion. UPS measurements confirmed the presence of the bonding (~6–8 eV), ionic hole (~3 eV), lone pair (~2 eV) below Fermi energy and the antibonding dipole states above the Fermi energy.

The order of bond formation is subject to the host atomic radius, lattice geometry, and the electronegativity difference between guest oxygen and host atoms. Figure 3.11 compares the STM images and bond configuration of O–Cu(001) and O–Rh(001) surfaces at 0.5 monolayer oxygen coverage and the ambient temperatures. For the Cu(100) surface, oxygen forms the first bond to a Cu (denoted 1) at the top layer and then the second bond to a Cu underneath (denoted 2), creating the missing-row type Cu(100)-($\sqrt{2} \times 2\sqrt{2}$)R45°-2O^{2-} geometry and dump bell-like, oppositely coupled pairing dipoles (denoted 3). For the Rh(100) surfaces, oxygen forms the first bond to a Rh in the second layer and then to one at the top layer, creating the clock and anti-clock wise Rh(100)-(2×2)p4g-2O^{2-} geometry with a Rhombi chain along the $\langle 11 \rangle$ direction. The M-O bond length varies with the coordination environment. The Cu-O bond of the top layer is 1.63 Å and the second Cu-O bond is 1.75 Å. The angle between the ionic bond is at 94°–103° and the angle between the lone pairs is 160°–135° towards the equilibrium. These processes of oxidation show the atomic scale anysotropy in the bond nature and energy, as well as charge distribution.

Thermal gravity investigations further verifies the geometric selectivity of diamond oxidation [134]. Diamond oxidizes easily at ~750 K through oxygen imping-

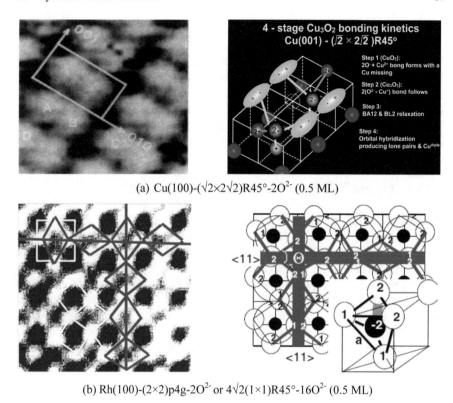

(a) Cu(100)-(√2×2√2)R45°-2O²⁻ (0.5 ML)

(b) Rh(100)-(2×2)p4g-2O²⁻ or 4√2(1×1)R45°-16O²⁻ (0.5 ML)

Fig. 3.11 STM images for the **a** O-Cu(100) surface [130] and VLEED derived four-stage bonding dynamics [131] and **b** Rh(100) [132] surfaces and the bond order and reconstruction patterns [133]. The lattice size and electronegativity resolved the order of M_2O bond formation to the fcc(100) surfaces at 0.5 ML coverage. The bright spots correspond to the polarization states of host atoms and the dark points are oxygen ions. Number 1 and 2 in panel labels the metal ions and 3 the dipoles. Number 1 in b denotes M^+ and 2 the dipoles ($R_{Cu} = 1.255$ Å, $\eta_{Cu} = 1.9$; $R_{Rh} = 1.342$ Å, $\eta_{Rh} = 2.3$). Reprinted with permission from [11]

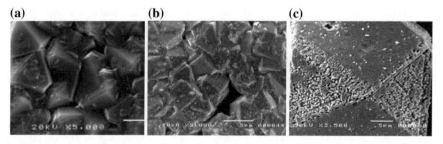

Fig. 3.12 Diamond **a** as-grown, **b** vacuum graphitization at 1100 K, and **c** (111) surface preferential oxidation at 750 K while the (110) surface remain stable unless at sites of defects. Reprinted with permission from [134]

ing into the densely-packed (111) planes of C_{3v} symmetry throughout the course of reaction, as shown in Fig. 3.12. In contrast, diamond graphitizes under vacuum or an Ar inert gas environment at about 1100 K without orientation preference. The diamond (111) plane is more favorable than the (110) surface for oxygen impinging into the bulk by bonding and re-bonding with lone pair production, leaving behind the weakly-interacting dipoles that are eroded away during the process of corrosion. The host lattice C_{3v} symmetry of the (111) is much more beneficial than the C_{2v} symmetrical lattice to the oxide tetrahedron bond formation.

3.8.1.2 Oxygen Floating and Impinging by Bond Switching

Because of the atomic scale M_2O anisotropy and the strong repulsion between its lone pairs, oxygen is "alive" and "flies" higher or lower in a way of "bond switching" rather than simple diffusion in the host as one could imagine. O diffuses into the bulk under certain condition and also floats up during epitaxial growth of metals on oxygen pre-recovered metal surfaces [130]. In oxidation and corrosion, oxygen breaks the adsorption barrier and the metal–metal bond to move into the bulk. The bulk oxidation can be seen with the naked eye as the oxide powders peeling off the master piece of metals, known as rusting. However, in epitaxial growth of metals onto oxygen pre-covered metal surfaces, oxygen atoms always float up to retaining at the surfaces.

STM observations [135, 136] revealed that the formation of an O/Cu surfactant structure on the surface due to migration of O initially located at the Ru surface in the temperature range of 350–450 K. Its surface coverage rises linearly with O pre-coverage up to 0.4 ML where it covers the surface completely. By increasing the coverage up to 0.5 ML, a drastic change in the morphology and density of the 2D islands occurs, which is accompanied by a change of the O/Cu surfactant structure. Work function measurements [137, 138] revealed that oxygen atom is always presented at the surface of the grown Cu films deposited on an oxygen pre-covered Ru(0001) surface. Under certain conditions (0.2–0.4 ML, 400 K), the work function, monitored during film deposition, oscillates with a period of one monolayer of copper epitaxial growth. Serving as a surfactant for a layer-wise Cu growth on the O–Ru(0001) surface, the floating oxygen periodically induces oxidized islands. For lower coverage, densely packed triangular islands are partially covered with an O/Cu surfactant structure. The O–Cu structure is locally ordered in a distorted hexagonal lattice, namely, with the hcp(0001) or fcc(111) features. The structure also consists of O–Cu–O strings inducing the observed corrugation.

Once the oxide structure is formed, elements of it float onto the top of the growing film and act as a surfactant layer for further Cu film growth. SIMS and secondary electron emission examination [139] revealed that the Cu layers deposited on the c(2 × 2)-O/Ni(001) substrate are always covered with an adsorbed layer of oxygen. Oxygen atoms segregate back to the surface during the epitaxial growth of Cu on the O–Cu(001) and Cu on the O–Cu(111) surface at about 400 K temperature [140, 141]. The co-adsorption of Ni and O on W(110) surface and subsequent annealing to 500–1000 K leads to segregation of Ni and O, with the formation of largely $\langle 111 \rangle$ oriented Ni crystallites along the surface normal [142]. Therefore, oxygen floating in the process of Cu/O/Ru(0001), Cu/O/Cu(111) and Cu/O/Cu(001) growth and Ni cluster on O–W(110) surface is driven by a common mechanism of bond switching.

VLEED investigation [143] revealed that annealing the O–Cu(001) surface at a 'dull red' (~450 K) temperature de-hybridizes the sp^3-orbital but after aging the sp^3-orbital rehybridized again. Hence, annealing at a certain temperature supplies forces that reverse the reaction by oxide bond breaking. At the critical temperature of 400 K, O^{2-} transits from de-hybridizing to hybridizing so the oxygen re-bonding occurs because annealing activates the bond breaking.

At the thermally activated state, oxygen can adjust itself towards a stable tetrahedron with two bonding and two lone pairs non-bonding orbitals. The lone pair induces dipoles that tend to be directed into the open end of a surface. The dipole forms periodically with the epitaxial layer growth, seen experimentally as the periodic change of the work function. Therefore, the strong repulsion between the dipoles and the repulsion between the lone pairs provide the driving force for oxygen to float up under a thermal activation in the process of homo- or hetero-epitaxial growth of metals [144]. With an external stimulus or under certain circumstances, the dipoles may escape from the bound by lone pairs, and then oxygen reactants have to re-bond to other atoms to form the stable tetrahedron, which is the case of bulk oxidation or rusting.

3.8.2 Hydration: Host Active

3.8.2.1 Solute Dissolution by H_2O Polarization

The driving force and the kinetic process of oxide-bond switching may infer the process of ionic crystal aqueous solvation. H_2O is the simplest oxide and molecular crystal with strong correlation and fluctuating. Unlike oxidation in which the oxygen atom undergoes bond and repulsive nonbonding lone pair formation with bond switching, a H_2O dipole hydrates an ion in a solid by approaching one of its four ends to the ion through Coulomb attraction, $Y^+:O^{2-}$ or $X^- \cdot H^+$. Meanwhile, H_2O polarization occurs associated with H_2O volume expansion or O:H–O elongation through H–O contraction and O:H expansion. The clustering and polarization of H_2O dipoles in turn weaken the local electric field and the $Y^+ \cdot X^-$ interaction of the solute. The H_2O volume expansion is subject to the O:H–O bond geometry conservation, and then some of the H_2O squeeze into the weakened $Y^+ \cdot X^-$.

Using the combination of STM and noncontact AFM, Jiang and colleagues [145] fabricated and resolved the individual Na^+ hydrates with 1–5 water molecules on a NaCl(001) surface at 5 K temperature. The $Na^+ + 3H_2O$ diffuses orders of magnitude faster than other Na^+ hydrates under field of the STM tip decorated with a Cl^- pulling the hydrates along a certain direction. The tip supplies a force overcoming the interaction between the hydrates and the NaCl substrate deposited on Au surface, as

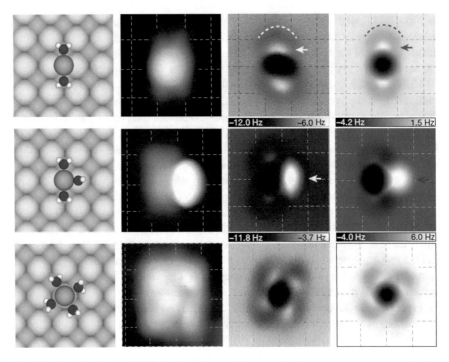

Fig. 3.13 Na·nH$_2$O (n = 2–4) hydration clusters. The n = 3 hydrate transports more easily than others on NaCl surface under the bias between the hydrate and STM tip decorated with a Cl$^-$ ion. From left column to right are the structure configurations, STM images, AFM images and simulations. Reprinted with permission from [145]

illustrated in Fig. 3.13. Ab initio calculations revealed that such high mobility arises from the existence of a peculiar metastable state, in which the three water molecules around the Na$^+$ can rotate collectively with an exceptionally low barrier. This scenario holds even at room temperature according to the classical MD simulations. These observations also evidence that hydration of ions can occur at extremely low temperature.

3.8.2.2 Salt Solution Lattice Geometry

One may assume both water and dissolved salt solute follow their own framework of lattice geometries, as typically shown in Fig. 3.14. H$_2$O molecules pinned to the ion and the undergo polarization and volume expansion without regular bond formation between solute and solvent, which weaken the electric field and the ionic bond, and eventually dissolve the solute into individual cations and anions packed in their original ways in the solution. The solution interlocks both geometries with each solute a cluster of water dipoles that screen the electric field. Ions occupy

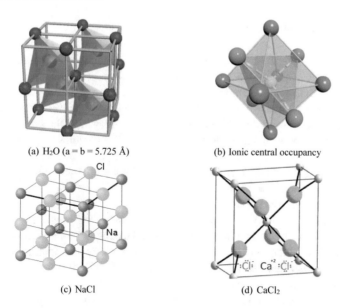

(a) H₂O (a = b = 5.725 Å)

(b) Ionic central occupancy

(c) NaCl

(d) CaCl₂

Fig. 3.14 Lattice geometry of liquid water and the typical YX and YX₂ type crystal cells. **a** Assembling the primary $2H_2O$ unit cells around the central oxygen ion forms the ideal diamond structure that has $8H_2O$ and 16 directional O:H–O bonds. This packing geometry and known mass density correlate the size and separation of molecular packed in water and in ice [6]. **b** illustrates the ionic eccentric occupancy of the interstatial vacancy and interacts with four nearest and six next-nearest oriented H2O molecules. **c –d** shows the structures of NaCl and CaCl₂ crystals ([146]). CsCl takes the centered simple cubic structure with one kind ion occupies the conor and the other the center

the interstatial hollow sites and interact with four nearest ($\sqrt{3}a/4$) and then six next-nearest (a/2) ortented water molecules. In contrat, H_3O^+ and OH^- replaces the central H_2O of the $H_2O·4H_2O$ motif.

The SIP ionic distance is much larger than its bulk as each of the ions is screened by its surrounding H_2O dipoles. Because the structure order of H_2O matrix, the solute ions may distort the crystal structure of the solvent water, the number of dipoles in a hydration shell follows a certain regulation that is subject to the structure of water. Increasing the solute concentration, the ionic distance will decrease, and the solute-solute interaction takes place.

For simplicity, one can mimic the $2CaCl_2$ structure to the $2H_2O$ unit cell with H^+ substitution for the Cl^- and O^{2-} for the Ca^+ without involvement of the lone pairs. In the $CaCl_2$ case, there will be twice number of Cl^- of the NaCl at the same solute concertation. The higher Cl^- density and the higher Ca^{2+} charge quantity shall differ its solution from those of NaCl and CsCl solutions.

3.8.2.3 Solute-Solute Interaction

As evidenced by Verlet and co-workers [147–152], injection of free electrons into water solvent does not influence the structure of the solvent but the electrons serve as probe for the local energetic environments. Photoelectron emission and the ultrafast PES detect their vertical bound energy and the lifetime of the electrons at different sites and sized water droplets. The hydrated electron may have the same function of Y^+ cation with inverse polarity on the hydrating H_2O molecules.

Besides the $H \leftrightarrow H$ anti-HB formation by replacing a ":" with H^+ upon acid HX solvation and the $O: \Leftrightarrow :O$ super-HB in base YOH solution, there will be X^- anions and Y^+ cations dispersed, respectively, in the acid and base solutions. The X^- and Y^+ perform the same as in the salt solutions.

Salt like HF forms the CIP dipoles, all other dissolved SIP ions dispersed into the solvent matrix in the interstatial hollow sites polarizing the neighbouring oriented H_2O molecules, which modulate the hydrogen network by polarization with solute-solute interactions that are subject to the solute concentration, ionic size, charge quantity and the extent of the hydration shell H_2O dipole screening. The number of screening H_2O dipoles in the hydration shell is subject to the H_2O geometry though slight distortion is workable. For the small alkali cations of $+1$ charge unit, its electric field may be fully screened, thus there is no interaction between the cation and other cations or anions.

Table 3.6 Comparison of oxidation and hydration dynamics

	Oxidation [10, 11]	Hydration [9, 21, 24]
Guest	Oxygen atom	Ionic or dipolar soluble crystal
Host	Solid state without specification	H_2O dipoles
Attribute	Guest active	Host active
Dynamics	Guest four-stage bond-nonbond formation and bond switching	Host pinning and clustering, polarization and solvent volume expansion, host dissociation and dispersion
Interaction energetics	Bond, nonbonding lone pair, anti-bond dipole, electron-hole	Electrostatic, O:H nonbond, anti-HB, super-HB, shielding, solute-solute interaction
Host characteristics	Reconstruction, relaxation, work function, property modulation	Surface stress, viscosity, conductivity, dielectrics, HB transition, HB network and property modulation
Detection method	Spatially and temporally resolved electron and phonon spectrometrics; STM/S, ZPS [153, 154], DPS [155], etc.	

Both oxidation and solvation are subject to the atomic/molecular under-coordination that shortens the bond and lengthens the nonbond. Electron lone pairs are of the key significance to both oxidation and hydration

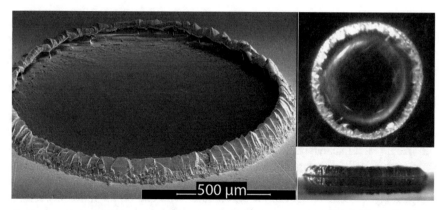

Fig. 3.15 SEM images of the deposited **a** open shell upon evaporation of water droplet on a NaCl crystal and **b** the top and side views of a water droplet after 33 s evaporating at 55 °C. Reprinted with permission from [156]

However, for the large halide X^- anions, the insufficient number of H_2O dipoles in the ordered hydration shell may partially screened, so anion-anion repulsion exists, and this repulsion increases with solution concentration. For the divalent YX_2 solution, the higher number of X and the higher charge quantity, $Y^{2+} \sim X^-$ attraction and the $X^- \leftrightarrow X^-$ will come into play. For larger molecular solvation, the solute grains are boundered by H^+ and lone pairs ":" interacting with their alike or unlike counterparts of solvent H_2O molecules. Such interactions distort the local structure of the hydration shells in addition the molecular dipoles.

Table 3.6 summarizes the difference between oxidation and hydration.

3.8.3 Undercoordination Enhanced Water-Salt Dissolubility

Water molecular undercoordination endows its stronger dissolubility of soluble substance because of the supersolidity that allows water molecules to impinge into the solute crystal. Figure 3.15 shows the ring-shape patterns formed by evaporating a water droplet deposited on a NaCl substrate via the coffee-stain effect [156] that is suggested resulting from outward flows inside the drop to compensate the enhanced loss of liquid due to a higher evaporation rate at the periphery [157, 158]. The evaporation of a water droplet left behind a deposit at the periphery of the dried drop along the pinned triple-phase line.

These patterns are formed in four stages. (i) At the drop deposition, the triple-phase line is pinned, an immediate dissolution of the substrate occurs, and a deposit starts to precipitate at the anchored triple phase line. (ii) During evaporation, the triple line leaves the initial perimeter of the drop and remains attached to the edge of the growing deposit. (iii) Approaching the end of the experiment, the drop surface becomes concave, and the drop surface in the center reaches the salt substrate, a new inner triple line then recedes outwards, until it reaches the deposit. (iv) At the very end of the evaporation process, the remaining liquid recedes inside the hollow shell to form the hollow rim like a coffee-stain.

Mailleur et al. [156] interpreted the phenomenon by taking into account of four basic ingredients. (i) Evaporation at the liquid-air interface. (ii) Dissolution at the solid-liquid interface. (iii) Diffusion and convection inside the liquid. (iv) Precipitation at the edge of the drop. Water evaporation and salt precipitation in the peripheral rim occurs simultaneously. The transfer of salt from the drop to the outer deposit shows, as for colloidal suspensions, that the divergence of the evaporation flux at the anchored triple line induces an outward capillary flow. Besides, the nonequilibrium morphology is found to depend on the contact angle and on the ratio between the drop radius and the shell thickness. Therefore, an interplay between drop wetting and substrate dissolution and growth leads to the unexpected surface morphology. They modeled the system by assuming that the evaporation flux is larger at the edge of the drop, and this inhomogeneity of evaporation drives an outward radial hydrodynamic flow. This process leads to a supersaturating in the triple-line region, which feeds the growth of the shell. This peripheral deposit exhibits a unique morphology with an open or closed hollow shell, in striking contrast with deposits from salty water evaporating on inert substrates [159, 160] or pure water evaporating on hydrosoluble solids [161, 162].

In addition to the simultaneous dynamics of water evaporation, dissolution, convection, diffusion, and precipitation, molecular undercoordination enhanced dissolubility and evaporability of the strongly polarized O:H–O bond at the rim could contribute significantly to the deposits. Molecular undercoordination lengthens the O:H and shortens the H–O to raise the polarity of the molecules. The polarity is an important indicator of the solubility of water. On the other hand, the undercoordinated O:H–O bond evaporates more easily as the evaporation is capped at the cohesive energy of the softened skin O:H nonbond about 0.1 eV or lower. The reduction of the O:H cohesive energy lowers the critical temperature of evaporation.

3.9 Summary

Briefly, solvation isolates a molecule from its crystallite or dissolves the molecule into counterions, or other forms of charge carriers, which can be viewed as aqueous charge injection, the dispersed charge interacts with the hydrogen bond network and modulates the solution properties in their respective ways. The solute-solvent molecular interaction proceeds through formation of the O:H vdW, the O:⇔:O super-HB and the H↔H anti-HB or through ionic or molecular dipolar polarization and screen shielding or their mixture. The O:⇔:O and the H↔H repulsion must co-exist with the O:H–O bond under stretching tension to stabilize the molecular interaction. The O:⇔:O has the same effect of mechanical compression that shortens the O:H nonbond and lengthens the H–O bond in a solution while the bond order deficiency

Table 3.7 Lewis and Hofmeister solutions and their functionalities[a,b,c]

Solution		Solvation dynamics	Solute capability	Solution property
Lewis acid (pH <7)	HX + H$_2$O ⇒	X⁻ + H$_3$O⁺ (H↔H anti-HB point breaker)	H$_3$O⁺ forms a NH$_3$-like tetrahedron with one lone pair, interacting with one of its four H$_2$O neighbors through the anti-HB that breaks the HB network point-by-point	Sour taste; capable of turning blue litmus red; corrosive, dilution, depressing viscosity and surface stress; relieving hypertension, etc.
Adduct (salt)	YX + H$_2$O	X⁻ + Y⁺ + H$_2$O (X⁻ and Y⁺ point polarizers)	Electric fields of the X⁻ and Y⁺ align, polarize and stretch the O:H–O bond, whose extent depends on the ion size, solute concentration and type separation and hydrating H$_2$O dipole shielding	Hofmeister series: stress and viscosity elevation; thermal stability; protein solubility; polarization; supersolid hydration shells; hypertension enhancement, etc.
Lewis base (pH ≥ 7) (alkali metals)	YOH + H$_2$O	OH⁻ + Y⁺ + H$_2$O (O:⇔:O super-HB point compressor)	HO⁻ forms an HF—like tetrahedron with three lone pairs. The O:⇔:O super HB point compressor shortens the neighboring O:H nonbond and lengthens the H–O bond	Solution greasiness and slipperiness; burning heat release (H–O softening)
Organic or complex molecules		X:H-Y, H↔H and X:⇔:Y formation and competition	Interface HB distortion	Hypertension mediation; DNA and proteins denatrization; T$_N$ modulation; etc.

[a]X = F, Cl, Br, I; Y = Li, Na, K, Rb, Cs

[b]The number of H$_2$O molecules in the hydration shell varies with the size and charge quantity of the Y⁺ and X⁻ ions

[c]Divalent salt, such as CaCl$_2$, dissolves into three independent ionic solutes CaCl$_2$ ⇒ Ca^{2+} + 2Cl⁻; Complex salt such as LiClO$_4$ dissolves into the LiClO$_4$ ⇒ Li⁺ + ClO$_4$⁻ with a greater anion symmetrically surrounded by 8 electron lone pairs [85]

of the solute OH$^-$ shortens its H–O bond. The H↔H has the same effect of heating that disrupts the solution network and surface stress.

The Y$^+$ and X$^-$ polarization has the same effect of molecular undercoordination on the H–O shortening and O:H lengthening associated with local supersolidity in the hydration shells. Screening the solute by the surrounding H$_2$O dipoles and the electrical repulsion between the alike solutes weaken the electric field in the hydration shell. For complex molecular solute, the anisotropic and short-range induction governs the relaxation of the solvent O:H–O bond and the performance of the solution. The spontaneous processes of bond formation and contraction absorb energy while bond dissociation and elongation emit energy. Solvent phonon frequency shifts disperse the quasisolid phase boundaries and the T_N and T_m of the solutions. Polarization transits the H–O phonon abundance-lifetime-stiffness and the surface stress and solution viscosity cooperatively.

Table 3.7 summarizes the solvation hydrogen-bonding dynamics, solute capabilities and solution properties for the monovalent YX salt, HX acidic, and YOH basic solutes. Solvation polarizes the O:H–O bond [24, 86, 87], creates the O:⇔:O super-HB [19, 105] or the H↔H anti-HB [18], which resolve the solute-solvent interactions in Lewis solutions. These interactions transit the O:H–O bonds from the mode of ordinary water to the hydrating in terms of the number fraction, bond stiffness, and molecular fluctuation order upon solvation [86, 87].

References

1. R. Qi, Q. Wang, P. Ren, General van der Waals potential for common organic molecules. Bioorg. Med. Chem. **24**(20), 4911–4919 (2016)
2. P.W. Atkins, *Physical Chemistry,* 4 edn. (Oxford University Press, 1990)
3. C. Vega, J.L.F. Abascal, M.M. Conde, J.L. Aragones, What ice can teach us about water interactions: a critical comparison of the performance of different water models. Faraday Discuss. **141**, 251–276 (2009)
4. V. Molinero, E.B. Moore, Water modeled as an intermediate element between carbon and silicon. J. Phys. Chem. B **113**(13), 4008–4016 (2009)
5. P.T. Kiss, A. Baranyai, Density maximum and polarizable models of water. J. Chem. Phys. **137**(8), 084506–084508 (2012)
6. Y. Huang, X. Zhang, Z. Ma, Y. Zhou, J. Zhou, W. Zheng, C.Q. Sun, Size, separation, structure order, and mass density of molecules packing in water and ice. Sci. Rep. **3**, 3005 (2013)
7. J. Alejandre, G.A. Chapela, H. Saint-Martin, N. Mendoza, A non-polarizable model of water that yields the dielectric constant and the density anomalies of the liquid: TIP4Q. Phys. Chem. Chem. Phys. **13**, 19728–19740 (2011)
8. M. Zhao, W.T. Zheng, J.C. Li, Z. Wen, M.X. Gu, C.Q. Sun, Atomistic origin, temperature dependence, and responsibilities of surface energetics: an extended broken-bond rule. Phys. Rev. B **75**(8), 085427 (2007)
9. C.Q. Sun, Y. Sun, The attribute of water: single notion, multiple myths. Springer Ser. Chem. Phys. **113** (2016)
10. C.Q. Sun, Relaxation of the chemical bond. Springer Ser. Chem. Phys. **108** (2014)
11. C.Q. Sun, Oxidation electronics: bond-band-barrier correlation and its applications. Prog. Mater Sci. **48**(6), 521–685 (2003)

12. W.T. Zheng, C.Q. Sun, Electronic process of nitriding: Mechanism and applications. Prog. Solid State Chem. **34**(1), 1–20 (2006)

13. Z. Zhang, D. Li, W. Jiang, Z. Wang, The electron density delocalization of hydrogen bond systems. Adv. Phys. X **3**(1), 1428915 (2018)

14. Y. Mo, J. Gao, Polarization and charge-transfer effects in aqueous solution via Ab initio QM/MM simulations. J. Phys. Chem. B **110**(7), 2976–2980 (2006)

15. C.Q. Sun, Dominance of broken bonds and nonbonding electrons at the nanoscale. Nanoscale **2**(10), 1930–1961 (2010)

16. Y.L. Huang, X. Zhang, Z.S. Ma, Y.C. Zhou, W.T. Zheng, J. Zhou, C.Q. Sun, Hydrogen-bond relaxation dynamics: resolving mysteries of water ice. Coord. Chem. Rev. **285**, 109–165 (2015)

17. J.R. Lane, CCSDTQ optimized geometry of water dimer. J. Chem. Theory Comput. **9**(1), 316–323 (2013)

18. X. Zhang, Y. Zhou, Y. Gong, Y. Huang, C. Sun, Resolving H(Cl, Br, I) capabilities of transforming solution hydrogen-bond and surface-stress. Chem. Phys. Lett. **678**, 233–240 (2017)

19. Y. Zhou, D. Wu, Y. Gong, Z. Ma, Y. Huang, X. Zhang, C.Q. Sun, Base-hydration-resolved hydrogen-bond networking dynamics: quantum point compression. J. Mol. Liq. **223**, 1277–1283 (2016)

20. R.A. Street, *Hydrogenated Amorphous Silicon* (Cambridge University Press, 1991)

21. Y. Huang, X. Zhang, Z. Ma, Y. Zhou, G. Zhou, C.Q. Sun, Hydrogen-bond asymmetric local potentials in compressed ice. J. Phys. Chem. B **117**(43), 13639–13645 (2013)

22. Y.L. Huang, X. Zhang, Z.S. Ma, G.H. Zhou, Y.Y. Gong, C.Q. Sun, Potential paths for the hydrogen-bond relaxing with $(H_2O)(N)$ cluster size. J. Phys. Chem. C **119**(29), 16962–16971 (2015)

23. X. Zhang, P. Sun, Y. Huang, T. Yan, Z. Ma, X. Liu, B. Zou, J. Zhou, W. Zheng, C.Q. Sun, Water's phase diagram: from the notion of thermodynamics to hydrogen-bond cooperativity. Prog. Solid State Chem. **43**, 71–81 (2015)

24. C.Q. Sun, J. Chen, Y. Gong, X. Zhang, Y. Huang, (H, Li)Br and LiOH solvation bonding dynamics: molecular nonbond interactions and solute extraordinary capabilities. J. Phys. Chem. B **122**(3), 1228–1238 (2018)

25. Y. Wang, H. Liu, J. Lv, L. Zhu, H. Wang, Y. Ma, High pressure partially ionic phase of water ice. Nat. Commun. **2**, 563 (2011)

26. C.Q. Sun, X. Zhang, W.T. Zheng, Hidden force opposing ice compression. Chem. Sci. **3**, 1455–1460 (2012)

27. N. Bjerrum, Structure and properties of ice. Nat. **115**(2989), 385–390 (1952)

28. M.d. Koning, A.l. Antonelli, A.J.R.d. Silva, and A. Fazzio, Orientational defects in ice Ih: an interpretation of electrical conductivity measurements. Phys. Rev. Lett. **96**(075501) (2006)

29. M. Millot, F. Coppari, J.R. Rygg, A. Correa Barrios, S. Hamel, D.C. Swift, and J.H. Eggert, Nanosecond X-ray diffraction of shock-compressed superionic water ice. Nat. **569**(7755), 251–255 (2019)

30. C.Q. Sun, X. Zhang, J. Zhou, Y. Huang, Y. Zhou, W. Zheng, Density, elasticity, and stability anomalies of water molecules with fewer than four neighbors. J. Phys. Chem. Lett. **4**, 2565–2570 (2013)

31. K.R. Wilson, R.D. Schaller, D.T. Co, R.J. Saykally, B.S. Rude, T. Catalano, J.D. Bozek, Surface relaxation in liquid water and methanol studied by x-ray absorption spectroscopy. J. Chem. Phys. **117**(16), 7738–7744 (2002)

32. F. Mallamace, C. Branca, M. Broccio, C. Corsaro, C.Y. Mou, S.H. Chen, The anomalous behavior of the density of water in the range 30 K < T < 373 K. Proc. Natl. Acad. Sci. U.S.A. **104**(47), 18387–18391 (2007)

33. C.Q. Sun, X. Zhang, X. Fu, W. Zheng, J.-L. Kuo, Y. Zhou, Z. Shen, J. Zhou, Density and phonon-stiffness anomalies of water and ice in the full temperature range. J. Phys. Chem. Lett. **4**, 3238–3244 (2013)

34. J. Harms, J.P. Toennies, F. Dalfovo, Density of superfluid helium droplets. Phys. Rev. B **58**(6), 3341 (1998)

35. J. Day, J. Beamish, Low-temperature shear modulus changes in solid He-4 and connection to supersolidity. Nature **450**(7171), 853–856 (2007)
36. C.Q. Sun, Size dependence of nanostructures: Impact of bond order deficiency. Prog. Solid State Chem. **35**(1), 1–159 (2007)
37. X. Zhang, Y. Huang, Z. Ma, Y. Zhou, W. Zheng, J. Zhou, C.Q. Sun, A common supersolid skin covering both water and ice. Phys. Chem. Chem. Phys. **16**(42), 22987–22994 (2014)
38. C. Medcraft, D. McNaughton, C.D. Thompson, D.R.T. Appadoo, S. Bauerecker, E.G. Robertson, Water ice nanoparticles: size and temperature effects on the mid-infrared spectrum. Phys. Chem. Chem. Phys. **15**(10), 3630–3639 (2013)
39. C. Medcraft, D. McNaughton, C.D. Thompson, D. Appadoo, S. Bauerecker, E.G. Robertson, Size and temperature dependence in the far-Ir spectra of water ice particles. Astrophys. J. **758**(1), 17 (2012)
40. X. Zhang, Y. Huang, P. Sun, X. Liu, Z. Ma, Y. Zhou, J. Zhou, W. Zheng, C.Q. Sun, Ice regelation: hydrogen-bond extraordinary recoverability and water quasisolid-phase-boundary dispersivity. Sci. Rep. **5**, 13655 (2015)
41. X. Zhang, Y. Huang, Z. Ma, Y. Zhou, J. Zhou, W. Zheng, Q. Jiang, C.Q. Sun, Hydrogen-bond memory and water-skin supersolidity resolving the Mpemba paradox. Phys. Chem. Chem. Phys. **16**(42), 22995–23002 (2014)
42. H. Sun, COMPASS: an ab initio force-field optimized for condensed-phase applications overview with details on alkane and benzene compounds. J. Phys. Chem. B **102**(38), 7338–7364 (1998)
43. K. Sotthewes, P. Bampoulis, H.J. Zandvliet, D. Lohse, B. Poelsema, Pressure induced melting of confined ice. ACS Nano **11**(12), 12723–12731 (2017)
44. H. Qiu, W. Guo, Electromelting of confined monolayer ice. Phys. Rev. Lett. **110**(19), 195701 (2013)
45. R. Moro, R. Rabinovitch, C. Xia, V.V. Kresin, Electric dipole moments of water clusters from a beam deflection measurement. Phys. Rev. Lett. **97**(12), 123401 (2006)
46. F.G. Alabarse, J. Haines, O. Cambon, C. Levelut, D. Bourgogne, A. Haidoux, D. Granier, B. Coasne, Freezing of water confined at the nanoscale. Phys. Rev. Lett. **109**(3), 035701 (2012)
47. B. Wang, W. Jiang, Y. Gao, B.K. Teo, Z. Wang, Chirality recognition in concerted proton transfer process for prismatic water clusters. Nano Res. **9**(9), 2782–2795 (2016)
48. H. Bhatt, A.K. Mishra, C. Murli, A.K. Verma, N. Garg, M.N. Deo, S.M. Sharma, Proton transfer aiding phase transitions in oxalic acid dihydrate under pressure. Phys. Chem. Chem. Phys. **18**(11), 8065–8074 (2016)
49. H. Bhatt, C. Murli, A.K. Mishra, A.K. Verma, N. Garg, M.N. Deo, R. Chitra, S.M. Sharma, Hydrogen bond symmetrization in glycinium oxalate under pressure. J. Phys. Chem. B **120**(4), 851–859 (2016)
50. Q. Zeng, T. Yan, K. Wang, Y. Gong, Y. Zhou, Y. Huang, C.Q. Sun, B. Zou, Compression icing of room-temperature NaX solutions (X = F, Cl, Br, I). Phys. Chem. Chem. Phys. **18**(20), 14046–14054 (2016)
51. Q. Zeng, C. Yao, K. Wang, C.Q. Sun, B. Zou, Room-temperature NaI/H_2O compression icing: solute–solute interactions. PCCP **19**, 26645–26650 (2017)
52. D. Kang, J. Dai, H. Sun, Y. Hou, J. Yuan, *Quantum simulation of thermally-driven phase transition and oxygen K-edge x-ray absorption of high-pressure ice* **3**, (2013). http://www.naturecom/srep/2013/131021/srep03005/metrics
53. K. Dong, S. Zhang, Q. Wang, A new class of ion-ion interaction: Z-bond. Sci. China Chem. **58**(3), 495–500 (2015)
54. F. Li, Z. Men, S. Li, S. Wang, Z. Li, C. Sun, Study of hydrogen bonding in ethanol-water binary solutions by Raman spectroscopy. Spectrochim. Acta Part A Mol. Biomol. Spectrosc. **189**, 621–624 (2018)
55. F.B. Li, Z.L. Li, S.H. Wang, S. Li, Z.W. Men, S.L. Ouyang, C.L. Sun, Structure of water molecules from Raman measurements of cooling different concentrations of NaOH solutions. Spectrochimica Acta Part a-Mol. Biomol. Spectrosc. **183**, 425–430 (2017)

56. J.P. Perdew, Y. Wang, Accurate and simple analytic representation of the electron-gas correlation-energy. Phys. Rev. B **45**(23), 13244–13249 (1992)
57. F. Ortmann, F. Bechstedt, W.G. Schmidt, Semiempirical van der Waals correction to the density functional description of solids and molecular structures. Phys. Rev. B **73**(20), 205101 (2006)
58. Y. Yoshimura, S.T. Stewart, M. Somayazulu, H. Mao, R.J. Hemley, High-pressure X-ray diffraction and Raman spectroscopy of ice VIII. J. Chem. Phys. **124**(2), 024502 (2006)
59. M. Erko, D. Wallacher, A. Hoell, T. Hauss, I. Zizak, O. Paris, Density minimum of confined water at low temperatures: a combined study by small-angle scattering of X-rays and neutrons. PCCP **14**(11), 3852–3858 (2012)
60. K. Liu, J.D. Cruzan, R.J. Saykally, Water clusters. Science **271**(5251), 929–933 (1996)
61. U. Bergmann, A. Di Cicco, P. Wernet, E. Principi, P. Glatzel, A. Nilsson, Nearest-neighbor oxygen distances in liquid water and ice observed by X-ray Raman based extended X-ray absorption fine structure. J. Chem. Phys. **127**(17), 174504 (2007)
62. K.R. Wilson, B.S. Rude, T. Catalano, R.D. Schaller, J.G. Tobin, D.T. Co, R.J. Saykally, X-ray spectroscopy of liquid water microjets. J. Phys. Chem. B **105**(17), 3346–3349 (2001)
63. A. Narten, W. Thiessen, L. Blum, Atom pair distribution functions of liquid water at 25 °C from neutron diffraction. Science **217**(4564), 1033–1034 (1982)
64. L. Fu, A. Bienenstock, S. Brennan, X-ray study of the structure of liquid water. J. Chem. Phys. **131**(23), 234702 (2009)
65. J.-L. Kuo, M.L. Klein, W.F. Kuhs, The effect of proton disorder on the structure of ice-Ih: a theoretical study. J. Chem. Phys. **123**(13), 134505 (2005)
66. A. Soper, Joint structure refinement of X-ray and neutron diffraction data on disordered materials: application to liquid water. J. Phys.: Condens. Matter **19**(33), 335206 (2007)
67. L.B. Skinner, C. Huang, D. Schlesinger, L.G. Pettersson, A. Nilsson, C.J. Benmore, Benchmark oxygen-oxygen pair-distribution function of ambient water from X-ray diffraction measurements with a wide Q-range. J. Chem. Phys. **138**(7), 074506 (2013)
68. K.T. Wikfeldt, M. Leetmaa, A. Mace, A. Nilsson, L.G.M. Pettersson, Oxygen-oxygen correlations in liquid water: addressing the discrepancy between diffraction and extended x-ray absorption fine-structure using a novel multiple-data set fitting technique. J. Chem. Phys. **132**(10), 104513 (2010)
69. X. Zhang, S. Chen, J. Li, Hydrogen-bond potential for ice VIII-X phase transition. Sci. Rep. **6**, 37161 (2016)
70. T.F. Kahan, J.P. Reid, D.J. Donaldson, Spectroscopic probes of the quasi-liquid layer on ice. J. Phys. Chem. A **111**(43), 11006–11012 (2007)
71. M.X. Gu, C.Q. Sun, Z. Chen, T.C.A. Yeung, S. Li, C.M. Tan, V. Nosik, Size, temperature, and bond nature dependence of elasticity and its derivatives on extensibility, Debye temperature, and heat capacity of nanostructures. Phys. Rev. B **75**(12), 125403 (2007)
72. F. Perakis, K. Amann-Winkel, F. Lehmkühler, M. Sprung, D. Mariedahl, J.A. Sellberg, H. Pathak, A. Späh, F. Cavalca, D. Schlesinger, A. Ricci, A. Jain, B. Massani, F. Aubree, C.J. Benmore, T. Loerting, G. Grübel, L.G.M. Pettersson, A. Nilsson, Diffusive dynamics during the high-to-low density transition in amorphous ice. Proc. Natl. Acad. Sci. **114**(31), 8193–8198 (2017)
73. J.A. Sellberg, C. Huang, T.A. McQueen, N.D. Loh, H. Laksmono, D. Schlesinger, R.G. Sierra, D. Nordlund, C.Y. Hampton, D. Starodub, D.P. DePonte, M. Beye, C. Chen, A.V. Martin, A. Barty, K.T. Wikfeldt, T.M. Weiss, C. Caronna, J. Feldkamp, L.B. Skinner, M.M. Seibert, M. Messerschmidt, G.J. Williams, S. Boutet, L.G. Pettersson, M.J. Bogan, A. Nilsson, Ultrafast X-ray probing of water structure below the homogeneous ice nucleation temperature. Nature **510**(7505), 381–384 (2014)
74. Y.R. Shen, V. Ostroverkhov, Sum-frequency vibrational spectroscopy on water interfaces: polar orientation of water molecules at interfaces. Chem. Rev. **106**(4), 1140–1154 (2006)
75. F. Mallamace, M. Broccio, C. Corsaro, A. Faraone, D. Majolino, V. Venuti, L. Liu, C.Y. Mou, S.H. Chen, Evidence of the existence of the low-density liquid phase in supercooled, confined water. Proc. Natl. Acad. Sci. U.S.A. **104**(2), 424–428 (2007)

76. B. Wang, W. Jiang, Y. Gao, Z. Zhang, C. Sun, F. Liu, Z. Wang, Energetics competition in centrally four-coordinated water clusters and Raman spectroscopic signature for hydrogen bonding. RSC Adv. **7**(19), 11680–11683 (2017)

77. Y. Shi, Z. Zhang, W. Jiang, Z. Wang, Theoretical study on electronic and vibrational properties of hydrogen bonds in glycine-water clusters. Chem. Phys. Lett. **684**, 53–59 (2017)

78. Y. Otsuki, T. Sugimoto, T. Ishiyama, A. Morita, K. Watanabe, Y. Matsumoto, Unveiling subsurface hydrogen-bond structure of hexagonal water ice. Phys. Rev. B **96**(11), 115405 (2017)

79. Y. Liu, J. Wu, Communication: long-range angular correlations in liquid water. J. Chem. Phys. **139**(4), 041103 (2013)

80. X.Z. Li, B. Walker, A. Michaelides, Quantum nature of the hydrogen bond. Proc. Natl. Acad. Sci. U.S.A. **108**(16), 6369–6373 (2011)

81. R.F. McGuire, F.A. Momany, H.A. Scheraga, Energy parameters in polypeptides. V. An empirical hydrogen bond potential function based on molecular orbital calculations. J. Phys. Chem. **76**, 375–393 (1972)

82. N. Kumagai, K. Kawamura, T. Yokokawa, An interatomic potential model for H_2O: applications to water and ice polymorphs. Mol. Simul. **12**, 177–186 (1994)

83. J. Sun, G. Niehues, H. Forbert, D. Decka, G. Schwaab, D. Marx, M. Havenith, Understanding THz spectra of aqueous solutions: glycine in light and heavy water. J. Am. Chem. Soc. **136**(13), 5031–5038 (2014)

84. K. Tielrooij, S. Van Der Post, J. Hunger, M. Bonn, H. Bakker, Anisotropic water reorientation around ions. J. Phys. Chem. B **115**(43), 12638–12647 (2011)

85. Y. Zhou, Yuan Zhong, X. Liu, Y. Huang, X. Zhang, C.Q. Sun, NaX solvation bonding dynamics: hydrogen bond and surface stress transition (X = HSO4, NO3, ClO4, SCN). J. Mol. Liq. **248**(432–438) (2017)

86. Y. Zhou, Y. Gong, Y. Huang, Z. Ma, X. Zhang, C.Q. Sun, Fraction and stiffness transition from the H–O vibrational mode of ordinary water to the HI, NaI, and NaOH hydration states. J. Mol. Liq. **244**, 415–421 (2017)

87. X. Zhang, Y. Xu, Y. Zhou, Y. Gong, Y. Huang, C.Q. Sun, HCl, KCl and KOH solvation resolved solute-solvent interactions and solution surface stress. Appl. Surf. Sci. **422**: 475–481 (2017)

88. G.C. Pimentel, A.L. McClellan, *The Hydrogen Bond* (ed. W.H. Freeman. San Francisco, CA, 1960), p. 475

89. L. Pauling, The Nature of the Chemical Bond. 3 edn. (Cornell University Press, Ithaca, NY 1960)

90. P. Banerjee, T. Chakraborty, Weak hydrogen bonds: insights from vibrational spectroscopic studies. Int. Rev. Phys. Chem. **37**(1), 83–123 (2018)

91. G.R. Desiraju, T. Steiner, *The weak hydrogen bond: in structural chemistry and biology*. Vol. 9. 2001: Oxford university press

92. P.A. Kollman, L.C. Allen, Theory of the hydrogen bond. Chem. Rev. **72**(3), 283–303 (1972)

93. E. Arunan, G.R. Desiraju, R.A. Klein, J. Sadlej, S. Scheiner, I. Alkorta, D.C. Clary, R.H. Crabtree, J.J. Dannenberg, P. Hobza, H.G. Kjaergaard, A.C. Legon, B. Mennucci, D.J. Nesbitt, Definition of the hydrogen bond (IUPAC recommendations 2011). Pure Appl. Chem. **83**(8), 1637–1641 (2011)

94. M.F. Perutz, The role of aromatic rings as hydrogen-bond acceptors in molecular recognition. Phil. Trans. R. Soc. Lond. A **345**(1674), 105–112 (1993)

95. G. Gilli, P. Gilli, *The Nature of the Hydrogen Bond: Outline of a Comprehensive Hydrogen Bond Theory*, vol. 23 (Oxford University Press, 2009)

96. E.A. Meyer, R.K. Castellano, F. Diederich, Interactions with aromatic rings in chemical and biological recognition. Angew. Chem. Int. Ed. **42**(11), 1210–1250 (2003)

97. J.L. Atwood, F. Hamada, K.D. Robinson, G.W. Orr, R.L. Vincent, X-ray diffraction evidence for aromatic π hydrogen bonding to water. Nature **349**(6311), 683 (1991)

98. W. Saenger, G. Jeffrey, *Hydrogen Bonding in Biological Structures* (Springer, Berlin, 1991)

99. O. Takahashi, Y. Kohno, M. Nishio, Relevance of weak hydrogen bonds in the conformation of organic compounds and bioconjugates: evidence from recent experimental data and high-level ab initio MO calculations. Chem. Rev. **110**(10), 6049–6076 (2010)
100. C.Q. Sun, Aqueous charge injection: solvation bonding dynamics, molecular nonbond interactions, and extraordinary solute capabilities. Int. Rev. Phys. Chem. **37**(3–4), 363–558 (2018)
101. D. Marx, M.E. Tuckerman, J. Hutter, M. Parrinello, The nature of the hydrated excess proton in water. Nature **397**(6720), 601–604 (1999)
102. Y. Gong, Y. Zhou, H. Wu, D. Wu, Y. Huang, C.Q. Sun, Raman spectroscopy of alkali halide hydration: hydrogen bond relaxation and polarization. J. Raman Spectrosc. **47**(11), 1351–1359 (2016)
103. P. Cotterill, The hydrogen embrittlement of metals. Prog. Mater Sci. **9**(4), 205–301 (1961)
104. D. Hollas, O. Svoboda, P. Slavíček, Fragmentation of HCl–water clusters upon ionization: non-adiabatic ab initio dynamics study. Chem. Phys. Lett. **622**, 80–85 (2015)
105. J. Chen, C. Yao, X. Liu, X. Zhang, C.Q. Sun, Y. Huang, H$_2$O$_2$ and HO- solvation dynamics: solute capabilities and solute-solvent molecular interactions. Chem. Sel. **2**(27), 8517–8523 (2017)
106. Y.Q. Fu, B. Yan, N.L. Loh, C.Q. Sun, P. Hing, Hydrogen embrittlement of titanium during microwave plasma assisted CVD diamond deposition. Surf. Eng. **16**(4), 349–354 (2000)
107. S. Huang, D. Chen, J. Song, D.L. McDowell, T. Zhu, Hydrogen embrittlement of grain boundaries in nickel: an atomistic study. NPJ Comput. Mater. **3**, 1 (2017)
108. L.J. Bartolotti, D. Rai, A.D. Kulkarni, S.P. Gejji, R.K. Pathak, Water clusters (H$_2$O)n [n = 9 − 20] in external electric fields: exotic OH stretching frequencies near breakdown. Comput. Theor. Chem. **1044**, 66–73 (2014)
109. V.M. Goldschmidt, Crystal structure and chemical correlation. Ber. Dtsch. Chem. Ges. **60**, 1263–1296 (1927)
110. M. Druchok, M. Holovko, Structural changes in water exposed to electric fields: a molecular dynamics study. J. Mol. Liq. **212**, 969–975 (2015)
111. Y. Zhou, Y. Huang, Z. Ma, Y. Gong, X. Zhang, Y. Sun, C.Q. Sun, Water molecular structure-order in the NaX hydration shells (X = F, Cl, Br, I). J. Mol. Liq. **221**, 788–797 (2016)
112. X. Zhang, T. Yan, Y. Huang, Z. Ma, X. Liu, B. Zou, C.Q. Sun, Mediating relaxation and polarization of hydrogen-bonds in water by NaCl salting and heating. Phys. Chem. Chem. Phys. **16**(45), 24666–24671 (2014)
113. A.W. Omta, M.F. Kropman, S. Woutersen, H.J. Bakker, Negligible effect of ions on the hydrogen-bond structure in liquid water. Science **301**(5631), 347–349 (2003)
114. K. Tielrooij, N. Garcia-Araez, M. Bonn, H. Bakker, Cooperativity in ion hydration. Science **328**(5981), 1006–1009 (2010)
115. M.R. Rahimpour, M.R. Dehnavi, F. Allahgholipour, D. Iranshahi, S.M. Jokar, Assessment and comparison of different catalytic coupling exothermic and endothermic reactions: a review. Appl. Energy **99**, 496–512 (2012)
116. R.C. Ramaswamy, P.A. Ramachandran, M.P. Duduković, Coupling exothermic and endothermic reactions in adiabatic reactors. Chem. Eng. Sci. **63**(6), 1654–1667 (2008)
117. N. Shahrin, Solubility and dissolution of drug product: a review. Int. J. Pharm. Life Sci. **2**(1), 33–41 (2013)
118. K. Haldrup, W. Gawelda, R. Abela, R. Alonso-Mori, U. Bergmann, A. Bordage, M. Cammarata, S.E. Canton, A.O. Dohn, T.B. Van Driel, Observing solvation dynamics with simultaneous femtosecond X-ray emission spectroscopy and X-ray scattering. J. Phys. Chem. B **120**(6), 1158–1168 (2016)
119. J.B. Rosenholm, Critical evaluation of dipolar, acid-base and charge interactions I. Electron displacement within and between molecules, liquids and semiconductors. Adv. Colloid Interface Sci. **247**, 264–304
120. J. Konicek, I. Wadsö, Thermochemical properties of some carboxylic acids, amines and N-substituted amides in aqueous solution. Acta Chem. Scand **25**(5), 1461–1551 (1971)
121. E.L. Ratkova, D.S. Palmer, M.V. Fedorov, Solvation thermodynamics of organic molecules by the molecular integral equation theory: approaching chemical accuracy. Chem. Rev. **115**(13), 6312–6356 (2015)

122. G. G., *Hydration thermodynamics of aliphatic alcohols*. PCCP, **1**(15): 3567–3576, (1999)
123. A.M. Ricks, A.D. Brathwaite, M.A. Duncan, IR spectroscopy of gas phase $V(CO_2)n+$ clusters: solvation-induced electron transfer and activation of CO_2. J. Phys. Chem. A **117**(45), 11490–11498 (2013)
124. M. Wohlgemuth, M. Miyazaki, M. Weiler, M. Sakai, O. Dopfer, M. Fujii, R. Mitrić, Solvation dynamics of a single water molecule probed by infrared spectra–theory meets experiment. Angew. Chem. Int. Ed. **53**(52), 14601–14604 (2014)
125. C. Velezvega, D.J. Mckay, T. Kurtzman, V. Aravamuthan, R.A. Pearlstein, J.S. Duca, Estimation of solvation entropy and enthalpy via analysis of water oxygen-hydrogen correlations. J. Chem. Theory Comput. **11**(11), 5090 (2015)
126. A. Zaichikov, M.A. Krest'yaninov, Structural and thermodynamic properties and intermolecular interactions in aqueous and acetonitrile solutions of aprotic amides. J. Struct. Chem. **54**(2), 336–344 (2013)
127. A. Magno, P. Gallo, Understanding the mechanisms of bioprotection: a comparative study of aqueous solutions of trehalose and maltose upon supercooling. J. Phys. Chem. Lett. **2**(9), 977–982 (2011)
128. K. Liu, C. Wang, J. Ma, G. Shi, X. Yao, H. Fang, Y. Song, J. Wang, Janus effect of antifreeze proteins on ice nucleation. Proc. Natl. Acad. Sci. U.S.A. **113**(51), 14739–14744 (2016)
129. X. Zhang, P. Sun, Y. Huang, Z. Ma, X. Liu, J. Zhou, W. Zheng, C.Q. Sun, Water nanodroplet thermodynamics: quasi-solid phase-boundary dispersivity. J. Phys. Chem. B **119**(16), 5265–5269 (2015)
130. F. Jensen, F. Besenbacher, E. Laegsgaard, I. Stensgaard, Dynamics of oxygen-induced reconstruction of Cu(100) studied by scanning tunneling microscopy. Phys. Rev. B **42**(14), 9206–9209 (1990)
131. C.Q. Sun, O–Cu(001): II. VLEED quantification of the four-stage Cu_3O_2 bonding kinetics. Surf. Rev. Lett. **8**(6), 703–734 (2001)
132. J.R. Mercer, P. Finetti, F.M. Leibsle, R. McGrath, V.R. Dhanak, A. Baraldi, K.C. Prince, R. Rosei, STM and SPA-LEED studies of O-induced structures on Rh(100) surfaces. Surf. Sci. **352**, 173–178 (1996)
133. C.Q. Sun, Electronic process of Cu(Ag, V, Rh)(001) surface oxidation: atomic valence evolution and bonding kinetics. Appl. Surf. Sci. **246**(1–3), 6–13 (2005)
134. Q.S. Chang, H. Xie, W. Zhang, H. Ye, P. Hing, Preferential oxidation of diamond {111}. J. Phys. D Appl. Phys. **33**(17), 2196 (2000)
135. H. Wolter, K. Meinel, C. Ammer, K. Wandelt, H. Neddermeyer, O-mediated layer growth of Cu on Ru (0001). J. Phys.: Condens. Matter **11**(1), 19 (1999)
136. K. Meinel, C. Ammer, M. Mitte, H. Wolter, H. Neddermeyer, Effects and structures of the O/Cu surfactant layer in O-mediated film growth of Cu on Ru (0 0 0 1). Prog. Surf. Sci. **67**(1–8), 183–203 (2001)
137. M. Schmidt, H. Wolter, M. Schick, K. Kalki, K. Wandelt, Compression phases in copper/oxygen coadsorption layers on a Ru (0001) surface. Surf. Sci. **287**, 983–987 (1993)
138. M. Schmidt, H. Wolter, K. Wandelt, Work-function oscillations during the surfactant induced layer-by-layer growth of copper on oxygen precovered Ru (0001). Surf. Sci. **307**, 507–513 (1994)
139. M. Karolewski, Determination of growth modes of Cu on O/Ni (1 0 0) and NiO (1 0 0) surfaces by SIMS and secondary electron emission measurements. Surf. Sci. **517**(1–3), 138–150 (2002)
140. W. Wulfhekel, N.N. Lipkin, J. Kliewer, G. Rosenfeld, L.C. Jorritsma, B. Poelsema, G. Comsa, Conventional and manipulated growth of Cu/Cu (111). Surf. Sci. **348**(3), 227–242 (1996)
141. M. Yata, H. Rouch, K. Nakamura, Kinetics of oxygen surfactant in Cu (001) homoepitaxial growth. Phys. Rev. B **56**(16), 10579 (1997)
142. J. Whitten, R. Gomer, Reactivity of Ni on oxygen-covered W (110) surfaces. J. Vac. Sci. Technol. A: Vac., Surf., Films **13**(5), 2540–2546 (1995)
143. C. Sun, Time-resolved VLEED from the O-Cu (001): atomic processes of oxidation. Vacuum **48**(6), 525–530 (1997)

144. J.W. Frenken, J. Van der Veen, G. Allan, Relation between surface relaxation and surface force constants in clean and oxygen-covered Ni (001). Phys. Rev. Lett. **51**(20), 1876 (1983)
145. J. Peng, D. Cao, Z. He, J. Guo, P. Hapala, R. Ma, B. Cheng, J. Chen, W.J. Xie, X.-Z. Li, P. Jelínek, L.-M. Xu, Y.Q. Gao, E.-G. Wang, Y. Jiang, The effect of hydration number on the interfacial transport of sodium ions. Nature **557**, 701–705 (2018)
146. M.A. Omar, *Elementary Solid State Physics: Principles and Applications* (Addison-Wesley, New York, 1993)
147. A. Bragg, J. Verlet, A. Kammrath, O. Cheshnovsky, D. Neumark, Hydrated electron dynamics: from clusters to bulk. Science **306**(5696), 669–671 (2004)
148. A.E. Bragg, J.R.R. Verlet, A. Kammrath, O. Cheshnovsky, D.M. Neumark, Electronic relaxation dynamics of water cluster anions. J. Am. Chem. Soc. **127**(43), 15283–15295 (2005)
149. A. Kammrath, J.R. Verlet, A.E. Bragg, G.B. Griffin, D.M. Neumark, Dynamics of charge-transfer-to-solvent precursor states in I-(water) n (n = 3–10) clusters studied with photoelectron imaging. The J. Phys. Chem. A **109**(50), 11475–11483 (2005)
150. A. Kammrath, G. Griffin, D. Neumark, J.R.R. Verlet, Photoelectron spectroscopy of large (water)[sub n][sup −] (n = 50–200) clusters at 4.7 eV. The J. Chem. Phys. **125**(7), 076101 (2006)
151. D. Sagar, C.D. Bain, J.R. Verlet, Hydrated electrons at the water/air interface. J. Am. Chem. Soc. **132**(20), 6917–6919 (2010)
152. J. Verlet, A. Bragg, A. Kammrath, O. Cheshnovsky, D. Neumark, Observation of large water-cluster anions with surface-bound excess electrons. Science **307**(5706), 93–96 (2005)
153. X.J. Liu, M.L. Bo, X. Zhang, L. Li, Y.G. Nie, H. TIan, Y. Sun, S. Xu, Y. Wang, W. Zheng, C.Q. Sun, Coordination-resolved electron spectrometrics. Chem. rev. **115**(14), 6746–6810 (2015)
154. C.Q. Sun, Atomic scale purification of electron spectroscopic information (US 2017 patent No. 9,625,397B2). 2017: United States
155. Y. Gong, Y. Zhou, C. Sun, *Phonon Spectrometrics of the Hydrogen Bond (O:H–O) Segmental Length and Energy Relaxation Under Excitation* (B.o. Intelligence, Editor., China, 2018)
156. A. Mailleur, C. Pirat, O. Pierre-Louis, J. Colombani, Hollow Rims from water drop evaporation on salt substrates. Phys. Rev. Lett. **121**, 124501 (2018)
157. R. Deegan, O. Bakajin, T. Dupont, G. Huber, T.W.S. Nagel, Capillary flow as the cause of ring stains from dried liquid drops. Nat. (Lond.) **389**, 827 (1997)
158. P. Sáenz, A. Wray, Z. Che, O. Matar, P. Valluri, J. Kim, K. Sefiane, Dynamics and universal scaling law in geometrically-controlled sessile drop evaporation. Nat. Commun. **8**, 14783 (2017)
159. N. Shahidzadeh-Bonn, S. Rafaï, D. Bonn, G. Wegdam, Salt crystallization during evaporation: impact of interfacial properties. Langmuir **24**, 8599 (2008)
160. N. Shahidzadeh, M. Schut, J. Desarnaud, M. Prat, D. Bonn, Salt stains from evaporating droplets. Sci. Rep. **5**, 10335 (2015)
161. A. Tay, D. Bendejacq, C. Monteux, F. Lequeux, How does water wet a hydrosoluble substrate? Soft Matter **7**, 6953 (2011)
162. J. Dupas, E. Verneuil, M. Ramaioli, L. Forny, L. Talini, F. Lequeux, Dynamic wetting on a thin film of soluble polymer: effects of nonlinearities in the sorption isotherm. Langmuir **29**, 12572 (2013)

Chapter 4
Lewis Acidic Solutions: H↔H Fragilization

Contents

Abstract Solvation dissolves the HX into an H^+ and an X^-. The H^+ bonds to a H_2O to form a firm H_3O^+ and a H↔H anti − HB point breaker. The H–O bond due H_3O^+ is 3% shorter and the associated O:H nonbond is 60% longer than normal. The H↔H compression shortens its nearest O:H nonbond by 11% and lengthens the H–O by 4%. The X^- point polarizer shortens the H–O bond and stiffens its phonon but relax the O:H nonbond oppositely in the supersolid hydration shell. The X^- solute capability of bond transition follows the $I > Br > Cl$ order in the form of $f_x(C) \propto 1 - \exp(-C/C_0)$ towards saturation because of the involvement of the $X^- \leftrightarrow X^-$ interaction that weakens the hydration-shell electric field at higher concentrations. However, the H^+ neither hops or tunnels freely nor polarize its neighbors, $f_H(C) = 0$. The H↔H has the same effect of heating on the surface stress and solution viscosity disruption.

Highlight

- $H^+ \leftrightarrow H^+$ anti-HB disrupts the solution network and surface stress without polarization.
- Anionic polarization shortens and stiffens the H–O bond but relaxes the O:H nonbond oppositely.
- DPS and DFT resolve H^+ and X^- solute capability of hydrating ordinary O:H–O bonds.
- Nonlinear fraction of bond-transition features the $X^- \leftrightarrow X^-$ repulsivity and the solution structure order.

© Springer Nature Singapore Pte Ltd. 2019   85
C. Q. Sun, *Solvation Dynamics*, Springer Series in Chemical
Physics 121, https://doi.org/10.1007/978-981-13-8441-7_4

4.1 Wonders of Excessive Protons

The H(F, Cl, Br, I) protonated aqueous solutions are ubiquitously important to fields varying from agriculture, biochemistry, to medical and pharmaceutical sciences. It remains however uncertain how the H^+ protons and the X^- anions interact with the solvent H_2O molecules despite intensive investigations made since the pioneer work done in 1900s by Arrhenius [1], Brønsted–Lowry [2, 3], and Lewis [4]. There are several acid-base theories describing the characteristics of the acids, bases, and adducts. Lewis [4] developed a theory of acids and bases that is based upon the sharing of electron pairs. A Lewis acid is a substance that can accept a pair of electrons from another atom to form a new bond. A Lewis base is a substance that can donate a pair of electrons to another atom to form a new bond. Typically, the acidic solution is corrosive, dilutive, and surface stress destructive [5–11]. Knowledge about the solute-solvent and solute-solute interactions is necessary to understanding the acidic solvation dynamics and their functionality of molecular crystals like drugs whose molecules are sided with excessive dangling H–O free radicals [12, 13] or lone pairs that functionalize unusually substance such as cells and proteins [14, 15].

The investigation of acid solvation is a long-standing issue with involvement of spectroscopies and theories from various perspectives. SFG probes the sublayer-resolved dipole orientation or the surface dielectrics, at the air-solution interface [16, 17]; time-resolved IR resolves the solute or water molecular diffusive motion dynamics in terms of phonon lifetime [18–21] that is related to the "structural diffusion" or "solute delocalization". Neutron scattering probes the density of states of phonons of the interested substance [22, 23].

Tremendous theoretical efforts have been made in past centuries focusing on the "proton transport dynamics" [20, 24–34] with less attendance of the solute capabilities of transiting the HB network and the solution properties. Figure 4.1 illustrates four outstanding models for explaining the motion dynamics of the excessive H^+/lone-pair in the protonated acidic/basic solutions.

Regarding the motion dynamics of protons and lone pairs, Grotthuss [35, 36] in 1900s firstly proposed that the "proton structural diffusion" or the "concerted random hopping" from one H_2O motif to another. The proton mobility was assumed high compared with the mobility of the H_2O molecules. This model applied to both acidic solvation with involvement of excessive protons and basic solvation with involvement of excessive lone pairs associated to the OH^- hydroxide. Subsequent development refined this concept by invoking proton thermal hopping [37], structural fluctuating [38], and quantum tunneling [39]. Although this widely-accepted scheme captures the concept of long-range charge translocation, numerous basic questions still remain regarding the rapidly evolving structure of an aqueous proton [40].

In the 1960s, Eigen [41] proposed an $H_9O_4^+$ complex model, as shown in Fig. 4.1a, suggesting that an H_3O^+ core is strongly O:H–O bonded to three H_2O molecules and leave the lone pair of the tetrahedrally-structured H_3O^+ free. The lone pairs of the neighboring three H_2O molecules oriented simultaneously in such a way that they are facing to the three H^+ protons to form the O:H–O bonds. According to the

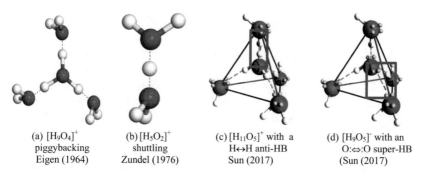

(a) $[H_9O_4]^+$ piggybacking Eigen (1964)	(b) $[H_5O_2]^+$ shuttling Zundel (1976)	(c) $[H_{11}O_5]^+$ with a H↔H anti-HB Sun (2017)	(d) $[H_9O_5]^-$ with an O:⇔:O super-HB (Sun (2017))

Fig. 4.1 Models for (**a, b**) proton and lone-pair motion and (**c, d**) O:H–O bond transition in Lewis solutions. (**a**) The $[H_9O_4]^+$ (or H_3O^+:$3H_2O$) complex of Eigen [41] and (**b**) the $[H_5O_2]^+$ (nonbonded H^+:$2H_2O$ [42] or covalently bonded H^+–$2H_2O$ [43] to the H_2O molecules) configuration of Zundel. The recently developed scheme of $2H_2O$ unit cell central replacement to form (**c**) a $[H_{11}O_5]^+$ (H_3O^+·$4H_2O$) motif for acidic solutions (**d**) a $[H_9O_5]^-$ (OH$^-$·$4H_2O$) motif for basic solutions, which transits an O:H–O bond into the H↔H anti-HB and an O:⇔:O super-HB, respectively [44, 45]

2N and the O:H–O configuration conservation for water ice, it is unlikely that the H_3O^+ only forms O:H–O bonds with its tetrahedral neighbors without considering its neighboring water molecular orientations in the solvent matrix. Water is a highly-ordered, strongly-correlated, and fluctuating crystal [44]. Charge X$^-$ and H$^+$ injection by acid solvation breaks the 2N conservation. It is forbidden for the H_3O^+ hydronium to form four O:H–O bonds with its three H$^+$ and one ":" to pairing up two alike and two unlike of its oriented neighbours, according to the 2N conservation [46].

Zundel [42] favors, however, a $[H_5O_2]^+$ complex in which the proton shuttles randomly between two H_2O molecules, as shown in Fig. 4.1b. A hydrogen atom frustrates in two positions off midway between two oxygen anions, by exchanging the location of the lone pairs [43]. This picture is the same to the transitional quantum tunneling of Bernal–Fowler proposed in 1932 [47] and the frustration mechanism of Pauling proposed in 1933 [48]. The shuttling transportation could be ruled out because of the ~4.0 eV H–O bond energy [46] and the requirement of at least 121.6 nm wavelength laser irradiation for the H–O bond dissociation in the gaseous phase [49, 50].

The last two models consider the broken of the 2N number and O:H–O conservation because of the excessive H$^+$/lone-pair [44, 51]. The central H_2O replacement with an H_3O^+ results in the H$^+$↔H$^+$ anti–HB point breaker for acid solution. The OH$^-$ substitution for the central H_2O turns out the O:⇔:O super–HB compressor in the basic solvation. The H$^+$ or the ":" is unlikely to stand alone or freely shuttling but tends to form firmly a H_3O^+ or a OH$^-$ tetrahedron to accommodate the excessive H$^+$/lone-pair. Eventually, the H_3O^+ or a OH$^-$ replacement results in the $[H_{11}O_5]^+$ motif for acidic solution and the $[H_9O_5]^-$ motif for basic solution because of the number variation of protons and lone pairs within the motifs.

Using the ab initio path integral approximations, Marx et al. [52, 53] suggest that the hydrated proton forms a fluxional defect in the hydration network, with both the $[H_9O_4]^+$ and the $[H_5O_2]^+$ complexes occurring only in the sense of 'limiting' or 'ideal" structures. The defect is delocalized over several hydrogen bonds owing to quantum fluctuations and the protons transfer collectively in ice. The manner of H^+ motion varies with temperature because of the involvement of thermal fluctuations and quantum effect. Moreover, ab initio MD calculations [54–57] suggest that "bond switch" occurs in the first H_3O^+ hydration shell. MD computations [58] suggest that HCl hydration make water clusters into smaller ones by fragmentation. On the other hand, quantum computations [59] suggest that the extra proton weakens the O–H bond strength in the H_3O^+ ion since the bond order of the O–H bond in the H_3O^+ ion is smaller than that in H_2O molecules, which causes a red shift of the O–H stretching mode in the H_3O^+ ion.

An ultrafast 2DIR spectroscopy investigation [20] suggested that the proton prefers Zundel motion manner accommodated by two H_2O at a time, $[H_5O_2]^+$, rather than the Eigen structure, $H_9O_4^+$. However, Wolke et al. [60] studied deuterated prototypical Eigen clusters, $[D_9O_4]^+$, bound to an increasingly basic series of hydrogen bond acceptors. By tracking the frequency of every D–O stretching vibration in the complex as the transferring hydrogen is incrementally pulled from the central H_3O^+ to a neighboring water molecule. Teschke and coworkers [61] measured the mobility of the protons under 4×10^6 V/m electric field to be five times larger than the reported mobility for protons in water, supporting the mechanism of Grotthuss. The hydrated excess proton spectral response through the mid-Raman at 1760 and 3200 cm^{-1} was attributed to the Zundel complex and the region at $\sim$2000 to $\sim$2600 cm^{-1} response is attributed to the Eigen complex, suggesting a core structure simultaneously with a Eigen-like and Zundel-like character, suggesting a rapid fluctuation between these two structures or a new specie.

The key difference among the first three explanations is the motion manners of the excessive H^+/lone-pair. The H^+ either hops freely from one H_2O molecule to another by switching the $H_2O \leftrightarrow H_3O^+$ on and off due to thermal and quantum fluctuation [52, 62] or firmly forms a H_3O^+ hydronium without restriction to its O:H–O bond orientations in the ordered water molecules. However, the presence of electron lone pairs on the O^{2-} anions is critical to molecular interactions. It is necessary to consider the location alteration between the lone pair and the proton in transportation.

The near ambient pressure XPS has detected a low surface coverage of adsorbed HCl at 253 K supersolid skin in both molecular and dissociated states [63], suggesting coexistence of both physisorption and chemisorption at the outermost ice surface and dissociation occurring upon solvation deeper in the interfacial region. Complementary XAS measurements confirm that the presence of Cl$^-$ ions induces significant changes to the HB network in the interfacial region. Being similar to HCl dissociation on warm ice, the concentrated HNO_3 solute shows lower dissociation ability at the surface and the fraction of its molecular forms increases with the concentration, according to the chemical shift in the XPS spectroscopy [64]. XPS measurements show that the dissociation rate at surface is 80% in the bulk solution and furthermore that dissociation occurs in the topmost solution layer. Therefore, acid solvation is subject to the coordination environment.

Nevertheless, little attention has been paid to the full-frequency scan of the Raman spectra or the correlation between the frequency shifts of the H–O and the O:H characteristic peaks –O:H–O bond segmental phonon cooperativity. It is unclear how the H^+ functionalizes the hydration network and properties of the acidic solutions [65–67]. One needs to link the solute motion dynamics to the solution performance upon acid being solvated.

In place of focusing on the proton/lone-pair motion manner and dynamics, Zhang et al. [44] examined the solute capabilities of transforming solution network HBs from the mode of the ordinary water to hydrating using the DPS distillation and DFT calculations. Observations verified that the H^+ or the ":" does not stand alone nor motion freely but firmly forms a H_3O^+ or a OH^- tetrahedron. HX acid solvation creates the halogenic X^- polarizer and the $H\leftrightarrow H$ point breaker to disrupt the solution network and the surface stress. Likewise, the central H_2O replacement of the $2H_2O$ unit cell with an OH^- for the base solvation [51] results in the $O:\Leftrightarrow:O$ super–HB point compressor. The screened X^- in acid and Y^+ in basic solutions serve as point polarizers associated with each a supersolid hydration shell. The combination of the point polarizer, point breaker or point compressor mediate the HB network and the performance of acidic and basic solutions. As it is seen, this premise is much more profoundly promising than focusing on the proton/lone-pair motion dynamics to the understanding of acid and basic solutions.

4.2 Phonon Spectrometric Resolution

4.2.1 Full-Frequency Raman Spectroscopy

Figure 4.2 compares the full-frequency Raman spectra for $HX/H_2O = 0.1$ molar ratio solutions and the concentrated HI/H_2O solutions. These spectra were plotted against the spectrum of deionized water collected under the same ambient conditions. Focusing on the evolution of the characteristic peaks for the O:H vibration centered at <200 cm^{-1} and the H–O stretching vibration at >3000 cm^{-1} would suffice. Figure 4.2a inset recaps the $H\leftrightarrow H$ anti–HB breaker formed between the central H_3O^+ and one of its H_2O neighbors. Inset b illustrates the X^- ionic polarization of the solvent O:H–O bonds. The X^- anion polarizes and stretches the H_2O molecules in the hydration shells, which is the same as the X^- does in salt solutions. The X^- is partially screened by the limited number of the ordered hydrating H_2O dipoles. The H_2O molecules must be subject to interacting with their neighbors. The superposition of the $H\leftrightarrow H$ point fragilization and the X^- point polarization dictate the properties of acidic solutions. However, the full-frequency Raman spectra could not resolve the effect of the $H\leftrightarrow H$ repulsive fragilization. Unlike the YX salt solvation, the H–O phonon abundance transition results from both Y^+ and X^- polarization, only the X^- polarization determines the phonon abundance transition into hydration of the HX solutions [68].

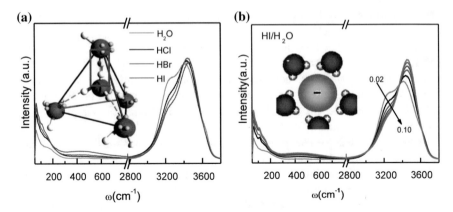

Fig. 4.2 Full-frequency Raman spectroscopy for (**a**) HX/H₂O at 0.1 molar concentration and (**b**) differently-concentrated HI/H₂O solutions. Inset a illustrates the processes of H↔H formation and (**b**) X⁻ ionic point polarization with hydrating molecular shielding. The H↔H does not polarize its neighboring O:H–O bond but disrupts the network locally. The X⁻ polarization shortens the H–O bond and stiffens its phonon but relaxes the O:H nonbond oppositely in the X⁻ hydration shell Reprinted with copyright permission from [44]

4.2.2 DPS Derivatives: Solute Capabilities

Figure 4.3 and Fig. 4.4 compare the similarity of the segmental ω_x DPS for the concentrated HX/H₂O solutions [44] and for the heated water [69]. The ω_L and the ω_H relax indeed cooperatively for all acidic solutions and the heated water. As the concentration increases from 0 to 0.1, X⁻ polarization transits the ω_L from 180 to 75 cm⁻¹ and the ω_H from 3200 to 3480 cm⁻¹ (ref [70]). H↔H repulsion reverts the ω_L from 75 to 110 cm⁻¹. The DPS also resolves the effect of the H↔H repulsion on the H–O bond elongation featured as a small hump below 3050 cm⁻¹. The H↔H repulsion lengthens its neighboring H–O bonds, as being the case of mechanical compression and O:⇔:O point compression in basic solutions [51]. The fraction of H–O bonds elongation by H↔H repulsion decreases when the X⁻ turns from Cl⁻ to Br⁻ and I⁻ as the stronger polarizability of I⁻ annihilates the effect of H↔H repulsion. The small spectral valley at 3650 cm⁻¹ results from the preferential skin occupation of X⁻ that strengthens the local electric field. The stronger electric field stiffens the dangling H–O bond but the X⁻ screening weakens the signal of detection.

Acid solvation shares the same effect of heating on the segmental length and phonon stiffness relaxation and surface stress depression. X⁻ polarization and H–O thermal contraction stiffen the ω_H, X⁻ polarization and O:H thermal expansion softens the ω_H. H↔H fragilization and thermal fluctuation lowers the surface stress.

The DPS also resolves the effect of the H↔H repulsion on the H–O bond elongation featured as a small hump below 3050 cm⁻¹. The H↔H repulsion lengthens its neighboring H–O bonds, as being the case of mechanical compression and O:⇔:O point compression in basic solution [51]. The extent of H↔H repulsion decreases

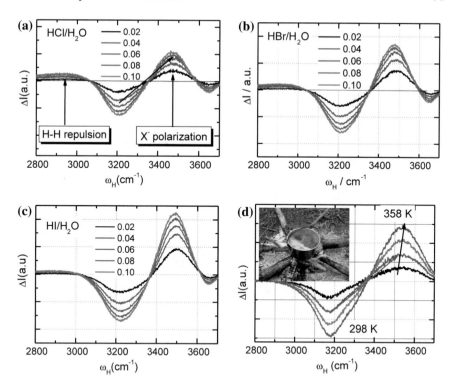

Fig. 4.3 Comparison of the ω_H DPS for the concentrated (**a**) HCl/H$_2$O, (**b**) HBr/H$_2$O, and (**c**) HI/H$_2$O solutions and (**d**) the heated water. The X$^-$ polarization transits the ω_H from 3200 to 3480 cm^{-1}. H$\leftrightarrow$H repulsion softens a fraction H–O bond to $\omega_H < 3050$ cm^{-1}. The < 3050 cm^{-1} feature attenuates when the X turns from Cl to I because of the stronger I$^-$ polarization, which overtones the effect of H$\leftrightarrow$H repulsion. The skin X$^-$ stronger polarization and its preferential occupancy results in the tiny valley at 3650 cm^{-1}. Reprinted with copyright permission from [44, 69]

when the X turns from Cl to I (see Fig. 4.3d) as the stronger polarizability of I annihilates the effect of H$\leftrightarrow$H repulsion. The small spectral valley at 3650 cm^{-1} results from the preference of X$^-$ skin occupation that strengthens the local electric field. The stronger electric field stiffens the dangling H–O bond but the X$^-$ screening weakens the signal of detection.

With the established database, $(d_H, E_H, \omega_H) = (0.10$ nm, 4.0 eV, 3200 cm$^{-1})_{water}$, one may evaluate the hydrated HB identities as $(0.095$ nm, 4.5 eV, 3500 cm$^{-1})_{solution}$, which indicates the H–O bonds in the X$^-$ supersolid hydration shell of the acidic solutions are indeed shorter and stiffer than those in the deionized water. For those featured at 3100 cm^{-1} and below, the H–O bond is longer and softer, being the same to the effect of mechanical compression and base solvation [71].

Figure 4.5 compares the solute capabilities of HB transition and surface stress disruption of H(Cl, Br, I)/H$_2$O solutions. Since the stronger H–O bond (~4.0 eV) only

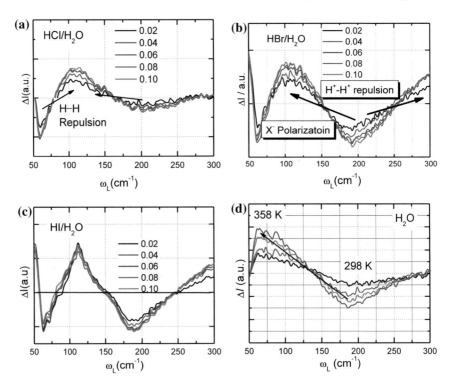

Fig. 4.4 Comparison of the ω_L DPS for the concentrated (**a**) HCl/H$_2$O, (**b**) HBr/H$_2$O, and (**c**) HI/H$_2$O solutions and (**d**) the heated water. The X$^-$ polarization transits the ω_L from 180 to 75 cm^{-1} but the H↔H repulsion reverts the ω_L partly to 120 and >250 cm^{-1}. Heating softens the ω_L to 80 cm^{-1}. Reprinted with copyright permission from [44, 51]

allows H$^+$ to form a H$_3$O$^+$ firmly, the H$^+$ polarization is absent, $f_H(C) = 0$. Therefore, $f_{HX}(C) \approx f_X(C)$. The nonlinear concentration trends $f_X(C) \propto 1 - \exp(-C/C_0)$ corresponds to the number fraction of bonds being polarized by X$^-$ in its closest hydration shells. The slope of the $f_X(C)$ is proportional to the number of bonds within the X$^-$ hydration shell. The pronounced X$^- \leftrightarrow$X$^-$ repulsion at higher concentrations, or smaller solute separations, weakens the ionic electronic field in the hydration shells, which indicates that the number inadequacy of the H$_2$O dipoles in the crystal-like semi-rigid hydration shells could not fully screen the X$^-$ solute. Such a trend becomes more pronounced with the X$^-$ size.

The H↔H fragilization disrupts the solvent network and the surface stress. Therefore, H↔H fragilization resolves the corrosive, dilutive, and skin-stress destructive nature of the acidic solutions. Results are consistent with MD calculations [58, 72] on the hydration shell molecular dynamics and the H$^+$ induced network fragmentation.

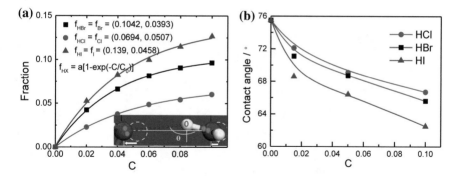

Fig. 4.5 Solute capability of (**a**) O:H–O bond transition and (**b**) surface stress disruption for the concentrated H(Cl, Br, I)/H_2O acidic solutions. The extent of stress disruption increases with the number and size of the X^- hydration shells. The $f_X(C) \propto 1 - \exp(-C/C_0)$ results from the $X^- \leftrightarrow X^-$ repulsion because of the insufficient number of the hydrating H_2O dipoles in the ordered hydration shell. Inset a illustrates the O:H–O elongation in the X^- hydration shells. Reprinted with copyright permission from [44]

4.2.3 Non-polarizability of Proton

One can assume that the I^- in the NaI and the HI has the same concentration dependence of hydration but Na^+ and H^+ are different. The ω_H DPS integrals for NaI in Fig. 4.6a and HI in Fig. 4.3c give rise to $f_{NaI}(C) = 0.4441 \times (1 - \exp(-C/0.1013)$, and $f_{HI}(C) = 0.1652 \times (1 - \exp(-C/0.0529)$ and their difference that is assumed $f_{Na}(C) \cong f_{NaI}(C) - f_{HI}(C)$, as shown in Fig. 4.6b (diamond). The following proves that the $f_H(C) \equiv 0$ [73].

The relative numbers of hydrating H_2O molecules in NaI and HI solutions meet the criterion. Letting $f_{NaI} = f_{Na} + f_I$ and $f_{HI} = f_H + f_I$, the ratio yields,

$$f_{NaI}/f_{HI} = (f_{Na} + f_I)/(f_H + f_I) = 1 + (f_{Na} - f_H)/(f_H + f_I)$$

reorganizing,

$$f_{Na} - f_H = (f_H + f_I)(f_{NaI}/f_{HI} - 1) = f_{NaI} - f_{HI} = f_{Na},$$

thus,

$$f_{Na} - f_H \equiv f_{Na} \text{ and } f_H \equiv 0.$$

The $f_{Na}(C)$ line in Fig. 4.6b (stars) increases linearly with the solute concentration, which is identical to the assumed $f_{NaI} - f_{HI}$. It is thus proven that $f_H(C) \equiv 0$. The H^+ does no polarize water molecules but bonds to a H_2O molecule to form the H_3O^- hydronium.

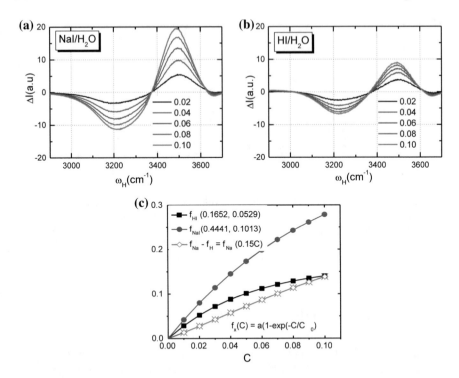

Fig. 4.6 The ω_H DPS for (**a**) the NaI/H$_2$O and (**b**) HI/H$_2$O solutions measured at 298 K [44, 74] and (**c**) their integrals $f_{NaI}(C)$ and $f_{HI}(C)$ and the assumed (diamond) and proven $f_{Na}(C) \cong 0.15C$. The $f_H(C) \equiv 0$ and $f_{Na}(C) - f_H(C) \equiv f_{Na}(C)$

4.3   DFT: Hydration Shell Configuration

Structural optimization with derivatives of the segmental bond lengths and vibration frequencies of the pure, H$^+$-contained, and X$^-$-contained water using both the (PW91) [75] and the dispersion-corrected OBS-PW [76] calculation methods for comparison. The OBS-PW method invokes the hydrogen bond interactions. All-electron calculations were performed based on the double numeric and polarization basis sets. The self-consistency threshold of total energy was set at 10^{-6} Hartree. The energy tolerance limit, force and displacement were set at 10^{-5} Hartree, 0.002 Hartree/Å and 0.005 Å, respectively. Harmonic vibrational frequencies were computed by diagonalizing the mass-weighted Hessian matrix [77]. The DFT calculation considers the structure under the temperature of 0 K.

Table 4.1 DFT derived bond strain ε (%) surround H^+ and X^- solutes. Reprinted with copyright permission from [44]

		OBS-PW		PW91	
		ε_{H-O}	$\varepsilon_{O:H}$	ε_{H-O}	$\varepsilon_{O:H}$
$H_2O–H^+$	Anti–HB	−3.08	63.4	−2.99	59.2
	Neighbor	4.38	−12.5	3.48	−10.3
$H_2O–Cl^-$	Cl·H–O	−0.91	24.1	−0.98	28.1
	Neighbor	−0.80	20.5	−0.65	19.2
$H_2O–Br^-$	Br·H–O	−1.05	31.6	−1.08	30.1
	Neighbor	−0.83	22.8	−0.73	22.9
$H_2O–I^-$	I·H–O	−1.09	40.5	−1.11	42.7
	Neighbor	−0.91	27.5	−0.76	29.8

*Neighbor—the O:H–O bonds in the second-nearest hydration shells as labeled in Fig. 4.7c

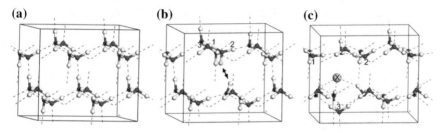

(a) **(b)** **(c)**

Fig. 4.7 DFT optimized cubic structure of (**a**) pure water, (**b**) H^+-contained (yellow) water with anti–HB (black arrow) and (**c**) X^--contained water with three nearest hydration neighbors. Number 1–3 denote the undercoordinated O^{2-} anions in the X^- hydration shell. Reprinted with copyright permission from [44]

4.3.1 Local Bond Length Relaxation

Table 4.1 features the calculated O:H and H–O segmental lengths associated with the H_3O^+ and the X^- solutes. ε_x is the solvation-induced strain at different locations. For pure water, each oxygen has two lone pairs and two protons, which form four identical O:H–O bonds with its tetrahedral neighbors, as shown in Fig. 4.7a. For the H_3O^+ induced bond relaxation, one must consider the H_3O^+ solute H–O bonds and the second nearest neighbors, as labeled in Fig. 4.7b. For the X^- hydration shell, three X·H–O bonds form the closest hydration shells with "·" representing for the Coulomb interaction between the X^- and the H^+, which is weaker than the regular O:H nonbond. The X^- hydration shell extends to the second O:H–O with inclusion of the under-coordinated O atoms, because of the X·H termination, as labeled in Fig. 4.7c.

Results in Table 4.1 show that both DFT schemes give the same trend of intrinsic O:H–O relaxation upon acid solvation. The H–O bond due H_3O^+ performs the same to the skin dangling H–O bond that is 3% shorter and its O:H counterpart is 60% longer than they are in pure water, which means that the H_3O^+ volume is smaller than a H_2O but a larger separation between its neighboring water molecules. However, the network disruption by H↔H repulsion shortens the next neighboring O:H–O bonds, which lengthens the H–O bond by 4% and shortens the O:H by about 11%, according to the calculations [44]. The O:H–O bond contraction reduces its dipole moment and depolarizes the local network, accordingly. The 4% elongation is responsible for the H–O vibration feature at <3100 cm^{-1} and the 11% contraction for the O:H transition of its vibration frequency from 75 to 110 cm^{-1}.

Results in Table 4.1 and Fig. 4.7c also indicate that the X^- anion attracts its three neighboring protons to form the X·H–O bonds in the closest hydration shells. The hydration shell extends to the subsequent O:H–O bonds oriented slightly towards the X^- charge center. Since the $X^- \cdot H^+$ interaction is weaker than the O^{2-}:H^+ nonbond and the number increase of the X^- with concentration also weakens the electric field in the hydration shell. Calculations confirm that the X^- solvation does shorten the H–O bond and lengthens the O:H and the X·H nonbond compared with the respective segment in pure water [44]. The X^- capability of relaxing the nearest X·H–O and the subsequent O:H–O follows the Hofmeister series order: I > Br > Cl. The H–O bond in the I^- hydration shell is 1.1% shorter and the I·H is 41% longer than they are in the bulk water. The O:H–O bond in the second hydration shell of the X^- anion becomes longer. The H–O bond contracts by 0.8% and the O:H expands by 28% in the second hydration shell, for instance.

Figure 4.7 compares the optimized H^+ and X^- solution structures with ordinary water. The H^+ forms an anti–HB with a neighboring water molecule, which results in two H–O radicals ending the continuous solvent network. The H↔H repulsion elongates and softens its neighboring H–O bond lined along the repulsion direction, as labeled in Fig. 4.7b. The X^- attracts three protons to form three X·H–O bonds in its closest hydration shells.

4.3.2 Vibrational Fine-Structure Features

Figure 4.8 shows the OBS-PW-derived H–O vibration spectra for the H_3O^+-contained and pure water. The DPS between H_3O^+-solvated and pure H_2O shows that the stretching and bending modes split into two components of blue and red shifting. The spectra include two parts –O:H stretching modes include contributions from the molecular translational and rotational motion, the bond bending torsional movement. The red shift is attributed to the nearest neighboring elongated O:H nonbonds of the H_3O^+ and the blue shift to the next nearest neighboring O:H elongation, which is consistent with the O:H length relaxation as shown in Table 4.1.

Results indicate that $H^+ \leftrightarrow H^+$ anti–HB induces a radical state and a depolarization state. For the stretching part, the pure water modes around 200 ~ 300 cm^{-1} shift to 110 cm^{-1} and below, resulting from the first nearest neighboring O:H nonbond elongation. The mode above 400 cm^{-1} corresponds to the neighboring O:H

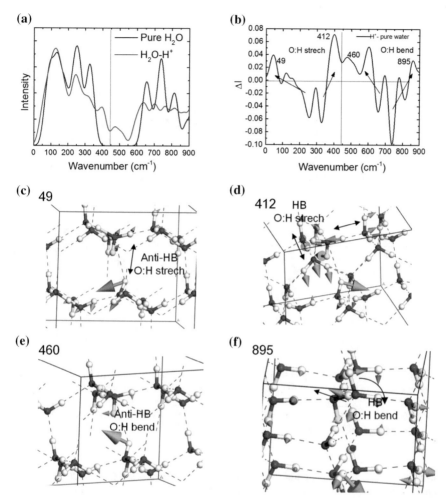

Fig. 4.8 **a** DFT-derived O:H vibrational spectra for the H_3O^+ and pure water and **b** the DPS deriva-tives [44]. DPS revealed that the O:H stretching, and the bending modes associated with the H↔H undergo a red shift but the next neighboring O:H modes undergo a blue shift. **c** The O:H stretching mode associated with the anti–HB is at 49 cm^{-1}; **d** the neighboring O:H depolarization mode is at 412 cm^{-1}; **e** the radical bending mode is at 406 cm^{-1}; **f** the neighboring O:H depolarization bending mode is at 895 cm^{-1}. Reprinted with copyright permission from [44]

depolarization and H↔H repulsion. The ∠O : H−O bending mode splits into two components as well. One shifts from 700 ~ 800 cm^{-1} to 600 cm^{-1} and below, and the other above 800 cm^{-1}, resulting from the first and the second neighboring O:H depolarization.

The O:H–O bond cooperative relaxation holds for both the H_3O^+ and the X^- hydration shells. Figure 4.9 shows the DFT-derived vibrational spectra for the H–O in the X^- hydration shell and pure water. The DPS indicates that blue shifts occur to

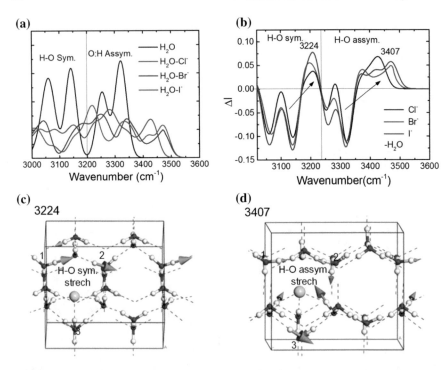

Fig. 4.9 a DFT-derived vibration spectra for the H–O bond in the X⁻ hydration shell and pure water and **b** their DPS. Blue shifts occur to the H–O symmetrical and asymmetrical stretching modes. **c** The H–O symmetric stretching mode in the X⁻ hydration shell shifts to 3200 cm⁻¹ and **d** The H–O asymmetric stretching mode shifts to 3400 cm⁻¹ in the X·H–O hydration shell. Reprinted with copyright permission from [44]

both the H–O symmetrical and asymmetrical stretching vibrations. The mode around 3200 cm⁻¹ characterizes the X⁻ terminated H–O symmetrical mode and the feature around 3400 cm⁻¹ characterizes the H–O asymmetric stretching mode in the X·H–O shell. As the X changes from Cl to I with increasing ionic radius, the X·H attraction turns to be weaker and the H–O bond becomes stronger. Thus, the characteristic H–O frequency increases following the Hofmeister series order: Cl⁻ < Br⁻ < I⁻.

4.4 Summary

In place of the uncertain conventions of proton motion, H⁺ ↔ H⁺ quantum fragilization and the X⁻ point polarization dictate the performance of the Lewis acidic solutions. Instead of drift diffusion, hopping and tunneling between water molecules, the proton bonds firmly to a H₂O molecule to form the H₃O⁺ hydronium, which derives the H↔H anti–HB point breaker that disrupts the solution network and surface stress. The H↔H anti–HB repulsion stiffens the polarized O:H phonon from 75 to $110\,cm^{-1}$ and softens a small fraction of the H–O phonon from 3200 to $3050\,cm^{-1}$ and below.

The X^- anion polarizes its surrounding molecules, shortens and stiffens the H–O bond and relaxes the O:H nonbond oppositely in its hydration shell. The central X·H–O and the next O:H–O bonds form the hydration shells, in which both H–O bonds undergo contraction. The three H–O bonds due H_3O^+ undergo contraction and the next H–O bond elongation. The phonon abundances of the H–O elongation due to H↔H repulsion and the H–O contraction due to X^- polarization vary with the X^- ionic size and electronegativity. The H_3O^+ associated neighboring H–O elongation follows the $Cl^- > Br^- > I^-$ order but the X^- associated H–O contraction follows its opposite. The nonlinear X^- capability of bond transition, $f_X(C)$, indicates that the number inadequacy of the hydrating H_2O dipoles in the hydration shell could not fully screen the solute electric field. The X^- is sensitive to interference of other X^- solutes.

Consistency in the DPS measurements and DFT calculations evidences the essentiality of the H↔H repulsion and X^- polarization in acidic solutions. It is essential to focus on the solute capabilities of transiting the hydrogen bonds and polarizing electrons from the modes of the ordinary water to hydrating and compressing by the H↔H anti-HB.

References

1. S. Arrhenius, *Development of the theory of electrolytic Dissociation.* Nobel Lecture, (1903)
2. J. Brönsted, Part III. Neutral salt and activity effects. The theory of acid and basic catalysis. Trans. Faraday Soc. **24**, 630–640 (1928)
3. T.M. Lowry, I.J. Faulkner, CCCXCIX.—Studies of dynamic isomerism. Part XX. Amphoteric solvents as catalysts for the mutarotation of the sugars. J. Chem. Soc. Trans. **127**, 2883–2887 (1925)
4. G.N. Lewis, Acids and bases. J. Franklin Inst. **226**(3), 293–313 (1938)
5. C.D. Cappa, J.D. Smith, K.R. Wilson, B.M. Messer, M.K. Gilles, R.C. Cohen, R.J. Saykally, Effects of alkali metal halide salts on the hydrogen bond network of liquid water. J. Phys. Chem. B **109**(15), 7046–7052 (2005)
6. W.J. Glover, B.J. Schwartz, Short-range electron correlation stabilizes noncavity solvation of the hydrated electron. J. Chem. Theory Comput. **12**(10), 5117–5131 (2016)
7. T. Iitaka, T. Ebisuzaki, Methane hydrate under high pressure. Phys. Rev. B **68**(17), 172105 (2003)
8. D. Liu, G. Ma, L.M. Levering, H.C. Allen, Vibrational spectroscopy of aqueous sodium halide solutions and air–liquid interfaces: observation of increased interfacial depth. J. Phys. Chem. B **108**(7), 2252–2260 (2004)
9. Y. Marcus, Effect of ions on the structure of water: structure making and breaking. Chem. Rev. **109**(3), 1346–1370 (2009)
10. J.D. Smith, R.J. Saykally, P.L. Geissler, The effects of dissolved halide anions on hydrogen bonding in liquid water. J. Am. Chem. Soc. **129**(45), 13847–13856 (2007)
11. J. Zhang, J.-L. Kuo, T. Iitaka, First principles molecular dynamics study of filled ice hydrogen hydrate. J. Chem. Phys. **137**(8), 084505 (2012)

12. B.L. Bhargava, Y. Yasaka, M.L. Klein, Computational studies of room temperature ionic liquid-water mixtures. Chem. Commun. **47**(22), 6228–6241 (2011)
13. S. Saita, Y. Kohno, N. Nakamura, H. Ohno, Ionic liquids showing phase separation with water prepared by mixing hydrophilic and polar amino acid ionic liquids. Chem. Commun. **49**(79), 8988–8990 (2013)
14. E.S. Stoyanov, I.V. Stoyanova, C.A. Reed, The unique nature of H+ in water. Chem. Sci. **2**(3), 462–472 (2011)
15. S. Heiles, R.J. Cooper, M.J. DiTucci, E.R. Williams, Hydration of guanidinium depends on its local environment. Chem. Sci. **6**(6), 3420–3429 (2015)
16. Y.R. Shen, V. Ostroverkhov, Sum-frequency vibrational spectroscopy on water interfaces: Polar orientation of water molecules at interfaces. Chem. Rev. **106**(4), 1140–1154 (2006)
17. H. Chen, W. Gan, B.H. Wu, D. Wu, Y. Guo, H.F. Wang, Determination of structure and energetics for Gibbs surface adsorption layers of binary liquid mixture 1. Acetone + water. J. Phy. Chem. B **109**(16), 8053–8063 (2005)
18. M.E. Tuckerman, D. Marx, M. Parrinello, The nature and transport mechanism of hydrated hydroxide ions in aqueous solution. Nature **417**(6892), 925–929 (2002)
19. S.T. van der Post, C.S. Hsieh, M. Okuno, Y. Nagata, H.J. Bakker, M. Bonn, J. Hunger, Strong frequency dependence of vibrational relaxation in bulk and surface water reveals sub-picosecond structural heterogeneity. Nat. Commun. **6**, 8384 (2015)
20. M. Thämer, L. De Marco, K. Ramasesha, A. Mandal, A. Tokmakoff, Ultrafast 2D IR spectroscopy of the excess proton in liquid water. Science **350**(6256), 78–82 (2015)
21. F. Dahms, R. Costard, E. Pines, B.P. Fingerhut, E.T. Nibbering, T. Elsaesser, The hydrated excess proton in the zundel cation H5 O2 (+): The role of ultrafast solvent fluctuations. Angew. Chem. Int. Ed. Engl. **55**(36), 10600–10605 (2016)
22. J.C. Li, A.I. Kolesnikov, Neutron spectroscopic investigation of dynamics of water ice. J. Mol. Liq. **100**(1), 1–39 (2002)
23. I. Michalarias, I. Beta, R. Ford, S. Ruffle, J.C. Li, Inelastic neutron scattering studies of water in DNA. Appl. Phys. A Mater. Sci. Process. **74**, s1242–s1244 (2002)
24. P.M. Kiefer, J.T. Hynes, Theoretical aspects of tunneling proton transfer reactions in a polar environment. J. Phys. Org. Chem. **23**(7), 632–646 (2010)
25. S. Daschakraborty, P.M. Kiefer, Y. Miller, Y. Motro, D. Pines, E. Pines, J.T. Hynes, Reaction mechanism for direct proton transfer from carbonic acid to a strong base in aqueous solution I: Acid and base coordinate and charge dynamics. J. Phys. Chem. B **120**(9), 2271–2280 (2016)
26. N.B.-M. Kalish, E. Shandalov, V. Kharlanov, D. Pines, E. Pines, Apparent stoichiometry of water in proton hydration and proton dehydration reactions in CH3CN/H2O solutions. J. Phys. Chem. A **115**(16), 4063–4075 (2011)
27. D. Borgis, G. Tarjus, H. Azzouz, An adiabatic dynamical simulation study of the Zundel polarization of strongly H-bonded complexes in solution. J. Chem. Phys. **97**(2), 1390–1400 (1992)
28. R. Vuilleumier, D. Borgis, Quantum dynamics of an excess proton in water using an extended empirical valence-bond Hamiltonian. J. Phys. Chem. B **102**(22), 4261–4264 (1998)
29. R. Vuilleumier, D. Borgis, Transport and spectroscopy of the hydrated proton: a molecular dynamics study. J. Chem. Phys. **111**(9), 4251–4266 (1999)
30. K. Ando, J.T. Hynes, Molecular mechanism of HCl acid ionization in water: Ab initio potential energy surfaces and Monte Carlo simulations. J. Phys. Chem. B **101**(49), 10464–10478 (1997)
31. K. Ando, J.T. Hynes, HF acid ionization in water: the first step. Faraday Discuss. **102**, 435–441 (1995)
32. D. Borgis, J.T. Hynes, Molecular-dynamics simulation for a model nonadiabatic proton transfer reaction in solution. J. Chem. Phys. **94**(5), 3619–3628 (1991)
33. M.I. Bernal-Uruchurtu, R. Hernández-Lamoneda, K.C. Janda, On the unusual properties of halogen bonds: A detailed ab initio study of X2 − (H2O) 1–5 clusters (X = Cl and Br). J. Phys. Chem. A **113**(19), 5496–5505 (2009)
34. H. Saint-Martin, J. Hernández-Cobos, M.I. Bernal-Uruchurtu, I. Ortega-Blake, H.J. Berendsen, A mobile charge densities in harmonic oscillators (MCDHO) molecular model for numerical simulations: the water–water interaction. J. Chem. Phys. **113**(24), 10899–10912 (2000)

35. C. de Grotthuss, *Sur la Décomposition de l'eau et des Corps Qu'elle Tient en Dissolution à l'aide de l'électricité.* Galvanique Ann Chim, **LVIII**: 54–74, (1806)
36. A. Hassanali, F. Giberti, J. Cuny, T.D. Kuhne, M. Parrinello, Proton transfer through the water gossamer. Proc. Natl. Acad. Sci. U.S.A. **110**(34), 13723–13728 (2013)
37. A.E. Stearn, H. Eyring, The deduction of reaction mechanisms from the theory of absolute rates. J. Chem. Phys. **5**(2), 113–124 (1937)
38. M.L. Huggins, Hydrogen bridges in ice and liquid water. J. Phys. Chem. **40**(6), 723–731 (1936)
39. G. Wannier, Die Beweglichkeit des Wasserstoff-und Hydroxylions in wäßriger Lösung. I. Annalen der Physik **416**(6), 545–568 (1935)
40. N. Agmon, The grotthuss mechanism. Chem. Phys. Lett. **244**(5), 456–462 (1995)
41. M. Eigen, Proton transfer, acid–base catalysis, and enzymatic hydrolysis. Part I: Elementary processes. Angew. Chem. Int. Ed. Engl. **3**(1), 1–19 (1964)
42. G. Zundel, P. Schuster, G. Zundel, C. Sandorfy, *The Hydrogen Bond.* Recent developments in theory and experiments, vol. 2, 1976
43. J.A. Fournier, W.B. Carpenter, N.H.C. Lewis, A. Tokmakoff, Broadband 2D IR spectroscopy reveals dominant asymmetric $H_5O_2^+$ proton hydration structures in acid solutions. Nat. Chem. **10**, 932–937 (2018)
44. X. Zhang, Y. Zhou, Y. Gong, Y. Huang, C. Sun, Resolving H(Cl, Br, I) capabilities of transforming solution hydrogen-bond and surface-stress. Chem. Phys. Lett. **678**, 233–240 (2017)
45. C.Q. Sun, J. Chen, X. Liu, X. Zhang, Y. Huang, (Li, Na, K)OH hydration bonding thermodynamics: Solution self-heating. Chem. Phys. Lett. **696**, 139–143 (2018)
46. C.Q. Sun, Y. Sun, *The Attribute of Water: Single Notion, Multiple Myths.* Springer Ser. Chem. Phys. vol. 113. (Springer, Heidelberg, 2016), 494pp
47. H.S. Frank, W.Y. Wen, Ion-solvent interaction. Structural aspects of ion-solvent interaction in aqueous solutions: a suggested picture of water structure. Discuss Faraday Soc. **24**, 133–140 (1957)
48. L. Pauling, The structure and entropy of ice and of other crystals with some randomness of atomic arrangement. J. Am. Chem. Soc. **57**, 2680–2684 (1935)
49. S.A. Harich, D.W.H. Hwang, X. Yang, J.J. Lin, X. Yang, R.N. Dixon, Photodissociation of H_2O at 121.6 nm: A state-to-state dynamical picture. J. Chem. phys. **113**(22), 10073–10090 (2000)
50. S.A. Harich, X. Yang, D.W. Hwang, J.J. Lin, X. Yang, R.N. Dixon, Photodissociation of D2O at 121.6 nm: A state-to-state dynamical picture. J. Chem. Phys. **114**(18), 7830–7837 (2001)
51. Q. Zeng, T. Yan, K. Wang, Y. Gong, Y. Zhou, Y. Huang, C.Q. Sun, B. Zou, Compression icing of room-temperature NaX solutions (X = F, Cl, Br, I). Phys. Chem. Chem. Phys. **18**(20), 14046–14054 (2016)
52. D. Marx, M.E. Tuckerman, J. Hutter, M. Parrinello, The nature of the hydrated excess proton in water. Nature **397**(6720), 601–604 (1999)
53. C. Drechsel-Grau, D. Marx, Collective proton transfer in ordinary ice: local environments, temperature dependence and deuteration effects. Phys. Chem. Chem. Phys. **19**(4), 2623–2635 (2017)
54. J.M. Heuft, E.J. Meijer, Density functional theory based molecular-dynamics study of aqueous chloride solvation. J. Chem. Phys. **119**(22), 11788–11791 (2003)
55. J.M. Heuft, E.J. Meijer, A density functional theory based study of the microscopic structure and dynamics of aqueous HCl solutions. Phys. Chem. Chem. Phys. **8**(26), 3116–3123 (2006)
56. S. Raugei, M.L. Klein, An ab initio study of water molecules in the bromide ion solvation shell. J. Chem. Phys. **116**(1), 196–202 (2002)
57. M. Tuckerman, K. Laasonen, M. Sprik, M. Parrinello, Ab initio molecular dynamics simulation of the solvation and transport of hydronium and hydroxyl ions in water. J. Chem. Phys. **103**(1), 150–161 (1995)
58. D. Hollas, O. Svoboda, P. Slavíček, Fragmentation of HCl–water clusters upon ionization: Non-adiabatic ab initio dynamics study. Chem. Phys. Lett. **622**, 80–85 (2015)
59. R. Shi, K. Li, Y. Su, L. Tang, X. Huang, L. Sai, J. Zhao, Revisit the landscape of protonated water clusters H + (H2O) n with n = 10–17: An ab initio global search. J. Chem. Phys. **148**(17), 174305 (2018)

60. C.T. Wolke, J.A. Fournier, L.C. Dzugan, M.R. Fagiani, T.T. Odbadrakh, H. Knorke, K.D. Jordan, A.B. McCoy, K.R. Asmis, M.A. Johnson, Spectroscopic snapshots of the proton-transfer mechanism in water. Science **354**(6316), 1131–1135 (2016)
61. O. Teschke, J. Roberto de Castro, J.F. Valente Filho, D.M. Soares, Hydrated excess proton raman spectral densities probed in floating water bridges. ACS Omega. **3**(10), 13977–13983 (2018)
62. E. Codorniu-Hernández, P.G. Kusalik, Probing the mechanisms of proton transfer in liquid water. Proc. Natl. Acad. Sci. **110**(34), 13697–13698 (2013)
63. X. Kong, A. Waldner, F. Orlando, L. Artiglia, T. Huthwelker, M. Ammann, T. Bartels-Rausch, Coexistence of physisorbed and solvated HCl at warm ice surfaces. J. Phys. Chem. Lett. **8**(19), 4757–4762 (2017)
64. T. Lewis, B. Winter, A.C. Stern, M.D. Baer, C.J. Mundy, D.J. Tobias, J.C. Hemminger, Does nitric acid dissociate at the aqueous solution surface? J. Phys. Chem. C **115**(43), 21183–21190 (2011)
65. K. Dong, S. Zhang, Hydrogen bonds: a structural insight into ionic liquids. Chem. A Eur. J. **18**(10), 2748–2761 (2012)
66. K. Dong, S. Zhang, Q. Wang, A new class of ion-ion interaction: Z-bond. Sci. China Chem. **58**(3), 495–500 (2015)
67. D.B. Wong, C.H. Giammanco, E.E. Fenn, M.D. Fayer, Dynamics of isolated water molecules in a sea of ions in a room temperature ionic liquid. J. of Phys. Chem. B **117**(2), 623–635 (2013)
68. X. Zhang, Y. Xu, Y. Zhou, Y. Gong, Y. Huang, C.Q. Sun, HCl, KCl and KOH solvation resolved solute-solvent interactions and solution surface stress. Appl. Surf. Sci. **422**, 475–481 (2017)
69. X. Zhang, T. Yan, Y. Huang, Z. Ma, X. Liu, B. Zou, C.Q. Sun, Mediating relaxation and polarization of hydrogen-bonds in water by NaCl salting and heating. Phys. Chem. Chem. Phys. **16**(45), 24666–24671 (2014)
70. Y. Gong, Y. Zhou, H. Wu, D. Wu, Y. Huang, C.Q. Sun, Raman spectroscopy of alkali halide hydration: hydrogen bond relaxation and polarization. J. Raman Spectrosc. **47**(11), 1351–1359 (2016)
71. Y. Zhou, D. Wu, Y. Gong, Z. Ma, Y. Huang, X. Zhang, C.Q. Sun, Base-hydration-resolved hydrogen-bond networking dynamics: Quantum point compression. J. Mol. Liq. **223**, 1277–1283 (2016)
72. M. Druchok, M. Holovko, Structural changes in water exposed to electric fields: A molecular dynamics study. J. Mol. Liq. **212**, 969–975 (2015)
73. C.Q. Sun, Perspective:Unprecedented O:⇔: O compression and H↔H fragilization in Lewis solutions. Phys. Chem. Chem. Phys. **21**, 2234–2250 (2019)
74. Y. Zhou, Y. Huang, Y. Gong, C.Q. Sun, O:H–O bond electrification in the aqueous YI solutions (Y = Na, K, Rb, Cs). Communicated, (2016)
75. J.P. Perdew, Y. Wang, Accurate and simple analytic representation of the electron-gas correlation-energy. Phy. Rev. B **45**(23), 13244–13249 (1992)
76. F. Ortmann, F. Bechstedt, W.G. Schmidt, Semiempirical van der Waals correction to the density functional description of solids and molecular structures. Phys. Rev. B **73**(20), 205101 (2006)
77. E.B. Wilson, J.C. Decius, P.C. Cross, *Molecular Vibrations* (Dover, New York, 1980)

Chapter 5
Lewis Basic and H$_2$O$_2$ Solutions: O:⇔:O Compression

Contents

Abstract The OH$^-$ and the H$_2$O$_2$ possess each two excessive pairs of electron lone pairs ":" that form an O:⇔:O super−HB upon solvation. The O:⇔:O compression shortens the O:H nonbond and stiffens its phonon but relaxes the H–O bond oppositely. The H–O bond elongation emits energy to heat up the solution. Bond-order-deficiency shortens the solute H–O bond and stiffens its phonon to 3550 cm^{-1} for H$_2$O$_2$ and 3610 cm^{-1} for OH$^-$. However, the O:⇔:O compression annihilates the weak cationic polarization. The H$_2$O$_2$ is less than the OH$^-$ capable of transiting the solvent H–O bonds and surface stress. The linear fraction coefficient f(C) suggests that the OH$^-$ be less sensitive to other solutes. The resultant of solvent exothermic H–O elongation by O:⇔:O compression and the solute endothermic H–O contraction by bond order deficiency heats up the solutions. Observations evidence not only the significance of the inter-lone-pair interaction but also the universality of the bond order-length-strength (BOLS) correlation to aqueous solutions.

Highlight

- The O:⇔:O super−HB has the same effect of compression on HBs relaxation and polarization.
- O:⇔:O compression elongates exothermically the solvent H–O bond, emitting heat.

© Springer Nature Singapore Pte Ltd. 2019   103
C. Q. Sun, *Solvation Dynamics*, Springer Series in Chemical
Physics 121, https://doi.org/10.1007/978-981-13-8441-7_5

- Bond-order-deficiency shortens and stiffens the solute H–O bond, absorbing energy.
- The H_2O_2 is less than the OH^- capable of transiting solute bond and surface stress.

5.1 Wonders of Excessive Electron Lone Pairs

Solvation of hydrogen peroxide (H_2O_2) and hydroxide (OH^-) in (Li, Na, K)OH bases forms important ingredients for bio- and organic–Chemistry such as cell functioning, signal processing, regulating [1–4], environmental catalysis [5], health care [6–8], electrochemical sensing [9], light-sensing [10], cancer therapy [11, 12], etc. Being less stable and easily explosive, H_2O_2 is becoming increasingly fashionable as an oxidant both in industry sectors and in academia. However, fine resolution and clarification of the solute-solvent molecular interactions and their consequence on the solution surface stress, solution temperature, solvation dynamics, critical temperature for phase transition, and network bond relaxation remain yet great challenge.

The OH^- mobility in solution has rarely been noted since the 1900's when Arrhenius [13], Brønsted–Lowry [14, 15], and Lewis [16] defined the base compounds solvation in terms of OH^- or electron lone pair acceptance or donation. The century-old notion regarded an OH^- as a H_2O molecule missing an H^+ proton. Such a "proton hole" transport mechanism followed the same H_3O^+ motion dynamics in acidic solutions. One can simply alter the excessive proton with the electron lone pair or simply reverse the H_3O^+ polarity to OH^- with alteration number of lone pairs from one to two.

The hydroxide was ever suggested highly active in solutions in comparison to the Y^+ and X^- ions that serve each as a point charge center of screened polarization [17–19]. The performance of the basic solution is in so far best explained in terms of "structural diffusion" or "solute delocalization" [20–23]—hydration complexes interconversion driven by the ionic solvation-shell fluctuations. First-principles calculations [22] suggested that an interplay between the hydration complexes and nuclear quantum effects determines the OH^- transport dynamics with little attention to the relaxation and mediation dynamics of the hydrogen bonding network.

The pump-probe spectroscopy investigation [24] of OH^- energy relaxation suggested that excitation for the OH^- solvation decays in 200 fs and this process is followed by a thermalization that becomes slower with the increasing of solute concentration. The thermalization is suggested to proceed by water molecular rotation, reorientation and diffusion [25]. Dynamic studies of YOH hydration in bulk water [1] and in water molecular clusters [26] revealed two processes of vibration phonon relaxations. One is the rapid processes on 200 ± 50 fs time scales and the other slower dynamics on 1–2 ps scales [23, 27]. The vibrational energy exchange between the bulk-like water and the hydrated water (or transition from the mode of ordinary water to the solvated state) takes place in ~200 fs.

DFT calculations suggest that the OH^- hydration shell contains $OH^- \cdot 3H_2O$ complex [28] and the hydrating molecular structure undergoes evolution by reorientation at heating [29]. The strong nonlinear coupling between intra- and inter-molecular vibrations and the non-adiabatic vibrational relaxation could be responsible for the rapid dynamics of phonon relaxation. The phonon frequency blue shift is associated with a longer phonon lifetime, a slower molecular dynamics, and a higher viscosity of the solution due to stronger polarization [30].

An addition of 7M-NaOH solute could fold the phonon lifetime and increase four-fold the viscosity of the solution from that of pure water [1]. On the other hand, NaOH hydration broadens the IR spectra between 800 and 3500 cm^{-1} with an association of redshift transiting the phonon abundance from above 3000 cm^{-1} to its below. This transition was attributed to the strong interactions between the OH$^-$ ion and its hydrating water molecules. In conjunction with harmonic vibrational analysis of OH$^-\cdot$(H$_2$O)$_{17}$ clusters, it was suggested that O–H stretching vibrations of aqueous hydroxides arise from coupling of multiple water molecules that solve the ions [26].

Likewise, the free and hydrogen-bonded OH$^-$ stretches the OH$^-\cdot$(H$_2$O)$_{4,5}$ cluster phonon band substantially in the frequency range 2650–3850 cm^{-1} with an estimation of cluster temperature of 170 K [31]. DFT calculations suggested that the water molecules can act either as single donor or acceptor, a double-proton-donor, or a double-donor-single-acceptor in both the first and the second solvation shell of the OH$^-$. Interconversion among the isomers appears to be rapid as manifested in the observed spectra dominated by the broad and congested absorptions.

FTIR spectroscopy of 1% HOD in varying concentrations of NaOD/D$_2$O with the D$_2$O background subtracted revealed that as the NaOD concentration increases, a new feature appears at 3600 cm^{-1} because of the OH$^-$ stretching, and a strong broadening appears at low frequency because of the HOD molecules bonded to the DO$^-$ ions [32]. The broadening of the IR spectra was explained in terms of Zundel model as proton shuttling motions between two oxygen atoms driven by the strong, rapidly fluctuating electrical fields of the surrounding polar solvent molecules [1, 33, 34].

Overwhelming contributions have been made to H$_2$O$_2$ solvation with a focus mainly on the production and application of the H$_2$O$_2$ solution for bleaching in hospitals and factories, for instance [35, 36]. H$_2$O$_2$ can bleach the textile waste water by using *in situ* production of hydrogen peroxide/hydroxyl radicals [37]. As an oxidant, H$_2$O$_2$ provides electrons to the fuel cells [38]. In the case of plants dealing with drought, H$_2$O$_2$ plays a key role in the process of inducing stomatal closure to defend cells coping with drought [39].

H$_2$O$_2$ is also presented in plants as a signaling molecule [40]. The presence of H$_2$O$_2$ in the somatic cells of animals is beneficial to organisms. In the heart regeneration process, fluorescence signal shows that H$_2$O$_2$ injures the Duox and Nox2 to produce H$_2$O$_2$, approaching the highest concentration of 30 μMol, which is distributed in the heart membrane and adjacent myocardium, serving as the active oxygen signal by degradation of redox-sensitive phosphatase Dusp6. This function lifts the inhibition of MAPK signaling pathways and enhances the pERK, thereby promoting myocardial proliferation, regeneration and inhibition of cardiac fibrosis [41].

However, little has yet been known about the H$_2$O$_2$ and OH$^-$ capabilities of functionalizing the solutions from the microscopic point of view. Zhou [2] and Chen [42] and coworkers examined H$_2$O$_2$ and OH$^-$ solvation dynamics and the solute capabilities of transforming the surface stress and solute bonding dynamics. The examination was compared with the capability of the OH$^-$ in KOH solution [2] and with the effect of mechanical compression of water on the O:H–O bond relaxation [43] using the combination of the DPS [44] and the contact angle detection. Uncovering the H–O vibration mode signatures of both the solvent O:H–O bonds and the solute H–O bonds, the DPS confirms the essentialities not only of the O:$\Leftrightarrow$:O super–HB compression [2] associated with excessive number of lone pairs in solutions but also the bond-order-deficiency induced solute H–O bond contraction [45].

5.2   Phonon Spectrometric Resolution

5.2.1   Solute and Solvent Discriminative HB Relaxation

YOH (or YOD) solvation broadens the main peak shifting to lower frequencies [26, 32, 46]. The solute type and concentration dependent full-frequency Raman spectra for YOH solvation shown in Fig. 5.1 confirmed that YOH solvation indeed softens the ω_H phonons, confirming the O:$\Leftrightarrow$:O compression that lengthens the neighboring H–O bond and softens its phonon. Besides, a sharp peak appears for the YOH solution at 3610 cm^{-1} that is identical to the dangling H–O bond at water surface. For the H$_2$O$_2$ solution in Fig. 5.2, a slightly broadened feature appears at 3550 cm^{-1}. The sharp peaks feature the less ordered H–O bond due the OH$^-$ and the H$_2$O$_2$ solutes. Besides, a sharp peak appears at 877 cm^{-1} due O–O vibration for the H$_2$O$_2$ molecules. The ω_x is less sensitive to the type of the alkali cations of the same concentration. These observations confirm that the O:$\Leftrightarrow$:O super–HB has the same effect of mechanical compression [42] and that the bond order-length-strength (BOLS) correlation [47] applies to aqueous solutions.

The DPS in Figs. 5.3, 5.4, 5.5 and 5.6 refined that OH$^-$ solvation shifts the ω_H partly to 3100 cm^{-1} and lower and the ω_L from below 220 cm^{-1} to higher besides the 3610 cm^{-1} sharp peak whose intensity is proportional to the solute concentration. The DPS in Fig. 5.6 shows that H$_2$O$_2$ solvation results in the same <3100 cm^{-1} feature and a concentration-resolved intensity of peak at 3550 cm^{-1}.

Observations justify that the O:$\Leftrightarrow$:O super–HB point compression (<3100 cm^{-1}; >220 cm^{-1}) has the same but much stronger effect of mechanically bulk compression (<3300 cm^{-1}; >200 cm^{-1}) at the critical pressure 1.33 GPa for the room-temperature water-ice transition [2]. The $\Delta\omega_L$ for the YOH solutions at lower concentrations duplicates the $\Delta\omega_L$ feature of the mechanically compressed water, because of the compression and polarization [48]. Compression lengthens the solvent H–O bond and softens its phonon but relaxes the O:H nonbond contrastingly. The strong effect of compression overweighs the effect of Y$^+$ polarization on the H–O phonon abundance transition. The excessive peak at 3610 cm^{-1} features the bond-order-deficiency induced H–O contraction of the due OH$^-$ solute, which is identical to the surface

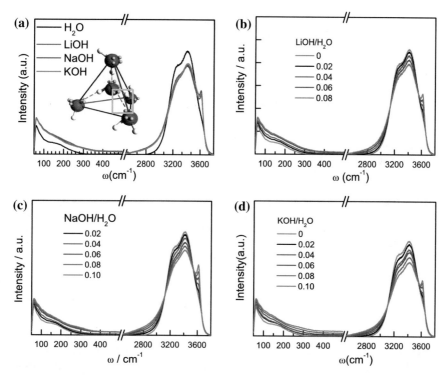

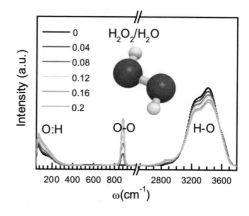

Fig. 5.1 Full-frequency Raman spectroscopy for **a** the (Li, Na, K)OH at 0.08 molar concentration and the concentrated **b** LiOH/H_2O, **c** NaOH/H_2O, and **d** KOH/H_2O solutions. Inset a illustrates the OH$^-$ substitution for the central H_2O in the $2H_2O$ unit cell with creation of the O:⇔:O super–HB point compressor as framed. Base solvation creates a sharp feature at 3610 cm^{-1} and flattens the main peak to lower frequencies. Reprinted with copyright permission from [42]

Fig. 5.2 Full-frequency Raman spectroscopy for the concentrated H_2O_2/H_2O solutions with a sharp peak at 877 cm^{-1} due O–O intramolecular vibration. Inset shows the molecular structure of H_2O_2 with two bonded protons and four lone pairs unindicated. Reprinted with copyright permission from [42]

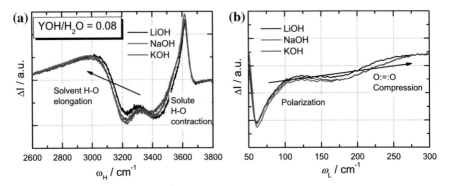

Fig. 5.3 The ω_x DPS for **a, b** YOH/H₂O = 0.08 solutions [27]. The ω_H transition at 3100 cm^{-1} for the basic solutions and at 3300 cm^{-1} for the compressed water to their below. The concentration dependent solute ω_H feature at 3610 cm^{-1} is identical to the H–O dangling bond vibration features. The ω_L DPS revealed features of compression above and polarization below the feature of bulk water at 200 cm^{-1}. Reprinted with copyright permission from [2, 27]

H–O dangling bond, shorter and stiffer than the skin H–O bond featured at 3450 cm^{-1} for ice and water, and bulk water at 3200 cm^{-1} [49].

The high broadness of the <3100 cm^{-1} hump shows the long distance of O:⇔:O super–HB compression forces acting on subsequent H₂O neighbors. The sharp feature at 3610 cm^{-1} for the YOH and 3550 cm^{-1} for the H₂O₂ indicates the strong localized nature of the solute H–O bond contraction. The DPS could resolve the bond order difference between the OH⁻ and the H₂O₂. The H–O and the O–O distances of the H₂O₂ are 0.95 and 1.475 Å and the ∠HOO is 94.8°. Compared with the single H–O bond of the OH⁻ solute, the H–O bond for the H₂O₂ is slightly longer and softer and the peak is slightly broader. The bond order for the H₂O₂ is slightly higher than that of the OH⁻. It is noted that the feature at 3610 cm^{-1} is extremely mechanically and thermal stable [50] because the shorter H–O is hardly deformed by a stimulus [51].

Observations justify that the O:⇔:O super–HB point compression (<3100 cm^{-1}; >220 cm^{-1}) has the same but much stronger effect of mechanically bulk compression (<3300 cm^{-1}; >200 cm^{-1}) at the critical pressure 1.33 GPa for the room-temperature water-ice transition [2], compared to the compression effect on water and ice shown in Figs. 5.5d and 5.6d. The $\Delta\omega_L$ for the YOH solutions at lower concentrations duplicates the $\Delta\omega_L$ feature of the mechanically compressed water, because of the compression and polarization [48]. Compression lengthens the solvent H–O bond and softens its phonon but relaxes the O:H nonbond contrastingly. The strong effect of compression overtones the effect of Y⁺ polarization. The excessive peak at 3610 cm^{-1} features the bond-order-deficiency induced H–O contraction of the due OH⁻ solute, which is identical to the surface H–O dangling bond of 10% shorter, shorter and stiffer than the skin H–O bond featured at 3450 cm^{-1} for ice and water, and bulk water at 3200 cm^{-1} [49].

Spectral feature consistence evidences the significance of the intermolecular O:⇔:O compression and polarization on the intramolecular H–O bond relaxation in

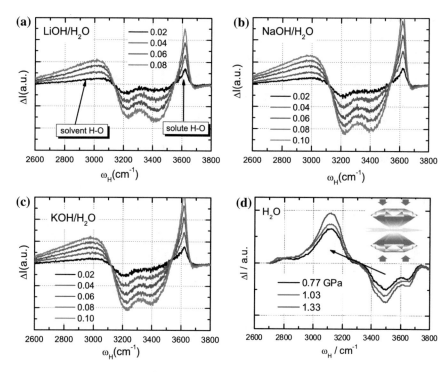

Fig. 5.4 Comparison of the DPS ω_H for the concentrated **a** LiOH/H$_2$O, **b** NaOH/H$_2$O, and **c** KOH/H$_2$O solutions and the **d** compressed water. All solutions share the same compression effect of O:$\Leftrightarrow$:O and mechanical pressure on the distant solvent H–O bond elongation (<3100 cm^{-1}), and the bond-order-deficiency induced solute H–O bond contraction (3610 cm^{-1}, being identical to the dangling H–O of 10% shorter than it is in bulk water). The O:$\Leftrightarrow$:O compression is much stronger than 1.33 GPa for room-temperature ice formation (d). Reprinted with copyright permission from [27]

the basic and in the H$_2$O$_2$ solutions due to the excessive electron lone pairs. The bond order-length-strength (BOLS) correlation [47] applies to aqueous solutions—bonds between undercoordinated atoms become shorter and stronger.

The broad hump at ω_H <3100 cm^{-1} shows the distant dispersion of the O:$\Leftrightarrow$:O compression forces acting on subsequent H$_2$O neighbors. The sharp feature at 3610 cm^{-1} for the YOH indicates the strong localized nature of the solute H–O bond contraction. These DPS H–O vibration peaks clarify the origin for the two ultrafast processes in terms of phonon lifetime [1, 26]. The longer 200 ± 50 fs lifetime features the slower energy dissipation of higher-frequency solute H–O bond vibrating at 3610 cm^{-1} and the other shorter time on 1–2 ps characterizes the elongated solvent H–O bond at lower-frequency of vibration <3100 cm^{-1} upon OH$^-$ solvation [27]. In the pump-probe ultrafast IR spectroscopy, the phonon population decay, or vibration energy dissipation, rate is proportional to the phonon frequency—the dynamics of higher frequency phonon is associated with a faster process.

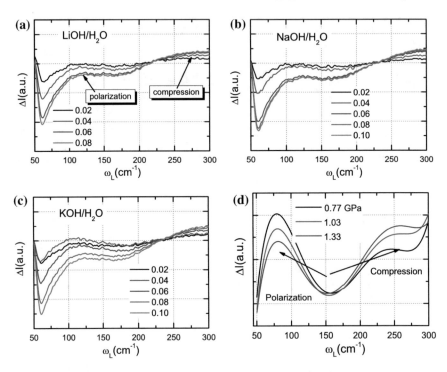

Fig. 5.5 Comparison of the DPS ω_L for the concentrated **a** LiOH/H$_2$O, **b** NaOH/H$_2$O, and **c** KOH/H$_2$O solutions and the **d** compressed water. All solutions share the same compression and polarization effect of O:⇔:O and mechanical pressure on the O:H phonons (~120 cm^{-1}, >250 cm^{-1}). Reprinted with copyright permission from [27]

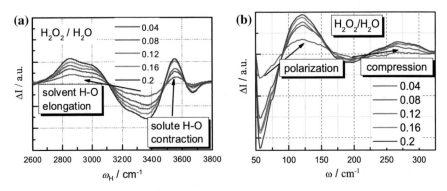

Fig. 5.6 Concentration dependence of the ω_x DPS for **a**, **b** H$_2$O$_2$/H$_2$O solutions shows the same effect of O:⇔:O compression on the solvent H–O bond elongation and O:H contraction (<3100 cm^{-1}; >220 cm^{-1}), and the bond-order-deficiency derived solute H–O bond contraction (3550 cm^{-1}). Reprinted with copyright permission from [27]

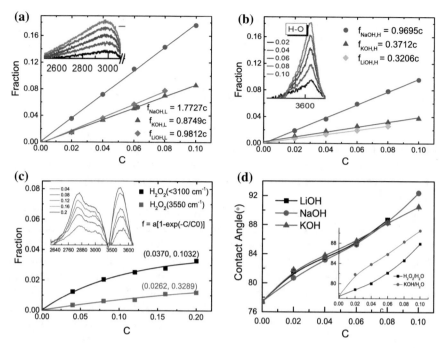

Fig. 5.7 Concentration dependence of the fraction coefficient for **a, b** (Li, Na, K)OH and **c** H_2O_2 solutions and the contact angles for **d** H_2O_2 compared with YOH solutions on glass substrate. H_2O_2 is less than the OH^- capable of transiting the solute H–O bonds and surface stress because of the effect of different bond order deficiencies between solute OH^- and H_2O_2. Reprinted with copyright permission from [42]

5.2.2 Hydrogen Bond Transition and Polarization

Integration of the ω_H DPS peaks gives rise to the phonon abundance transition from the ordinary water to the hydrating, or the number fractions of the solvent H–O bonds (<3100 cm^{-1}) and solute H–O bonds (3610 cm^{-1}) transition to the elongated and the shortened states. Figure 5.7a, b show the f(C) for the OH^- transiting the ordinary H–O bond to the elongated solvent H–O bonds (<3100 cm^{-1}) and the solute H–O phonons (3610 cm^{-1}) in YOH solutions. Although their ionic polarization was annihilated, the Y^+ cation resolves the fraction of bond transition. The f(C) for the H_2O_2 solute, in Fig. 5.7c, shows the nonlinear form of $f_{3550}(C) < f_{3100}(C)$ towards saturation, which shows the presence of the solute-solute repulsion that weakens the $O:\Leftrightarrow:O$ compression. However, the linear $f_{355-0}(C)$ indicates that the effect of solute bond-order-deficiency is independent of the solute concentration.

Figure 5.7d shows that the YOH solutions have the similar polarization effect on the surface stress. YOH solvation constructs the surface stress through polarization by both cation and $O:\Leftrightarrow:O$ polarization. In contrast, H_2O_2 is less than the OH^- capable of transiting the solvent H–O bonds, the solute H–O bonds, and the surface stress. Compared with the $H \leftrightarrow H$ in acid solutions, the $O:\Leftrightarrow:O$ repulsivity is much greater as the two pairs of lone pairs packed closely.

5.2.3 Solute Undercoordination and $O:\Leftrightarrow:O$ Compression

Why does the solvent H–O bond turn to be longer and the solute H–O bond shorter? Besides the formation of the $O:\Leftrightarrow:O$ super–HB point compressor between the OH^- tetrahedron or the H_2O_2 twin tetrahedra and one of their neighboring O^{2-} anion [2], the bond-order deficiency induced bond contraction comes into play as well in the aqueous solution [47].

For a H_2O_2 twin tetrahedra with an O–O bond, there are two H^+ and four lone pairs interacting with six neighboring H_2O molecules through three H^+ and three lone pairs of its neighboring H_2O. There are two excessive lone pairs that form an $O:\Leftrightarrow:O$ super–HB with one of its six neighboring water molecules. One needs to note that the electron lone pairs of the $O:\Leftrightarrow:O$ repel each other not only inserting compressive force locally but serve as a charge center of polarization. From the perspective of $O:\Leftrightarrow:O$ formation, H_2O_2 is the same to the OH^- in reacting to their coordinating water neighbors but the H–O bond order is slightly greater than the OH^- and the lone pairs distributed anisotropically surrounding the H_2O_2 molecules.

On the other hand, atomic undercoordination shortens and stiffens the remaining bonds between undercoordinated atoms [47]. Bond contraction raises the local density of charge and energy; bond stiffening deepens the local potential and cause quantum entrapment of binding energy of core electrons. The locally densely entrapped electrons in turn polarize the valence charge. These sequential entities have formed the bond order-length-strength (BOLS) correlation and nonbonding electron polarization (NEP) notion that has been the subject of previous studies [47]. The BOLS-NEP is responsible for the performance of topological defects, surfaces, and nanostructures of various shapes [47]. For instance, the bond contracts by 30% for a monatomic chain, or the edge of graphene by 18% for the three coordinated monolayer graphene. For the undercoordinated water molecules, the H–O covalent bond follows the BOLS notion and the entire O:H–O bond follows the NEP as well [30]. Theoretical computations confirmed that the dangling H–O bond is shorter and stiffer than its partner pointing inward the bulk water of the same H_2O molecule located in the outmost shell of molecular clusters [52] and nanobubbles [53]. The BOLS holds true to molecular liquids and crystals as well.

Furthermore, H–O bond dissociation occurs neither to the H_2O_2 nor to the OH^- upon solvation as the H–O bond has 4.0 eV energy or greater under the ambient conditions. Experimental observations revealed that one can only break the H–O bond of water monomer using a 121.6 nm UV laser excitation [54, 55], unlike Na^+ that can be dissolved from NaCl at 5 K temperature under vacuum conditions without needing excitation [56].

It is now clear why the solutions share the same ω_H DPS signatures of mechanical compression and the same feature of surface dangling bonds. The red shift of the

Table 5.1 Vibration frequency, length and energy of the H–O bond under various conditions

H–O bond	ω_H (cm^{-1})	d_H (Å)	E_H (eV)	References
Dangling	3610	≪0.84	5.11	[2, 30, 55]
OH$^-$ solute	3610			
H$_2$O$_2$ solute	3550	<0.84	>4.66	[42]
Water skin	3450	0.84	4.66	[53]
Bulk water (4 °C)	3200	1.00	3.97	[48]
Compressed water	<3300	>1.00	<3.97	[2]
OH$^-$ and H$_2$O$_2$ solution matrix	<3100	≫1.00	2.70	[2]

solvent H–O phonon from 3200 to < 3100 cm^{-1} evidences clearly that the O:⇔:O compression is much greater than the upper limit of 1.33 GPa above which ice forms at room temperature [2]. The OH$^-$ single H–O bond is exactly the case of dangling bond featured at 3610 cm^{-1}, which is shorter and stiffer than the twin H–O bonds due a H$_2$O$_2$ featured at 3550 cm^{-1}, the skin H–O bond at 3450 cm^{-1}, and the bulk at 3200 cm^{-1}, demonstrating that the BOLS notion applies to aqueous solutions without exception. Table 5.1 summarizes the length, energy and vibration frequencies of the coordination-resolved H–O bond under different coordination conditions.

5.3 Heat Emission upon Solvation

5.3.1 Factions of H–O Bond Transition

Exothermic and endothermic reactions are of great importance to both basic and engineering sciences [28, 57, 58], as well as the efficiency of drug functioning [59]. However, how the OH$^-$ hydration releases energy to heat up its solution remains as an issue of amazing. Rosenholm [60] and Ratkova and co-workers [61] reviewed recently the thermodynamic chemistry occurred in liquids, solids, and semiconductor materials and suggested that the concurrent understandings are mainly within the framework of classical thermodynamics in terms of enthalpy [61, 62] and Gibbs free energy [63]. The heat generation at reaction is mainly attributed to the solute-solvent electron transportation [60, 64], molecular interactions [65], water molecular motion dynamics [66], and water H–O correlation [67]. The inter- and intra-molecular cooperative interactions govern the path, ultimate outcome, and efficiency of aqueous thermic solvation [68].

From the perspective of O:H–O cooperativity [30] and the solute-solvent O:⇔:O super–HB compression [27], one can correlate the solution temperature T(C) and the fraction f(C) of bonds being elongated by O:⇔:O repulsion. According to Pauling [69], chemical bonds are the ingredient of energy storage. Energy emission or

Table 5.2 Thermodynamics of base solvation (reprinted with permission from [70])

YOH + H$_2$O → Y$^+$ + OH$^-$ + H$_2$O + (heat ↑)	Three pairs of lone pairs ":" attached to the tetrahedrally coordinated OH$^-$ hydroxide, forming an O:⇔:O compressor
H$_2$O$_2$ + H$_2$O → H$_2$O$_2$ + H$_2$O + (heat ↑)	Four pairs of ":" pertained to the H$_2$O$_2$ hydrogen peroxide, forming an O:⇔:O compressor

O:⇔:O repulsion shortens the O:H and lengths the solvent H–O emitting energy to heat up the solutions

Table 5.3 Thermodynamics of YOH solvation [70]

Heat absorption Q$_a$	Solvent H–O bond thermal contraction [71]	Significant
	Undercoordinated solute H–O bond contraction [72]	Significant
	Hydrating H–O bond contraction by Y$^+$ polarization [18]	Negligible because YX solvation and polarization change little the solution temperatures
Heat emission Q$_e$	YOH dissolution into Y$^+$ and OH$^-$ [27]	
	O:⇔:O compressed solvent H–O bond elongation [27]	Significant
	Y$^+$ polarization elongated O:H nonbond [18]	Negligible O:H energy dissipation
Heat dissipation Q$_{dis}$	Molecular motion and structure fluctuation	With little contribution to energy absorption or emission
	Heat loss due to the non-isothermal calorimetric detection	Cause error of tolerance

absorption proceeds by bond relaxation—the equilibrium atomic distance and binding energy change under non-conservative perturbation [47] such as aqueous solvation or mechanical compression. Table 5.2 summarizes the thermodynamic processes involved in YOH and H$_2$O$_2$ solvation. The combination of O:H–O and the O:⇔:O repulsion govern the solute-solvent interaction [70].

Energy absorption, emission and dissipation are involved in solvation. Bond dissociation and elongation emit energy, but bond formation and bond contraction absorb energy, leading to the respective exothermic and endothermic reaction. Molecular motion and structure fluctuation dissipate energy within the range of O:H cohesive energy at 0.1 eV, which contribute little to the change of solution temperature. Table 5.3 suggests the possible thermodynamics involved in the YOH solvation under the non-adiabatic calorimetric detection and their significance of energy change. H$_2$O$_2$ solvation is not subject to YOH molecular dissolution but only H–O relaxation and molecular motion and structure fluctuation.

One can readily correlate the solution temperature T(C) to the f(C) by the calorimetric detection. The detection was conducted using a regular thermometer to record the solution temperature in a glass beaker under the ambient temperature. The solution was stirred using a magnetic bar rotating in the beaker in 5 Hz frequency. For concept proofing, it is reasonable to omit the effects of non-adiabatic detection and the stirring friction on solution temperature, as these artefacts only influence the accuracy of detection but not the nature and trend of observations.

Figure 5.8a, b show the T(C) for the solution self-heating on solvation. The T(C) $\propto$ C for the YOH solutions but deviates at higher H_2O_2 higher concentrations. The f(C) for the H_2O_2 in Fig. 5.8c shows saturation trends because of the solute-solute interactions. Figure 5.8d compares the time spent to heating the sample to the highest temperatures. LiOH is more easily dissolved than the rest two. This thermal relaxation time in second or minute scale is completely different from that used in the time-resolved infrared spectroscopy for spectral signal decay at the fs ~ ps level with involvement of the water molecular and solute drift motion [73–75]. Furthermore, the lower capability of the H_2O_2 than YOH of raising the T(C), the solution surface stress and the f(C) of H–O bonds transiting from water to the solvated states further confirms the effect of bond-order-deficiency.

5.3.2 Solvation Bonding Exo-thermodynamics

One may estimate the energy exchange during solvation. According to the proposed thermal dynamics in Table 5.3. The total energy should conserve [76]:

$$\sum_3 Q_{e,i}(C) - \sum_3 Q_{a,j}(C) - \sum_2 Q_{dis,l}(C) = 0$$

Only three processes are significant. The energy Q_0 heating up a unit mass solution (m = 1) from T_i to T_f by increasing the solute concentration up to C_M equals ($h_0 = 4.18$ J(g·K)$^{-1} = 0.00039$ eV(bond·K)$^{-1}$ is the specific heat for liquid water):

$$\int_0^{Q_0} dq = h_0 \int_0^1 dm \int_0^{C_M} \frac{dt(C)}{dC} dC = h_0 T(C_M) \tag{5.1}$$

The energy difference between exothermic solvent H–O elongation Q_e by O:$\Leftrightarrow$:O compression and the endothermic solute H–O contraction Q_a by bond-order-deficiency heats up the solution (h_e and h_a are the energy emission and absorption per bond),

$$\int_0^{Q_e} dq_e - \int_0^{Q_a} dq_a = \left[h_e \int_0^{f_e} dm_e - h_a \int_0^{f_a} dm_a \right] \int_{T_i}^{T_f} dt = [h_e f_e - h_a f_a] T(C_M) \tag{5.2}$$

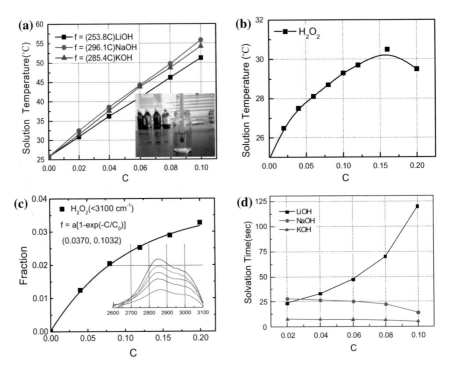

Fig. 5.8 Solution temperatures for the concentrated **a** YOH and **b** H₂O₂ solutions in (**b**) mol/L and **c** molar concentration. **d** The thermal relaxation time required to reach the highest possible solution temperature for OH⁻ solutions. Reprinted with copyright permission from [70]

Table 5.4 Derivative of energy emission from the H–O bond elongation by O:⇔:O compression

YOH	C_M	f_e/C	f_a/C	T_M/C	h_e/h_0	Q_e (eV/bond)	Q_e/E_L
LiOH	0.08	0.981	0.321	253.8	1.515	0.150	1.58
NaOH	0.1	1.773	0.970	296.1	1.245	0.144	1.52
KOH	0.1	0.875	0.371	285.4	1.984	0.221	2.33

One can obtain the heat emission from H–O elongation,

$$Q_e = h_e T(C_M) = \frac{h_0 T(C_M)}{f_e - f_a h_a / h_e} \approx \frac{h_0 T(C_M)}{f_e - f_a}, \ (with \ h_e \approx h_a) \quad (5.3)$$

One may approximate that the h_a is compatible to the h_e for estimation. Table 5.4 tabulates the outcome showing that the H–O elongation by O:⇔:O compression emits at least 150% the O:H cohesive energy of 0.1 eV at room temperature [72]. It is thus verified that the energy remnant of the solvent H–O exothermic elongation and the solute H–O endothermic contraction heats up the solution, which has little to do with energy dissipation of molecular motion or structure fluctuation.

One can also estimate the energy emission from H–O bond elongation with the documented values of $(d_H, E_H, \omega_H) = (1.0\,\text{Å}, 4.0\,\text{eV}, 3200\,\text{cm}^{-1})$ for the mode of bulk water [77], and for basic hydration $(1.10 \sim 1.05\,\text{Å}, E_2, 2500 \sim 3000\,\text{cm}^{-1})$ using the $\omega^2 \propto E/d^2$ relation:

$$\Delta E = 2E\left(\frac{\Delta d}{d} + \frac{\Delta\omega}{\omega}\right) = 2 \times 4\left(\frac{0.10 \sim 0.05}{1} + \frac{2500 \sim 3000}{3200}\right)$$

$$= \begin{cases} -0.95\,(2500\,\text{cm}^{-1}) \\ -0.1\,(3000\,\text{cm}^{-1}) \end{cases} \tag{5.4}$$

The H–O bond elongation losses its cohesive energy from 4.0 by $0.1 \sim 1.0$ eV depending on the frequency shift. Although the 0.15 eV energy emission may be underestimated, the exothermic H–O bond elongation by $O:\Leftrightarrow:O$ compression is certainly the intrinsic and primary source for heating up the YOH solution. The elongated H–O bond emits >150% the O:H cohesive energy at 0.095 eV that caps the energy dissipation by molecular motion and thermal fluctuation. Therefore, the intramolecular H–O bond relaxation driven by the bond-order-deficiency and inter-molecular $O:\Leftrightarrow:O$ repulsion governs the hydration bonding thermodynamics and the performance of the solutions.

5.4 DFT: Solute Occupancy and Anisotropic Polarizability

5.4.1 DFT Examination of Theoretical Expectations

The geometry optimizations and phonon spectra computation for pure water and water containing Y^+ cation and OH^- hydroxide were carried out by repeating the DFT iteration of Chap. 4. Using a $(2a)^3$ cube that accommodates eight H_2O molecules distributed in four of the eight partitions in terms of the $2H_2O$ motif (see Fig. 3.14a, b). The cation takes up one of the six face-centered position or one of the eight vertices of the cube to form the $Y^+ \cdot 4H_2O$ motif without any regular bond formation. The maximal occupancy is eight at the ratio of $N_{ion}/N_{H2O} = 1/4$. The ideal distance between the Y^+ and the nearest O is $d_{Y-O} = d_{O-O} = 2.695\,\text{Å} = \sqrt{3}a/2$ and the distance to the six next-nearest O neighbors is $d_{Y-O} = a$ ($a = 2d_{O-O}/\sqrt{3} = 3.112\,\text{Å}$ at 4°C).

As the oriented H_2O molecule has already formed O:H–O bonds with its four H_2O neighbors, only one lone pair or a H–O bond points directly away from the central Y^+ and the center diameter of the rest three ':' or H^+ faces to the Y^+, with either ':' or H^+ interaction dominance. Being different from the convention of Y^+ having three nearest H_2O neighbors, each pair of the four neighboring oriented H_2O molecules are subject to $Y^+ \leftrightarrow H^+$ repulsion and $Y^+:O$ attraction. The $Y^+:O$ attraction shortens itself but lengthens the opposing H–O bond, while the $Y^+ \leftrightarrow H$ repulsion does the opposite because of the O:H–O cooperativity. The O:H–O configuration and orientation conserve, which allows the oriented H_2O molecules only rotates slightly under the ionic electric field. The cation prefers the eccentric interstitial hollow site to form the Y^+·$(4H_2O + H_2O)$ motif and polarizes its surroundings anisotropically. For demonstration purpose, focus was given on the cationic polarization of the nearest $4H_2O$ molecules by neglecting the six next-nearest neighbors. The entities of $H↔Y^+$:O interactions, O:⇔:O compression, polarization, and the solute H–O contraction distort the local bonding network and disperse the phonon frequencies, but the phonon spectroscopy could hardly resolve the subtle yet important information.

5.4.2   DFT Outcomes: Network Distortion

Table 5.5 summarizes the PW91 and OBS-PW derived O:H–O length relaxations surrounding the Y^+ and OH^- solutes. The strain ε represents the relative length change with respect to those scales for pure water. The two DFT schemes revealed the same trend and nature of bond relaxation despite the accuracy due to possible artifacts in the calculation codes.

For the LiOH solvation instance, the Li^+:O attraction relaxes the Li^+:O–H:O segments by −33.70, 2.99, and −6.91%, respectively; the $Li^+↔H$ repulsion relaxes the $Li^+↔H$–O:H by 41.73, −3.76, and 2.30%, in the sequence order, with respect to their references to the central vacancy and the O:H–O bond of pure water. The O:⇔:O repulsion relaxes the O:⇔:O:H–O segments by 63.1, −20.8, and 3.8%, in the sequence order. Bond order deficiency shortens the solute H–O bond by 1.83% and lengthens. The cation dislocates eccentrically by ΔY^+ varying from 0.74 (for Li^+) to 0.86 Å (K^+).

5.4.3   DFT: Phonon Band Dispersion

Fig. 5.9a, b show the PW91-derived O:H–O vibration spectra for the 1/64 concentrated Y^+/H_2O solution and Fig. 5.9c, d the DPS upon all the spectral peak areas being area normalized. The DPS below 1000 cm^{-1} in Fig. 5.9c displays two major components: the O:H vibration mode below ~500 cm^{-1} consisting of molecular translational motion and the ∠H:O:H bending and polarization (<200 cm^{-1}). The ∠O:H–O bending vibration mode shifts from 750 to its above, respectively.

The $Y^+↔H^+$ repulsion shifts upward while the Y^+ polarization shift downward of these low-frequency vibration features in Fig. 5.9. For the stretching component, the O:H stretching mode around 300 cm^{-1} shifts to 253 cm^{-1} due to the longer and weaker O:H by polarization. The O:H phonon frequency shifts from 380 to around 439 cm^{-1} corresponding to shorter and stiffer O:H of the Y^+:O. Both redshift and blueshift happen to the O:H stretching and the bending vibration parts, which is in accordance with the O:H performance within the first hydration shell. The broad fluctuation around the main peaks demonstrates the weak O:H–O bonds transition

Table 5.5 DFT-derived O:H–O segmental cooperative and discriminative relaxations in the vicinal hydration shells of the Y^+ and the OH^- solute and the O:H–O due solute. The strain ε is in %.*

		OBS-PW					PW91				
		$\varepsilon_{Y:H2O}$	ε_{H-O}	$\varepsilon_{O:H}$	ε_{H-O}	ΔY^+	$\varepsilon_{Y:H2O}$	ε_{H-O}	$\varepsilon_{O:H}$	ε_{H-O}	ΔY^+
$Li^+ \cdot H_2O$	$Li^+:O_{h1}-H_{h1}:O_{h2}-H_{h2}$	-33.79	3.12	-7.81	1.29	1.06	-33.60	2.86	-6.00	1.14	0.74
	$Li^+ \leftrightarrow H_{h1}-O_{h1}:H_{h2}-O_{h2}$	42.72	-3.93	2.67	-0.28		40.74	-3.58	1.92	-0.69	
$Na^+ \cdot H_2O$	$Na^+:O_{h1}-H_{h1}:O_{h2}-H_{h2}$	-19.10	1.89	-2.72	0.81	1.09	-17.60	2.35	-3.70	0.87	0.78
	$Na^+ \leftrightarrow H_{h1}-O_{h1}:H_{h2}-O_{h2}$	60.49	-4.52	4.56	-0.27		46.10	-3.77	5.87	-1.33	
$K^+ \cdot H_2O$	$K^+:O_{h1}-H_{h1}:O_{h2}-H_{h2}$	-4.86	1.87	-2.29	0.54	1.18	-8.18	1.86	-1.34	0.56	0.86
	$K^+ \leftrightarrow H_{h1}-O_{h1}:H_{h2}-O_{h2}$	63.63	-5.22	5.24	0.79		50.83	-4.42	5.96	-1.71	
$OH^- \cdot H_2O$	$O:\Leftrightarrow:O$	–	–	63.54	–	–	–	–	62.61	–	
	Solute O:H–O	–	-1.90	3.83	–		–	-1.76	2.24	–	
	Solvent $(O:\Leftrightarrow:)O:H-O$	–	3.81	-21.04	–		–	3.74	-20.49	–	

[a] The segmental length $d_{H-O} = 1.0004$ Å and $d_{O:H} = 1.6946$ Å for H_2O at 4 °C [77]. [b] The lattice $a = 3.1119$ Å
Reprinted with permission from [78]

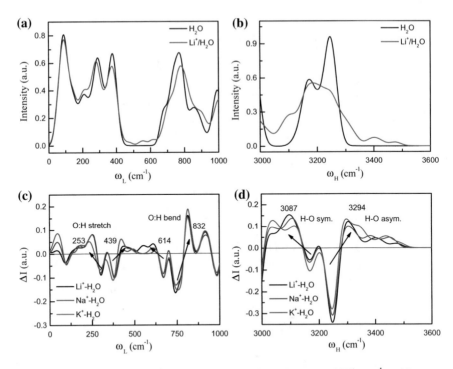

Fig. 5.9 Cation-dispersed vibration spectra in the regime of **a** $\omega_L \leq 1000$ cm^{-1} and **b** $\omega_H \geq$ 3000 cm^{-1} for Li$^+$/H$_2$O solution and the **c**, **d** DPS for (Li, Na, K)$^+$/H$_2$O solutions

beyond the first hydration shell. Peak dispersion over the range of 450–600 cm^{-1} is ascribed to the suppression of the O:H stretching and bending modes due to the incomplete shielding of Y$^+$ ionic polarization that extends beyond the first hydration shell.

Fig. 5.9d shows the Y$^+$-dispersed ω_H DPS. The H–O asymmetric stretching mode that describes the simultaneous motion of H away or close to the oxygen, but the H–O symmetric mode is an alternation—one goes away and the other closer. The H–O bond mode shifting from around 3169 to 3087 cm^{-1} is attributable to the H–O symmetric stretching on account of the elongation and softening of H–O of Y$^+$:OH$_2$. The H–O phonon frequency shifts from 3250 to 3294 cm^{-1} due to the H–O asymmetric stretching vibration. Thus, the Y$^+$:O attraction and Y$^+$↔H repulsion disperse both the solvent H–O and the O:H stretching modes into additional two satellites. The cation capability of phonon frequency dispersion follows the Li$^+$ > Na$^+$ > K$^+$ order, according to the calculated DPS.

Fig. 5.10a, b show the OH$^-$ dispersed vibration spectra for the segmented O:H–O bonds from 50 to 600 cm^{-1} (ω_L) and from 3000 to 3600 cm^{-1} (ω_H) and Fig. 5.10c, d the corresponding DPS. The O:⇔:O compression disperses the O:H stretching mode from 75 to 125 cm^{-1} and 375–429 cm^{-1}, which conforms to the neighboring shorter and stiffer O:H nonbonds. Meanwhile, the O:⇔:O repulsion disperses the

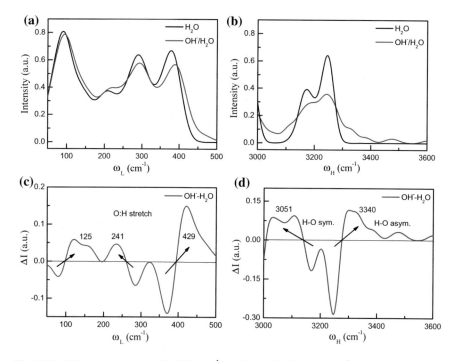

Fig. 5.10 OH$^-$-dispersed **a** ω_L (50–600 cm^{-1}) and **b** ω_H (3100–3600 cm^{-1}) for the 1/64 concentrated OH$^-$/H$_2$O solutions and **c, d** the respective DPS

H–O phonon frequency from 3170 cm^{-1} to its below. The O:$\Leftrightarrow$:O polarization transits the H–O characteristic mode to 3246–3340 cm^{-1}, which also conforms to the contraction and enhancement of solute H–O bond. The O:$\Leftrightarrow$:O polarization shifts the O:H characteristic phonon frequency from 285 to 241 cm^{-1} as the polarization lengthens and softens the O:H nonbond. Hence, strong O:$\Leftrightarrow$:O compression and polarization have dual functionalities on the solvent O:H–O bonds.

It is noted that the O:$\Leftrightarrow$:O and H$^+\leftrightarrow$Y$^+$ repulsion has the similar effect on the ω_H dispersion because of the polarization and repulsion nature. The O:H phonon is more sensitive to these two types of interactions. The computational spectral peaks in panel (b) should correspond respectively to the bulk at 3150 cm^{-1} and the skin component at 3450 cm^{-1} in reality [72]. The tiny feature at 3600 cm^{-1} corresponds to the solute OH$^-$ bond contraction.

Fig. 5.11a, b show the Y$^+$ and OH$^-$ joined effect on the vibration spectra dispersion for the segmental O:H nonbond from 50 to 300 cm^{-1} (ω_L) and for the H–O bond from 3000 to 3800 cm^{-1} (ω_H). One may note that the cation has two choices of its occupancy at lower concentrations with respect to the O:$\Leftrightarrow$:O super-HB—one is sitting at the surface normal along, and the other is aside the O:$\Leftrightarrow$:O direction. Computational results resolve slight difference between the spectra dispersed by the cation positions. Compared with the experimental DPS in Fig. 5.3 for the

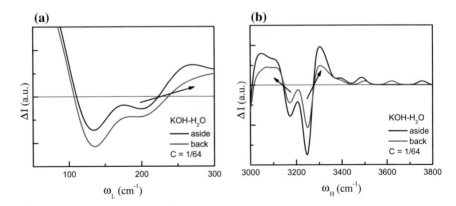

Fig. 5.11 Y$^+$ and OH$^-$ combination dispersed DPS profiles of **a** ω_L (50–300 cm^{-1}) and **b** ω_H (3000–3800 cm^{-1}) for the 1/64 concentrated solution and the spectroscopic measurements (refer to the measured spectra shown in Fig. 5.3 for comparison).

(Li, Na, K)OH solutions [72], the calculated DPS produces the major features of experimental observations despite the frequency offset due to the potential function used in calculation.

The discrepancy in the phonon frequency of stretching vibration mode between theoretical and experimental suggests the essentiality of choosing proper potential functions in computational approach. The solute concentration difference between calculation (1/64) and experiment (1/50) could unlikely be the cause of weakening the 3610 cm^{-1} mode intensity.

O:$\Leftrightarrow$:O and H$^+\leftrightarrow$ Y$^+$ compression shifts the ω_L from the bulk component 200 cm^{-1} to its above and polarization reverts part of the bulk component to its below. The compression lengthens the solvent H–O bond to create the feature below 3200 cm$^-$ and the polarization generates the feature slightly above. The feature of solute H–O contraction is less significant compared to the spectroscopic measurements.

To conclude, DFT examination revealed the detailed information on the local bond relaxation and the corresponding spectroscopic responses, despite possible artifacts in the computational codes, which confirmed the following:

(1) The solvent prefers the crystal-like structure of oriented H$_2$O molecules instead of the random or amorphous state of freely moving molecules.

(2) The cation Y$^+$ prefers eccentrically the interstitial hollow site interacting with neighboring oriented H$_2$O molecules through H$^+\leftrightarrow$Y$^+$:O interactions, which distort the local bonding network anisotropically.

(3) Instead of shuttling freely between adjacent H$_2$O molecules, as conventionally thought, the excessive ':' turns one O:H–O bond into the O:$\Leftrightarrow$:O super-HB by altering the YOH into the OH$^-$ hydroxide that replaces a H$_2$O. The O:$\Leftrightarrow$:O compresses and polarizes its neighboring O:H–O bonds.

(4) The repulsive Y$^+\leftrightarrow$H$^+$ and the O:$\Leftrightarrow$:O expand but the attractive Y$^+$:O contracts. The Y$^+$:O attraction and the O:$\Leftrightarrow$:O repulsion lengthen the opposing H–O bond but shorten the subsequent O:H cooperatively; the Y$^+\leftrightarrow$H repulsion and the solute bond-order deficiency do their H–O bonds and the subsequent

O:H nonbonds contrastingly. The $Y^+ \leftrightarrow H$ repulsion and $Y^+:O$ attraction disperse the H–O and O:H phonon band each into two satellites.

(5)   The Y^+ anisotropic polarization, $O:\Leftrightarrow:O$ compression, and the solute OH^- bond contraction distort the network and disperse the O:H–O vibration frequencies accordingly, agreeing with spectroscopic detection.

5.5   Summary

It is thus confirmed the significance of the inter-lone-pair repulsion and the bond order-deficiency to the YOH and H_2O_2 hydration bonding dynamics. The solvent H–O elongation by $O:\Leftrightarrow:O$ compression and the solute H–O contraction by bond-order-deficiency, and the associated polarization govern the performance of the basic and the H_2O_2 solutions in terms of phonon frequency shift, surface stress, the exo-thermodynamics, and the solution temperature. Although the Y^+ polarization was annihilated by compression, different kinds of the Y^+ resolve the OH^- capability of bond transition.

The $O:\Leftrightarrow:O$ point compressor has the same effect to the applied pressure on the O:H–O bond network relaxation. The compression lengthens the solvent H–O bond and transits their phonons from above 3100 cm^{-1} to the below in a long-range manner while relaxes the O:H nonbonds oppositely. The bond order-deficiency shortens the solute H–O bonds giving rise to the localized phonons at 3610 for OH^- and 3550 cm^{-1} for H_2O_2, discriminating the extent of bond order deficiency in both cases.

The energy difference between the solvent H–O bond exothermic elongation and the solute H–O endothermic contraction heats up the solutions. Both the cation polarization and the $O:\Leftrightarrow:O$ compression raise the surface stress of the basic solutions. The OH^- is more than the H_2O_2 capable of transiting the ordinary O:H–O bonds, raising the surface stress and solution temperature. Findings prove not only the essentiality of expanding the solvation investigation from the perspective of molecular motion dynamics to the hydration bonding and nonbonding thermodynamics but also the power of the DPS strategy and the developed strategies resolving the fraction and stiffness transition and the related energetics. DFT optimization confirms that the cation Y^+ sits eccentrically the interstitial hollow site to form the $Y^+ \cdot 4H_2O$ motif and the excessive electron lone pair ':' is attached to the OH^- to form the $OH^- \cdot 4H_2O$ in the YOH basic solutions. The Y^+ interacts with its oriented H_2O neighbors through $Y^+:O$ attraction and $Y^+ \leftrightarrow H^+$ repulsion. The ':' turns one O:H-O bond into the $O:\Leftrightarrow:O$ by replacing the center H_2O of the $H_2O \cdot 4H_2O$ with the OH^-. The repulsive $Y^+ \leftrightarrow H^+$ and $O:\Leftrightarrow:O$ expand but the attractive $Y^+:O$ contracts. The Y^+ anisotropic polarization, $O:\Leftrightarrow:O$ compression and polarization, and the solute OH^- bond contraction distorts discriminatively the solute local bond network and disperses the O:H–O phonon frequencies, as probed spectroscopically.

References

1. M. Thämer, L. De Marco, K. Ramasesha, A. Mandal, A. Tokmakoff, Ultrafast 2D IR spectroscopy of the excess proton in liquid water. Science **350**(6256), 78–82 (2015)
2. Q. Zeng, T. Yan, K. Wang, Y. Gong, Y. Zhou, Y. Huang, C.Q. Sun, B. Zou, Compression icing of room-temperature NaX solutions (X = F, Cl, Br, I). Phys. Chem. Chem. Phys. **18**(20), 14046–14054 (2016)
3. H.S. Marinho, C. Real, L. Cyrne, H. Soares, F. Antunes, Hydrogen peroxide sensing, signaling and regulation of transcription factors. Redox Biol. **2**, 535–562 (2014)
4. H. Satooka, M. Hara-Chikuma, Aquaporin-3 controls breast cancer cell migration by regulating hydrogen peroxide transport and its downstream cell signaling. Mol. Cell. Biol. **36**(7), 1206–1218 (2016)
5. A.H. Gemeay, I.A. Mansour, R.G. El-Sharkawy, A.B. Zaki, Catalytic effect of supported metal ion complexes on the induced oxidative degradation of pyrocatechol violet by hydrogen peroxide. J. Colloid Interface Sci. **263**(1), 228–236 (2003)
6. G. Strukul, *Catalytic Oxidations with Hydrogen Peroxide as Oxidant*, Vol. 9. (Springer Science & Business Media, 2013)
7. J.-G. Kim, S.-J. Park, J.S.S. Damsté, S. Schouten, W.I.C. Rijpstra, M.-Y. Jung, S.-J. Kim, J.-H. Gwak, H. Hong, O.-J. Si, Hydrogen peroxide detoxification is a key mechanism for growth of ammonia-oxidizing archaea. Proc. Natl. Acad. Sci. **113**(28), 7888–7893 (2016)
8. I.J. Amanna, H.P. Raue, M.K. Slifka, Development of a new hydrogen peroxide-based vaccine platform. Nat. Med. **18**(6), 974–979 (2012)
9. A. Noorbakhsh, M. Khakpoor, M. Rafieniya, E. Sharifi, M. Mehrasa, Highly sensitive electrochemical hydrogen peroxide sensor based on iron oxide-reduced graphene oxide-chitosan modified with DNA-celestine blue. Electroanalysis **29**(4), (2017)
10. K. Bodvard, K. Peeters, F. Roger, N. Romanov, A. Igbaria, N. Welkenhuysen, G. Palais, W. Reiter, M.B. Toledano, M. Kall, M. Molin, Light-sensing via hydrogen peroxide and a peroxiredoxin. Nat. Commun. **8**, 14791 (2017)
11. W. Fan, N. Lu, P. Huang, Y. Liu, Z. Yang, S. Wang, G. Yu, Y. Liu, J. Hu, Q. He, Glucose-responsive sequential generation of hydrogen peroxide and nitric oxide for synergistic cancer starving-like/gas therapy. Angew. Chem. Int. Ed. **56**(5), 1229–1233 (2017)
12. G. Vilema-Enriquez, A. Arroyo, M. Grijalva, R.I. Amador-Zafra, J. Camacho, Molecular and cellular effects of hydrogen peroxide on human lung cancer cells: potential therapeutic implications. Oxidative Med. Cell. Longevity **2016**, 1908164 (2016)
13. S. Arrhenius, *Development of the theory of electrolytic Dissociation.* Nobel Lecture, (1903)
14. J. Brönsted, Part III. Neutral salt and activity effects. The theory of acid and basic catalysis. Trans. Faraday Soc. **24**, 630–640 (1928)
15. T.M. Lowry, I.J. Faulkner, CCCXCIX.—Studies of dynamic isomerism. Part XX. Amphoteric solvents as catalysts for the mutarotation of the sugars. J. Chem. Soc.Trans. **127**, 2883–2887 (1925)
16. G.N. Lewis, Acids and bases. J. Franklin Inst. **226**(3), 293–313 (1938)
17. Y. Gong, Y. Zhou, H. Wu, D. Wu, Y. Huang, C.Q. Sun, Raman spectroscopy of alkali halide hydration: hydrogen bond relaxation and polarization. J. Raman Spectrosc. **47**(11), 1351–1359 (2016)
18. Y. Zhou, Y. Huang, Z. Ma, Y. Gong, X. Zhang, Y. Sun, C.Q. Sun, Water molecular structure-order in the NaX hydration shells (X = F, Cl, Br, I). J. Mol. Liq. **221**, 788–797 (2016)
19. X. Zhang, T. Yan, Y. Huang, Z. Ma, X. Liu, B. Zou, C.Q. Sun, Mediating relaxation and polarization of hydrogen-bonds in water by NaCl salting and heating. Phys. Chem. Chem. Phys. **16**(45), 24666–24671 (2014)
20. Y.R. Shen, V. Ostroverkhov, Sum-frequency vibrational spectroscopy on water interfaces: polar orientation of water molecules at interfaces. Chem. Rev. **106**(4), 1140–1154 (2006)
21. H. Chen, W. Gan, B.-h. Wu, D. Wu, Y. Guo, H.-f. Wang, Determination of structure and energetics for Gibbs surface adsorption layers of binary liquid mixture 1. Acetone + water. J. Phys. Chem. B **109**(16), 8053–8063 (2005)

22. M.E. Tuckerman, D. Marx, M. Parrinello, The nature and transport mechanism of hydrated hydroxide ions in aqueous solution. Nature **417**(6892), 925–929 (2002)
23. S.T. van der Post, C.S. Hsieh, M. Okuno, Y. Nagata, H.J. Bakker, M. Bonn, J. Hunger, Strong frequency dependence of vibrational relaxation in bulk and surface water reveals sub-picosecond structural heterogeneity. Nat. Commun. **6**, 8384 (2015)
24. L. Liu, J. Hunger, H.J. Bakker, Energy relaxation dynamics of the hydration complex of hydroxide. J. Phys. Chem. A **115**(51), 14593–14598 (2011)
25. J. Hunger, L. Liu, K.-J. Tielrooij, M. Bonn, H. Bakker, Vibrational and orientational dynamics of water in aqueous hydroxide solutions. J. Chem. Phys. **135**(12), 124517 (2011)
26. A. Mandal, K. Ramasesha, L. De Marco, A. Tokmakoff, Collective vibrations of water-solvated hydroxide ions investigated with broadband 2DIR spectroscopy. J. Chem. Phys. **140**(20), 204508 (2014)
27. Y. Zhou, D. Wu, Y. Gong, Z. Ma, Y. Huang, X. Zhang, C.Q. Sun, Base-hydration-resolved hydrogen-bond networking dynamics: Quantum point compression. J. Mol. Liq. **223**, 1277–1283 (2016)
28. W.H. Robertson, E.G. Diken, E.A. Price, J.-W. Shin, M.A. Johnson, Spectroscopic determination of the OH− solvation shell in the OH− ·(H$_2$O) n clusters. Science **299**(5611), 1367–1372 (2003)
29. R.-J. Lin, Q.C. Nguyen, Y.-S. Ong, K. Takahashi, J.-L. Kuo, Temperature dependent structural variations of OH–(H 2 O) n, n = 4–7: effects on vibrational and photoelectron spectra. Phys. Chem. Chem. Phys. **17**(29), 19162–19172 (2015)
30. C.Q. Sun, Y. Sun, The Attribute of Water: Single Notion, Multiple Myths. Springer Ser. Chem. Phys. **113**. (2016). Heidelberg: Springer-Verlag. 494 pp
31. C. Chaudhuri, Y.-S. Wang, J. Jiang, Y. Lee, H.-C. Chang, G. Niedner-Schatteburg, Infrared spectra and isomeric structures of hydroxide ion-water clusters OH-(H$_2$O) 1-5: a comparison with H3O (H$_2$O) 1-5. Mol. Phys. **99**(14), 1161–1173 (2001)
32. S.T. Roberts, P.B. Petersen, K. Ramasesha, A. Tokmakoff, I.S. Ufimtsev, T.J. Martinez, Observation of a Zundel-like transition state during proton transfer in aqueous hydroxide solutions. Proc. Natl. Acad. Sci. **106**(36), 15154–15159 (2009)
33. F. Dahms, R. Costard, E. Pines, B.P. Fingerhut, E.T. Nibbering, T. Elsaesser, The hydrated excess proton in the Zundel cation H$_5$O$_2$ (+): The role of ultrafast solvent fluctuations. Angew. Chem. Int. Ed. Engl. **55**(36), 10600–10605 (2016)
34. C.T. Wolke, J.A. Fournier, L.C. Dzugan, M.R. Fagiani, T.T. Odbadrakh, H. Knorke, K.D. Jordan, A.B. McCoy, K.R. Asmis, M.A. Johnson, Spectroscopic snapshots of the proton-transfer mechanism in water. Science **354**(6316), 1131–1135 (2016)
35. H. Sies, Hydrogen peroxide as a central redox signaling molecule in physiological oxidative stress: Oxidative eustress. Redox Biol. **11**, 613–619 (2017)
36. S. Fukuzumi, Y. Yamada, Hydrogen peroxide used as a solar fuel in one-compartment fuel cells. Chemelectrochem **3**(12), 1978–1989 (2016)
37. A. Asghar, A.A. Abdul Raman, W.M.A. Wan Daud, Advanced oxidation processes for in-situ production of hydrogen peroxide/hydroxyl radical for textile wastewater treatment: a review. J. Cleaner Prod. **87**, 826–838 (2015)
38. L. An, T. Zhao, X. Yan, X. Zhou, P. Tan, The dual role of hydrogen peroxide in fuel cells. Sci. Bull. **60**(1), 55–64 (2015)
39. Z.-M. Pei, Y. Murata, G. Benning, S. Thomine, B. Klusener, G.J. Allen, E. Grill, J.I. Schroeder, Calcium channels activated by hydrogen peroxide mediate abscisic acid signalling in guard cells. Nature **406**(6797), 731–734 (2000)
40. S.J. Neill, R. Desikan, A. Clarke, R.D. Hurst, J.T. Hancock, Hydrogen peroxide and nitric oxide as signalling molecules in plants. J. Exp. Bot. **53**(372), 1237–1247 (2002)
41. P. Han, X.H. Zhou, N. Chang, C.L. Xiao, S. Yan, H. Ren, X.Z. Yang, M.L. Zhang, Q. Wu, B. Tang, J.P. Diao, X. Zhu, C. Zhang, C.Y. Li, H. Cheng, J.W. Xiong, Hydrogen peroxide primes heart regeneration with a derepression mechanism. Cell Res. **24**(9), 1091–1107 (2014)
42. J. Chen, C. Yao, X. Liu, X. Zhang, C.Q. Sun, Y. Huang, H2O2 and HO- solvation dynamics: solute capabilities and solute-solvent molecular interactions. Chem. Sel. **2**(27), 8517–8523 (2017)

43. X. Zhang, P. Sun, Y. Huang, T. Yan, Z. Ma, X. Liu, B. Zou, J. Zhou, W. Zheng, C.Q. Sun, Water's phase diagram: from the notion of thermodynamics to hydrogen-bond cooperativity. Prog. Solid State Chem. **43**, 71–81 (2015)
44. C.Q. Sun, *Atomic Scale Purification of Electron Spectroscopic Information (US 2017 patent No. 9,625,397B2)* (United States, 2017)
45. C.Q. Sun, X. Zhang, J. Zhou, Y. Huang, Y. Zhou, W. Zheng, Density, elasticity, and stability anomalies of water molecules with fewer than four neighbors. J. Phys. Chem. Lett. **4**, 2565–2570 (2013)
46. Y. Crespo, A. Hassanali, Characterizing the local solvation environment of OH− in water clusters with AIMD. J. Chem. Phys. **144**(7), 074304 (2016)
47. C.Q. Sun, *Relaxation of the Chemical Bond*. Springer Ser. Chem. Phys. **108**, (2014). Heidelberg: Springer-Verlag. 807 pp
48. C.Q. Sun, X. Zhang, W.T. Zheng, Hidden force opposing ice compression. Chem. Sci. **3**, 1455–1460 (2012)
49. X. Zhang, Y. Huang, Z. Ma, Y. Zhou, W. Zheng, J. Zhou, C.Q. Sun, A common supersolid skin covering both water and ice. Phys. Chem. Chem. Phys. **16**(42), 22987–22994 (2014)
50. F. Li, Y. Wang, C. Sun, Z. Li, Z. Men, Spectra study hydrogen bonds dynamics of water molecules at NaOH solutions. J. Mol. Liq. **277**, 58–62 (2019)
51. Q. Zeng, C. Yao, K. Wang, C.Q. Sun, B. Zou, Room-temperature NaI/H_2O compression Icing: solute–solute interactions. PCCP **19**, 26645–26650 (2017)
52. B. Wang, W. Jiang, Y. Gao, Z. Zhang, C. Sun, F. Liu, Z. Wang, Energetics competition in centrally four-coordinated water clusters and Raman spectroscopic signature for hydrogen bonding. RSC Adv. **7**(19), 11680–11683 (2017)
53. X. Zhang, X. Liu, Y. Zhong, Z. Zhou, Y. Huang, C.Q. Sun, Nanobubble Skin Supersolidity. Langmuir **32**(43), 11321–11327 (2016)
54. S.A. Harich, X. Yang, D.W. Hwang, J.J. Lin, X. Yang, R.N. Dixon, Photodissociation of D_2O at 121.6 nm: a state-to-state dynamical picture. J. Chem. Phys. **114**(18), 7830–7837 (2001)
55. S.A. Harich, D.W.H. Hwang, X. Yang, J.J. Lin, X. Yang, R.N. Dixon, Photodissociation of H_2O at 121.6 nm: a state-to-state dynamical picture. J. Chem. Phys. **113**(22), 10073–10090 (2000)
56. J. Peng, J. Guo, R. Ma, X. Meng, Y. Jiang, Atomic-scale imaging of the dissolution of NaCl islands by water at low temperature. J. Phys.: Condens. Matter **29**(10), 104001 (2017)
57. M.R. Rahimpour, M.R. Dehnavi, F. Allahgholipour, D. Iranshahi, S.M. Jokar, Assessment and comparison of different catalytic coupling exothermic and endothermic reactions: a review. Appl. Energy **99**, 496–512 (2012)
58. R.C. Ramaswamy, P.A. Ramachandran, M.P. Duduković, Coupling exothermic and endothermic reactions in adiabatic reactors. Chem. Eng. Sci. **63**(6), 1654–1667 (2008)
59. N. Shahrin, Solubility and dissolution of drug product: a review. Int. J. Pharma. Life Sci. **2**(1), 33–41 (2013)
60. J.B. Rosenholm, Critical evaluation of dipolar, acid-base and charge interactions I. Electron displacement within and between molecules, liquids and semiconductors. Adv. Colloid Interface Sci. **247**, 264–304 (2017)
61. E.L. Ratkova, D.S. Palmer, M.V. Fedorov, Solvation thermodynamics of organic molecules by the molecular integral equation theory: approaching chemical accuracy. Chem. Rev. **115**(13), 6312–6356 (2015)
62. J. Konicek, I. Wadso, Thermochemical properties of some carboxylic acids, amines and N-substituted amides in aqueous solution. Acta Chem. Scand. **25**(5), 1461–1551 (1971)
63. G. Graziano, Hydration thermodynamics of aliphatic alcohols. PCCP **1**(15), 3567–3576 (1999)
64. A.M. Ricks, A.D. Brathwaite, M.A. Duncan, IR spectroscopy of gas phase V(CO_2)n + clusters: solvation-induced electron transfer and activation of CO_2. J. Phys. Chem. A **117**(45), 11490–11498 (2013)
65. A.M. Zaichikov, M.A. Krest'yaninov, Structural and thermodynamic properties and intermolecular interactions in aqueous and acetonitrile solutions of aprotic amides. J. Struct. Chem. **54**(2), 336–344 (2013)

66. M. Wohlgemuth, M. Miyazaki, M. Weiler, M. Sakai, O. Dopfer, M. Fujii, R. Mitrić, Solvation dynamics of a single water molecule probed by infrared spectra–theory meets experiment. Angew. Chem. Int. Ed. **53**(52), 14601–14604 (2014)

67. C. Velezvega, D.J. Mckay, T. Kurtzman, V. Aravamuthan, R.A. Pearlstein, J.S. Duca, Estimation of solvation entropy and enthalpy via analysis of water oxygen-hydrogen correlations. J. Chem. Theory Comput. **11**(11), 5090 (2015)

68. K. Haldrup, W. Gawelda, R. Abela, R. Alonso-Mori, U. Bergmann, A. Bordage, M. Cammarata, S.E. Canton, A.O. Dohn, T.B. Van Driel, Observing solvation dynamics with simultaneous femtosecond X-ray emission spectroscopy and X-ray scattering. J. Phys. Chem. B **120**(6), 1158–1168 (2016)

69. L. Pauling, *The Nature of the Chemical Bond.* 3 ed. (Cornell University press, Ithaca, NY, 1960)

70. C.Q. Sun, J. Chen, X. Liu, X. Zhang, Y. Huang, (Li, Na, K)OH hydration bondin thermodynamics: Solution self-heating. Chem. Phys. Lett. **696**, 139–143 (2018)

71. C.Q. Sun, X. Zhang, X. Fu, W. Zheng, J.-L. Kuo, Y. Zhou, Z. Shen, J. Zhou, Density and phonon-stiffness anomalies of water and ice in the full temperature range. J. Phys. Chem. Lett. **4**, 3238–3244 (2013)

72. Y.L. Huang, X. Zhang, Z.S. Ma, Y.C. Zhou, W.T. Zheng, J. Zhou, C.Q. Sun, Hydrogen-bond relaxation dynamics: resolving mysteries of water ice. Coord. Chem. Rev. **285**, 109–165 (2015)

73. T. Brinzer, E.J. Berquist, Z. Ren, 任哲, S. Dutta, C.A. Johnson, C.S. Krisher, D.S. Lambrecht, S. Garrett-Roe, Ultrafast vibrational spectroscopy (2D-IR) of CO_2 in ionic liquids: Carbon capture from carbon dioxide's point of view. J. Chem. Phys. **142**(21), 212425, (2015)

74. Z. Ren, A.S. Ivanova, D. Couchot-Vore, S. Garrett-Roe, Ultrafast structure and dynamics in ionic liquids: 2D-IR spectroscopy probes the molecular origin of viscosity. J. Phys. Chem. lett. **5**(9), 1541–1546 (2014)

75. Q. Zhang, T. Wu, C. Chen, S. Mukamel, W. Zhuang, Molecular mechanism of water reorientational slowing down in concentrated ionic solutions, in *Proceedings of the National Academy of Sciences*: 201707453, (2017)

76. C.Q. Sun, Perspective:Unprecedented O:⇔:O compression and H↔H fragilization in Lewis solutions. Phys. Chem. Chem. Phys. **21**, 2234–2250 (2019)

77. Y. Huang, X. Zhang, Z. Ma, Y. Zhou, J. Zhou, W. Zheng, C.Q. Sun, Size, separation, structure order, and mass density of molecules packing in water and ice. Sci. Rep. **3**, 3005 (2013)

78. Y.H. Siyan Gao, X. Zhang, C.Q. Sun, *Unexpected Solute Occupancy and Anisotropic Polarizability in Lewis Basic Solutions.* Communicated, (2019)

Chapter 6
Hofmeister Salt Solutions: Screened Polarization

Contents

Abstract Water dissolves salt into ions and then hydrates the ions in an aqueous solution. Hydration of ions deforms the hydrogen bonding network and triggers the solution with what the pure water never shows such as conductivity, molecular diffusivity, thermal stability, surface stress, solubility, and viscosity, having enormous impact to many branches in biochemistry, chemistry, physics, and energy and environmental industry sectors. However, regulations for the solute-solute-solvent interactions are still open for exploration. From the perspective of the screened ionic polarization and O:H–O bond relaxation, this chapter is focused on understanding the hydration dynamics of Hofmeister ions in the typical YI, NaX, ZX$_2$, and NaT salt solutions (Y = Li, Na, K, Rb, Cs; X = F, Cl, Br, I; Z = Mg, Ca, Ba, Sr; T = ClO$_4$, NO$_3$, HSO$_4$, SCN). Phonon spectrometric analysis turned out the f(C) fraction of bond transition from the mode of deionized water to the hydrating.

© Springer Nature Singapore Pte Ltd. 2019
C. Q. Sun, *Solvation Dynamics*, Springer Series in Chemical Physics 121, https://doi.org/10.1007/978-981-13-8441-7_6

The linear $f(C) \propto C$ form features the invariant hydration volume of small cations that are fully-screened by their hydration H_2O dipoles. The nonlinear $f(C) \propto 1 - \exp(-C/C_0)$ form describes that the number insufficiency of the ordered hydrating H_2O diploes partially screens the anions. Molecular anions show stronger yet shorter electric field of dipoles. The screened ionic polarization, inter-solute interaction, and O:H–O bond transition unify the solution conductivity, surface stress, viscosity, and critical energies for phase transition.

Highlight

- Ionic polarization shortens the H–O bond and stiffens its phonon but relaxes the O:H contrastingly.
- Anion-anion interaction limits a $f_X(C) \propto 1 - \exp(-C/C_0)$ fraction of O:H–O bonds to be polarized.
- Cations form invariantly sized hydration volumes without being interfered by any other solutes.
- Polarization reconciles skin stress, conductivity, supersolidity, and viscosity.

6.1 Wonders of Salt Solvation

6.1.1 Background

Injection of ions by salt solvation distorts the hydrogen bond (O:H–O or HB with ':' being the electron lone pair) network and triggers the solution with unexpected properties compared with the deionized water that serves as the solvent. Solvation dynamics has been amusing the community with debating mechanisms despite overwhelming contributions made since 1888 when Frank Hofmeister [1] discovered the ionic series order on protein solubility and the solution surface stress. Salt solvation is ubiquitously important to innumerable reactions occurring to living and nonliving systems, which has profound impact to many biological, chemical, and environmental issues [2–4] such as DNA engineering [5], disease curing and health caring [2, 6]. Salt solvation changes the solution viscoelasticity [7, 8], thermal and mechanical stability [9], critical energies in terms of temperatures and pressures for phase transition [10, 11], and even reduces the friction when the salt solutions are used as lubricants [12–14]. Ionic hydration forms a stiffener and cylindrical volume coupling a cation and a pair of point defects in the adjacent graphite-oxide layers. The stiffer '(−) ~ (+) ~ (−)' vertical hydration volume enlarges the graphene-oxide layer separation from 0.33 up to 1.5 nm, which could be promising for sea water purification [15, 16].

The anionic abilities in precipitating certain egg proteins follow the Hofmeister series order and it is solute concentration dependent [17–22]. The phenomenon was explained by means of water structure order maker that shortens the O:H distance and water breaker that does it otherwise [23–25], ionic specification and affinity matching [8, 26], quantum dispersion [8, 27], skin induction [28], length scale of interaction [29], and many more. Hydrogen bond quantum dispersion was believed important in ligand-water interaction [30] and cellobiose-water interactions [31].

Ions could be categorized as having the 'salting in' and 'salting out' effect on solubility of a protein [32]. The 'salting in' refers to the stronger ionic solubility of proteins that are surrounded by ions, which lowers the electrostatic energy of the protein to raise the proteins' solubility, and vice versa. It is expected that the ionic size, electronegativity, and polarity determine the ions acting on the surrounding water molecules leading to salting in or salting out proteins in aqueous solutions.

Compared with the monovalent ions of the same centration C, the Z^{2+} and $2X^-$ in a divalent salt solution "hardens" drinking water and they are harmful to living creatures [33, 34]. From the perspective of classical thermodynamics, namely, the density, isothermal compressibility, and surface tension, a coarse-grained electrolyte solution model was proposed, being capable of reproducing some trends on how salts influence the diffusivity and viscosity of the solutions [35, 36].

Salt solvation differs the local physical–chemical properties of the hydration volume from those of unsalted water. A neutron diffraction investigation of nickel chloride hydration as a function of solute concentration, pressure, and temperature revealed that heating can reduce the Ni^{2+} and Cl^- hydration volume [37]. Monte Carlo simulation and neutron diffraction [34] showed that the electrostatic field from the divalent cations in $(Mg, Ca)Cl_2$ solutions strengthens their interactions with solvent water molecules to form a more rigid and stable supersolid, first hydration shell [38, 39], compared with the monovalent ionic solutions.

Molecular dynamics (MD) computations and light-scattering investigation suggested that the negatively charged carboxylic acid side-chain groups preferring pairing with smaller to the bigger cations [40, 41]. This phenomenon was attributed to the different proteins' surface sensitivity to the Na^+ or K^+ cations, because the Na^+ 'poisons' the proteins more readily than the K^+, which might be a clue why the cytosol is rich in K^+ but poor in Na^+ [42]. Similarly, the positively charged side-chain groups of the basic amino acids pair are more efficiently interacting with smaller cations, than larger anions, resulting in an reverse Hofmeister ordering at these sites [43, 44]. Interactions with these charged groups can be comparable to—or even overwhelm—those with the backbone, which would then cause the Hofmeister reversal for the peptide or protein.

The ionic solute–water dipolar interaction affects water's reorientation dynamics and the H–O (ω_H) and D–O (ω_D) stretching vibrational frequencies in the H_2O + D_2O mixture solvent. For instance, Smith et al. [45] probed the phonon spectra of the H_2O + D_2O mixture with and without the solvation of 1 mol K(F, Cl, Br, I) and uncovered that larger anions with lower electronegativity could stiffen the ω_H more than the contrast. The F^- has negligible effect on the ω_H phonon frequency shift.

From 2DIR spectroscopy and MD simulations, Park and coworkers [46] found that 5% NaBr addition stiffens the D–O stretching vibration frequency from 2509 to 2539 cm^{-1} in the HOD + H_2O solutions and the phonon frequency shifts with the number ratio of N_{H2O}/N_{NaBr} = 8, 16, and 32. Higher solute concentration induces more significant ω_{D-O} peak shift with weaken screened by the hydrating H_2O dipoles or other anions. The ω_{D-O} blueshift raises its lifetime. NaBr hydration not only causes the ω_{D-O} blueshift but also slows the molecular dynamics—or in another term, the rate of phonon energy dissipation varies inversely with the phonon frequency—high frequency phonons have longer lifetime. An increase of NaBr concentration also raises the solution viscosity, which is accompanied by lowering the translational motion dynamics of water molecules or the order of the molecular fluctuation dynamics. Salt such as (Na, Li)Cl [47, 48], Na(Br, ClO_4) [46], and $Mg(ClO_4)_2$ [49–51] solvation also shifts positively the ω_H of the hydroxyl group (OH^- or OD^-).

DFT optimizations [52] suggested that within the closest hydration shell of an I^- anion, the $I \cdot H$ distance is about 40% longer and the H–O is some ~1% shorter than the O:H and H–O respectively in bulk water. However, the O:H in the hydration shell is 29% longer and the H–O is 0.9% shorter. The X^- capability of inducing $X^- \cdot H–O$ and O:H–O relaxations obeys the $I^- > Br^- > Cl^-$ Hofmeister series order.

The elastic second harmonic scattering from >0.05 M (Cs, Li, K, Na, Rb)(Cl, NH_4) and (Ca, Mg, Sr)Cl_2 solutions showed the ionic specificity in both the local electronic anisotropy and in the H_2O molecule nanoscopic orientation ordering [53]. Cations with higher valences and smaller sizes affect more than anions the charge distribution around ions and the structural orientation of water molecules in the extended hydration shells. As it has been confirmed that ions in salt [11, 19, 54, 55], basic [11], and acidic [56, 57] solutions serves each as a charge center to align, cluster, stretch and polarize its neighboring H_2O molecules to form a supersolid hydration shell. Within the hydration volume, the H–O bond turns to be stiffer and the O:H nonbond softer.

Extensive attention has also been paid to the ionic effect in complex salts [58–61]. A combination of MD simulations and ultrafast IR measurements [58] suggested that the rotational mobility of the SCN^- in KSCN solutions is relatively higher. 2DIR investigation [50] revealed that the ω_H band shifts positively as the $NaClO_4$ molar concentration is increased from 5 to 10%. An interplay of the liquid jet NEX-AFS, MD simulations, and ab initio electronic structure calculations [62] elucidated that $MgCl_2$ salt ions affect the electronic properties of water molecules in the close vicinity—short-range.

Salt solvation lowers the critical temperature for homogeneous ice nucleation, T_N, substantially. Reducing the $H_2O/LiCl$ molecular number ratio from 100 to 7, the aqueous solution performs the same in lowering the T_N of the solution from 248 to 190 K [48]. The O:H nonbond energy, E_L, is related to the T_N and the H–O bond energy, E_H, to the melting temperature T_m [63], according to Einstein's relation of $\Theta_{Dx} \propto \omega_x \propto (E/d^2)_x$ with subscript x = L and H for the O:H and H–O interactions. Θ_{Dx} is the Debye temperature.

Salt solvation raises the solution viscosity. The viscosity is an important macroscopic parameter that is often used to classify water-soluble salts into structure order making or structure breaking. In an aqueous solution, the solute molecules can be taken as Brownian particles to move with a drift diffusivity $D(\eta, R, T)$ of Stokes-

Einstein relation [64]. The solution viscosity η changes with the solute concentration C, in terms of the Jones–Dale notion [65, 66],

$$\Delta\eta(C)/\eta(0) = AC^{1/2} + BC.$$

The prefactor A is related to the inter-solute ionic interactions and keeps a constant for the alkali halide solutions. The B describes the solute-solvent molecular interactions. The $\eta(0)$ is the reference of water. However, it has been [38] clarified that the cationic polarization governs the linear term whilst the interanion repulsion and anionic polarization dictates the nonlinear term of the Jones–Dale description. For the basic and acid solutions, divalent and complex molecular salt solutions, the viscosity doe not follow the notion of Jones–Dale, showing the $\partial^2\eta/\partial C^2 > 0$, originating from the dominance of polarization and solute-solute repulsion.

The diffusivity and viscosity are correlated to the H–O phonon relaxation time $\tau(\omega)$ obtained from the 2DFR spectroscopy,

$$\frac{I(\omega, t)}{I(\omega, 0)} = \exp[-t/\tau(\omega)]$$

One can gain the $\tau(\omega)$ by recording and simulating the decay profile in the 2DIR experiments. The $\tau(\omega)$ varies not only with the exciting frequency but also the solute concentration, solute type, temperature, and molecular coordination environment [67]. Cooling or increasing solute concentration could prolong the $\tau(\omega)$. Figure 6.1a shows that the faster H–O bond stretching vibration corresponds to a longer relaxation time or slower molecular motion dynamics. The H–O dangling bond whose vibration is featured at $(3700 \text{ cm}^{-1}, 850 \text{ fs})$ in the SFG (at 3610 cm^{-1} of Raman spectrum) diffuses slower than those in the skin $(3450 \text{ cm}^{-1}, 500 \text{ fs})$ and in the bulk $(3200 \text{ cm}^{-1}, 250 \text{ fs})$. Alternatively, the lifetime can be related to the rate of vibration energy dissipation—higher vibration energy requires a longer time for dissipation [39].

Figure 6.2 correlates the relaxation time with the viscosity of SCN^- and CO_2 solutions obtained from SFG measurements [68, 69]. One can modulate the solution viscosity by changing solution temperature or solute concentration. Indeed, the relaxation time is proportional to the solution viscosity, which confirms that the H–O phonon blueshift is associated with slower molecular dynamics, or the energy dissipation rate is inversely proportional to the phonon frequency. However, liquid heating and acid solving do not follow this trend because of the thermal fluctuation and network disruption, see Fig. 6.2a, though the H–O phonon blueshift retains [56, 58].

Using a polarization selective IR pump-probe spectroscopy, Wei et al. [70] examined the cation effects on the D–O phonon relaxation dynamics of water molecules in (Li, Na)ClO$_4$ aqueous solutions. They found that D–O peak splits into two components and the viscosity increases dramatically, with $\partial^2\eta/\partial C^2 > 0$. The Li$^+$ has a much weaker effect than the Na$^+$ on the population peak at 2630 cm^{-1}, and thus its effect on the D–O bond relaxation dynamics is hardly resolvable.

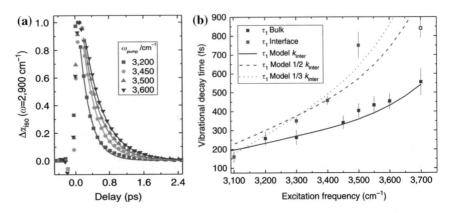

Fig. 6.1 a Normalized infrared pump-probe data for H-bonded OH groups in bulk H_2O. The delay traces are taken at $\omega_{probe} = 2900\ cm^{-1}$ with the pump frequencies centered at $\omega_{pump} = 3200, 3450, 3500$ and $3600\ cm^{-1}$. **b** H–O vibrational relaxation time constants of bulk (blue) and interfacial (red) H_2O depend functionally on the H–O stretching excitation frequency. The open-red symbol corresponds to the vibrational relaxation time of the free H–O radicals. Reprinted with copyright permission from [67]

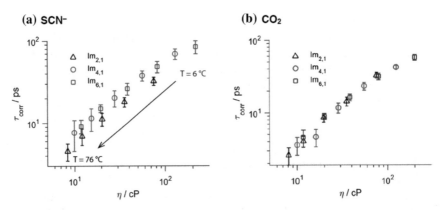

Fig. 6.2 Correlation between the H–O phonon relaxation time and the viscosity of **a** SCN^- and **b** CO_2 solutions. Solution cooling and solute concentration increasing raise the relaxation time and viscosity because of polarization effect. Reprinted from copyright permission from [68, 69]

In contrast, an ultrafast pump-probe spectroscopy examination of the ionic effect on the phonon relaxation time of water molecules in $Mg(ClO_4)_2$, $NaClO_4$, and Na_2SO_4 solutions suggested that an addition of ions does not affect the rotational dynamics of the free water molecules outside the first hydration shells or enhance or breakdown the HB network in liquid water [71]. However, supported by neutron diffraction [72] and viscosity measurements [73], Tielrooij et al. [74] showed that a certain kind of salts do have an effect on the dynamics of water molecules of longer distance away from the ionic solute of multiple shells. A terahertz and femtosecond IR spectroscopic investigation revealed that the Mg, Li, Na, and Cs cations and S,

Cl, I anions can interact strongly with H_2O molecules well beyond the first hydration shell of water molecules.

An IR examination [75] of the $\Delta\omega_{D\text{-}O}$ in the CsCl, (K, Cs)F, and (Na, K)I, solutions of $0.04D_2O + 0.96H_2O$ solvent correlated the $\Delta\omega_{D\text{-}O}$ to the solute capability of structure making ($B > 0$) and breaking ($B < 0$) in Jones–Dale notion. The $\Delta\omega_{D\text{-}O}$ redshift at $2500\ cm^{-1}$ for (Na, K)I and CsCl was attributed to the structure order breaking but the $\Delta\omega_{D\text{-}O}$ blueshift for (K, Cs)F to structure order making. Slight $\Delta\omega_D$ redshift does happen only at higher solute concentrations, arising from the partially-screened anion-anion interaction that weakens the local electric field of the anion and reduces its hydration volume.

The high elasticity, solubility, and viscosity of the supersolid skin of the deionized water and the ionic hydration shells have been attributed to the electrostatic polarization by molecular undercoordination or by the ionic hydration [2, 39]. The supersolid phase is associated with slower molecular motion, shorter H–O bond, longer lifetime, longer O:H; lower T_N, and higher T_m. The skin supersolid stems the hydrophobicity of water, slipperiness of ice, and Mpemba effect—warm water cools faster because of the lower mass density and high thermal diffusivity [76, 77].

To examine the microscopic mechanism of solution viscosity, Omta et al. [71] examined the effects of ions on the orientation correlation time of water molecules in $Mg(ClO_4)_2$, $NaClO_4$, and Na_2SO_4 solutions by means of femtosecond pump-probe spectroscopy. They suggested that an addition of ions had no influence on the rotational dynamics of water molecules outside the first ionic hydration shells and that the presence of ions does not lead to an enhancement or a breakdown of the HB network in liquid water.

However, supported by neutron diffraction [72] and viscosity measurements [73], Tielrooij et al. [74] showed that a certain kind of salts can change water reorientation dynamics at a longer distance or multiple shells. A joint terahertz and femtosecond IR spectroscopic study of water dynamics around different, specifically Mg, Li, Na, and Cs cations and S, Cl, I anions, revealed that the effect of ions and counterions on water can be strongly interdependent and non-additive, and in certain cases extends well beyond the first solvation shell of water molecules directly surrounding the ion.

Nickolov and Miller [75] examined the $\Delta\omega_D$ in (K, Cs)F, (Na, K)I, and CsCl solutions of 4 wt% D_2O mixed in H_2O solvent using FTIR at the full concentration range. They correlated the $\Delta\omega_D$ to the solute characteristics of structure making ($B > 0$) and breaking ($B < 0$) in Jones–Dale notion. They attributed the $\Delta\omega_D$ redshift at $2500\ cm^{-1}$ for (Na, K)I and CsCl to structure breaking but the $\Delta\omega_D$ blueshift for (K, Cs)F to structure making.

It has been intensively demonstrated [2] that the high elasticity, solubility, and viscosity of the supersolid skin of the ordinary water and the ionic hydration shells arise from polarization induced by molecular undercoordination or by the ionic electrostatic field. In the supersolid phase, molecular motion is slow, the H–O bond is shorter and its phonon relaxation time is longer, the O:H is longer; the T_N is lower, and the T_m is higher. The supersolid nature determines the hydrophobicity of water skin, slipperiness of ice, and the high thermal conductivity for thermal diffusion dominating the Mpemba effect—warm water cools faster [76, 77].

6.1.2 Known Mechanisms

Hofmeister [1] made a heuristic attempt at interpreting his observations, based on the theory of electrolytic dissociation. Hofmeister extended his studies to additional proteins and colloidal particles and tried to connect the observed ordering of ions with their strength of hydration, denoted as the water-absorbing effects of salts. Hofmeister's explanation for the ionic ordering was eventually framed into the theory of structure-making and structure-breaking ions during 1930–1950s [23–25].

Numerous hypothetic theories have been developed on the solute-solvent interactions and solvation dynamics. Collins [8, 26], a biologist at Maryland, proposed the "ion-water affinity matching", assuming that cations and anions of comparable hydration free energies tend to pair in water. Salt solvation effect not only depends on the type of ions, but also shows its cooperativity with other ions as a collection. Ion pairing influences the capability of aqueous solutions on activity coefficients, dissolving proteins, and varying surface stress. It is currently believed that Hofmeister effect results from the versatile abilities of ions to replace water molecules at the surfaces. The different surface potential and stress could be used to explain how ions affect protein's stability through indirect interactions [73].

One can hardly explain the specific ion properties of electrolyte solutions if based solely on electrostatics without considering quantum mechanical effect or specific ion dispersion with other ions and water molecules [27]. Collins's rule could be extended the dispersion interactions to the ionic interactions in calculations, which could explain some puzzles of electrolyte solutions [8]: the solvation energy differences among ions of the same size, the small repulsion of I^- at the air/water interface, and the ionic affinity of large ions for each other in water embodied in Collins's rules.

Conversely, Liu et al. [28] identified that the Hofmeister effect is strong in the Ca^{2+} and Na^+ exchange on a charged surface over a broad range of ionic strengths. They disputed that their findings result from a strong force of ion-surface induction than the classical induction, dispersion, hydration, or ionic size. The strength of this induction force was about 104 times that of the classical induction, which could be comparable to the Coulomb interaction force. The presence of the observed strong non-classical induction force suggested that energies of electrons in the deeper bands of ions and atoms at the interface might be important but underestimated, and it is just the underestimated energies determine the Hofmeister effects. Yes, indeed, the ions and H–O bond at a surface perform differently from they do in the bulk because of bond contraction, quantum entrapment and polarization effect [39].

Hofmeister [1] interpreted his observations from the framework of electrolytic dissociation and tried to connect the observed ions ordering with their strength of hydration, denoted as the water-absorbing effects of salts. Hofmeister's effort was eventually framed into the theory of ionic structure-making and structure-breaking during the 1930–1950s [23–25].

However, it is hard to tell a simple story about the effect of ionic hydration on solution properties using any single concept under a given condition. Zangi et al. [78] showed in 2006, ions of high charge quantity (q) could induce salting-out by

Table 6.1 Attributes of ion structure-making and structure-breaking[a]

	Structure maker (kosmotropes)	Structure breaker (chaotropes)
Solute	Small size and high-q	Large size and low-q
Functionality	• Enhance O:H order • Strengthen hydration • Salting-out- 'steal' water from proteins • Precipitate proteins and prevent unfolding • Strengthen the hydrophobic interaction • Commonly used in protein purification through ammonium sulfate precipitation	• Weaken O:H bonding • Disrupt water structure order and denature DNA • Increases the unfolding or denaturation of protein • Weaken the hydrophobic effect • Interact more strongly with the unfolded form of a protein than with its native form
Examples	F^-; Mg_2^+	I^-; NH_4^+; SCN^-

[a]Ions produce long—range effects on the structure of water, leading to changes in water's ability of letting proteins fall out of, or stay dissolved in, a solution

promoting the hydrophobic interactions to aggregate the proteins. Conversely, the low-q ions could have either a salting-out or a salting-in effect, depending on the ionic concentrations. Without considering the nature of the protein surface itself or solvent-solute-protein interactions, it could be impossible to rationalize the ordering of ions, or exceptions to the Hofmeister behavior [79]. Table 6.1 features the attributes of structure-maker and structure-breaker.

6.1.3 Challenge and Objectives

Currently, the majority approaches is focused on the solute motion dynamics, hydration length scale, water structure making or breaking, salting in or out and molecular relaxation lifetime by taking an H_2O molecule as the basic structure element. However, solute capabilities, solute-solute and solute-solvent interactions and their effect on the HB network transition and charge polarization is a high challenge. Correlating the H–O bond and O:H nonbond interactions, transition of HB length and phonon abundance-stiffness, H–O bond energy dissipation, molecular motion dynamics, solute diffusivity, solution viscosity, surface stress, and critical energies (pressures and temperatures) for phase transitions is necessary. Most importantly, discriminating the effects of Hofmeister ions on the hydrogen bonding network distortion and solute properties is the foremost task of the community [80].

Further progress in understanding ion-specific effects in biological systems might require researchers to go beyond the concept of separate anionic and cationic Hofmeister series [22]. What matters the performance of salt solutions is not only the behavior of individual ions at the protein surface but, to varying extents, also interactions between the solute ions themselves, both near the protein and in the bulk aqueous solution. Such effects become operational at high salt concentrations and are distinct from non-specific electrostatic interactions.

In fact, the nature of the chemical bond bridges the structure and properties of a molecule and a crystal [81] and the formation, dissociation and relaxation of the chemical bond and nonbond and their associated energetics, localization, entrapment, and polarization of electrons in various energy levels govern the performance of a substance [82]. Bond length and energy relaxation dictate the phonon frequency shift and the electron polarization and the solute-solvent-solute interactions dictate the solution conductivity, hydrophobicity, surface stress, viscosity, and solubility of the solution [83, 84]. The joint effect of bond relaxation and charge polarization dictates the performance of water, ice and aqueous solutions [81, 82].

It is the opinion of the practitioner that one shall transit from the solvation molecular dynamics to hydration bonding dynamics. Intermolecular O:H, H↔H, and O:⇔:O interactions, and the screened ionic polarization in association of the O:H–O cooperative relaxation would be essential. One should treat solvent water as a highly-ordered crystal rather than the randomly ordered structure or protons being freely motion. Consideration of the anomalous performance of water solvent from the perspective of ion electrification is necessary. Focusing on the O:H–O segmental relaxation and the associated polarization in the first hydration subshell would be suffice [3, 4, 20]. Without knowing the attribute of water properly, it could be very hard to understand correctly the solvation dynamics.

Progress in the NaX [19], YI [54], complex NaT [52], and divalent ZX_2 [85] hydration dynamics has confirmed consistently our proposals of solvation charge injection. Salt hydration creates Hofmeister ions that form each a point or a dipolar charge center to align, cluster, stretch, and polarize water molecules to form supersolid hydration volume. H–O contraction and O:H elongation take place with an association of strong polarization in the supersolid hydration volume [76, 86] associated with ω_H blueshift and ω_L redshift. The supersolidity due to electric polarization by charge injection and molecular undercoordination is elastic, and thermo-mechanically stable. Water clusters, skins of water and ice, droplets and nanobubbles perform similarly to the ionic hydration volume. Encouragingly, this way of approach has enabled discrimination of the solute functionalities on the hydrogen bonding network relaxation and property change of salt solutions [38].

With the aid of DPS and XAS spectrometrics, this chapter deals with the length, manner, strength of ionic field interacting with the hydrating O:H–O bonds in the hydration volume, and the effect on the network deformation by bond relaxation and transition from the mode of water to hydrating. A systematic examination of the hydration dynamics of Hofmeister ions in the typical YI, NaX, ZX_2, and NaT salt solutions (Y = Li, Na, K, Rb, Cs; X = F, Cl, Br, I; Z = Mg, Ca, Ba, Sr; T = ClO_4, NO_3, HSO_4, SCN) has led to discrimination of functionalities of these ions. Progress evidences the essentiality of the framework of solvation charge injection, O:H–O bond cooperativity, and the power of the DPS and XAS spectrometrics. Developed strategies and knowledge about factors dictating the properties of the solutions such as conductivity, surface stress, supersolidity, viscosity and the capability of phase boundary dispersion would supply important impact to related fields such as solute-protein interface, colloid-matrix interfaces, water distillation and pollution control, and energy and environment science.

6.2 Ionic Polarization and O:H–O Bond Transition

6.2.1 Na(F, Cl, Br, I) Solvation

6.2.1.1 DPS Derivatives: Fraction of Bond Transition

Examination of Na(F, Cl, Br, I) solvation aims to discriminate the functionalities of X^- anions one from another on the O:H–O phonon frequency shift and bond transition as the Na^+ is assumed to perform identically in these NaX solutions. Figure 6.3 shows the full-frequency Raman spectra for the NaX/H_2O solutions of 0.06 molar concentration and for the concentrated NaI/H_2O solutions. These spectra are compared with the spectrum collected from deionized H_2O probed at the same 298 K temperature and ambient pressure. Salt hydration stiffens the ω_H and softens the ω_L cooperatively and the fraction of phonons transited from the referential mode of ordinary water to the hydrating states increases with solute concentration. The spectral comparison clarified that no vibration feature is resolved for possible solute-solvent bond formation.

The ω_H hydration mode overlaps the skin H–O modes of water and ice featured at 3450 cm^{-1}, showing that water skin and the solute hydration shells share the same H–O bond length and energy though ionic polarization and molecular undercoordination are irrelevant degrees of freedom. The O:H mode shifts from some 200 to

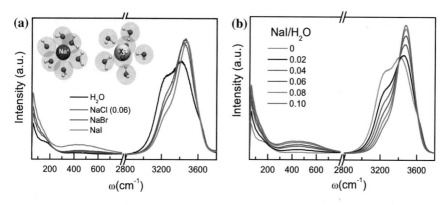

Fig. 6.3 Full-frequency Raman spectra for the **a** solute-type dependent NaX/H_2O solutions of 0.06 molar concentration and the **b** concentrated NaI/H_2O solutions. Inset a illustrates the effect of ionic electrification of the separated ion pairs on the aligning and clustering of water molecules and H_2O dipolar shielding in the hydration shells. Reprinted with copyright permission from [19]

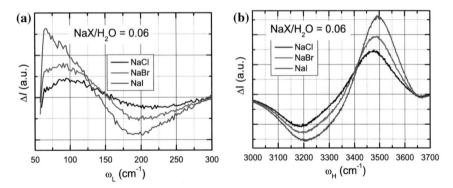

Fig. 6.4 Solute-type-resolved **a** ω_L and **b** ω_H DPS for the 0.06 NaX/H$_2$O solutions with reference to deionized water show the transition of the bond stiffness ($\Delta\omega_x$), phonon abundance (peak area), and fluctuation order (line width) of the O:H–O bond from ordinary water (200; 3200 cm^{-1}) to the hydration shells (75; ~3500 cm^{-1}). The bond stiffness and the fluctuation order vary with the type of X solute. The capability of solute electrification follows the I$^-$ > Br$^-$ > Cl$^-$ Hofmeister series order. Reprinted with copyright permission from [19]

~75 cm^{-1} by ionic polarization. These observations evidence that the intermolecular O:H interaction cooperates strongly with the H–O intramolecular bond relaxation, and one shall consider them inclusively.

The ω_x DPS for the NaX/H$_2$O of the same 0.06 molar concentration in Fig. 6.4 and for the concentrated NaI/H$_2$O solutions in Fig. 6.5 compare the X$^-$ capability of transiting the number fraction (abundance), bond stiffness (frequency shift), and fluctuation order (FWHM) of the solvent O:H–O bonds. The X$^-$ solutes follow the I$^-$ > Br$^-$ > Cl$^-$ capability order of polarization. The solute capability of polarization increases as one switches the X from Cl to I, in terms of the phonon stiffness and phonon abundance. The X$^-$ of larger radius and less electronegative effects more significantly on the O:H–O bonds transition. The X$^-$ and Na$^+$ have the same effect of ionic polarization that shortens the H–O bond and stiffens its ω_H phonon. The O:H nonbond always relaxes in the opposite direction to the H–O bond under perturbation. The $\Delta\omega_x$ recovers slightly at higher concentration, which suggests weakening of the electric field in the hydration shells. However, Raman spectroscopy can hardly discriminate the effect of Y$^+$ from the effect of X$^-$ on contributing to phonon relaxation at this stage.

6.2.1.2 Bond Transition and Polarization

Figure 6.6 compares the fraction coefficients for bond transition and the contact angles between solution and glass substrate for the concentrated NaX/H$_2$O solutions, and the fraction coefficients derived from $f_{Na}(C) = f_{NaX}(C) - f_{HX}(C)$ [87]. See Chap. 4, for proving $f_{HX}(C) = f_X(C)$ and $f_H(C) = 0$. Slight deviation from the linear dependence (bending up) of Na$^+$ in NaCl and NaBr solutions suggests the presence

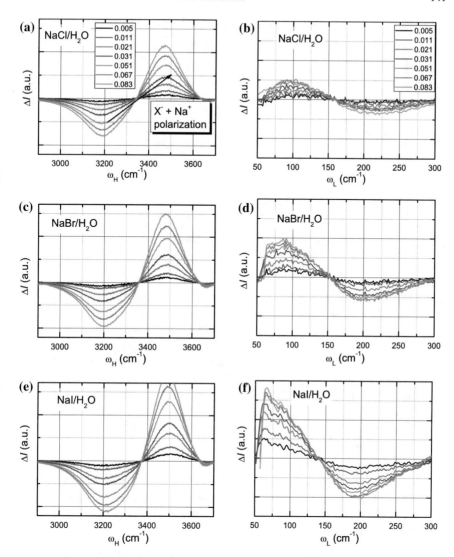

Fig. 6.5 The ω_x DPS for the concentrated **a, b** NaCl/H$_2$O, **c, d** NaBr/H$_2$O, and **e, f** NaI/H$_2$O solutions. NaF hydration is not included due to its limited solubility of 0.022. The $\Delta\omega_x$ recovers slightly at higher concentration due to weakening of the electric field in the hydration shells by anion-anion repulsion. Reprinted with copyright permission from [19]

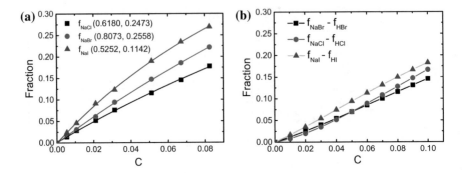

Fig. 6.6 The fraction coefficients for **a** NaX and **b** Na$^+$ estimated from $f_{Na}(C) = f_{NaX}(C) - f_{HX}(C)$ and **c** the contact angles between NaX solutions and glass substrate. $f_X(C) \cong f_{HX}(C)$ and $f_H(C) \cong 0$ is applied [87]. The $f_{YX}(C)$ follows the I$^-$ > Br$^-$ > Cl$^-$ > F$^-$ Hofmeister capability series order. The $f_{Na}(C)$ varies its curvature slightly with the X$^-$ anions: Cl$^-$ > Br$^-$ > I$^-$ with zero curvature because of the strongest I$^-$ polarizability. Reprinted with copyright permission from [19]

of the Na$^+$ − X$^-$ attraction that enhances the local electric field of Na$^+$ solute though the estimation may be subject to precision. The nonlinearity of the $f_X(C) = a[1 - \exp(-C/C_0)]$ features the involvement of the X$^- \leftrightarrow$ X$^-$ repulsion that weakens the X$^-$ capability of polarization. The extent of polarization is proportional to the solute concentration despite the slope at concentration higher than 0.03. The capability of bond transition and polarization for the halogenic solutes varies with the solute type in the same I$^-$ > Br$^-$ > Cl$^-$ > F$^-$ ~ 0 (saturates at 0.022) Hofmeister capability series order [88].

The spectral analysis revealed the solute type and concentration dependent capabilities of the halogenic solutes:

(1) Compared to the Raman spectrum collected from deionized water under the same conditions, no extra features are associated with NaX solutions, which implies that cations and anions remain as isolated charge centers without forming covalent or ionic bonds between counter-ions or between solute and solvent water molecules.

(2) Salt hydration has the same, but even stronger, effect of skin molecular undercoordination on the O:H−O bond relaxation and polarization. Ionic polarization lengthens the O:H nonbond and softens its phonon from 200 to 75 cm^{-1}, and meanwhile shortens the H−O bond and stiffens phonon from 3200 to 3500 cm^{-1}, by weakening the O−O repulsion [39].

(3) The solute capability of H−O phonon abundance and stiffness transition, and polarization, from the mode of water to its hydrating changes with solute type and solute concentration. Interanion repulsion comes into play at higher concentration to weaken the local electric field and solute capability of bond transition. The quasi-linear $f_{Na}(C)$ form suggests its invariant electric field and constant hydration volume size because of the hydrating H$_2$O dipolar shielding.

(4) The effect of polarization for a larger ion with lower electronegativity is more pronounced than others. The effect of NaF addition is too small to be resolved spectrometrically. This tiny spectral difference evidences that the Na$^+$ and F$^-$

stay closely to form the contact ionic pair, which explains why the maximal solubility of NaF is only 0.022 ($n_{H2O}/n_{NaF} \approx 45/1$).

(5) The effect of X^- polarization on the phonon relaxation follows the $X(R/\eta) = I(2.2/2.5) > Br(1.96/2.8) > Cl(1.81/3.0) > F(1.33/4.0) \approx 0$ Hofmeister capability series order with R being the X^- ionic radius and η the electronegativity. The discrepancy between the $f_X(C)$ and the $f_Y(C)$ indicates the number inadequacy of the hydrating H_2O dipoles to screen the X^- from being interfered by other X^-, which becomes more significant as the R increases.

6.2.2 (Li, Na, K, Rb, Cs)(Cl, I) Solvation

6.2.2.1 DPS of (Na, K, Rb, Cs)I

One can also discriminate the Y^+ capabilities of bond transition and polarization one from another by spectrometrically examining the Y type and concentration resolved YI solutions as the I^- is assumed to perform identically in their respective solutions. Figure 6.7 shows the full-frequency Raman spectra of aqueous (Na, K, Rb, Cs)I solutions of 0.036 molar concentration. Irrespective of the cation type, the ω_H blueshift and the ω_L redshift are consistent with that for the Na(Cl, Br, I) solutions, which confirms the O:H–O cooperativity upon the I-based salt being solvated though the cation-induced shift is less significant than that induced by anions. For the ω_H mode, the skin component with phonon frequencies centered at ~ 3450 cm^{-1} become more pronounced, indicating the cationic hydration shell formation. Following the order of $Y(R/\eta)$: Li$^+$(0.78 Å/1.0), Na$^+$(0.98/0.9), K$^+$(1.33/0.8), Rb$^+$(1.49/0.8), Cs$^+$(1.65/0.8) [89], the ionic size and electronegativity of the alkali cations determine their capabilities of polarization.

Fig. 6.7 Full-frequency Raman spectra of 0.036 molar ratio (Na, K, Rb, Cs)I/H_2O solutions at 298 K. The ω_x DPS profiles show that the O:H–O relaxation is less sensitive to the type of cations. Reprinted with copyright permission from [54]

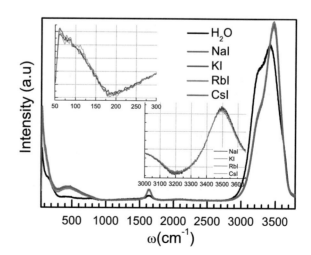

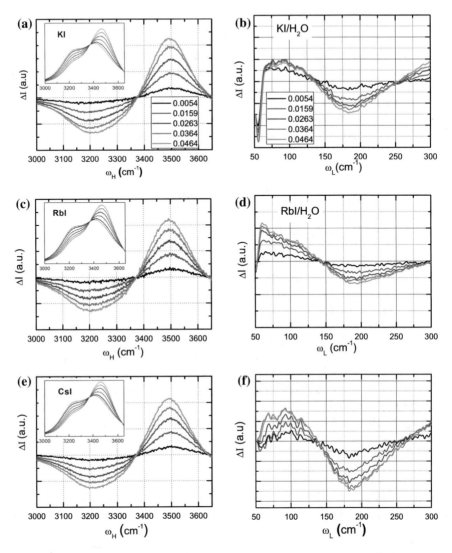

Fig. 6.8 Concentration dependent ω_x DPS for **a, b** KI/H$_2$O, **c, d** RbI/H$_2$O, **e, f** CsI/H$_2$O solutions. Reprinted with copyright permission from [90]

Repeating the same iteration of the ω_H DPS peaks results in the fraction coefficient for the concentrated YI/H$_2$O solutions, which feature the Y$^+$ cation capability of O:H–O bond transition as shown ion Fig. 6.8. One can hardly tell the difference between Y$^+$ cations because of their compatible 0.8–0.9 Å ionic radii and almost identical electronegativity.

6.2.2.2   DPS of (Li, Na, K, Rb, Cs)Cl

Likewise, Fig. 6.9 shows the full-scan of the Raman spectra Figs. 6.8 (for NaCl/H₂O), 6.10 and 6.11 show the ω_H and ω_L DPS for (Cl, K, Rb, Cs)Cl/H₂O solutions without vibration features of solute-solvent bond formation.

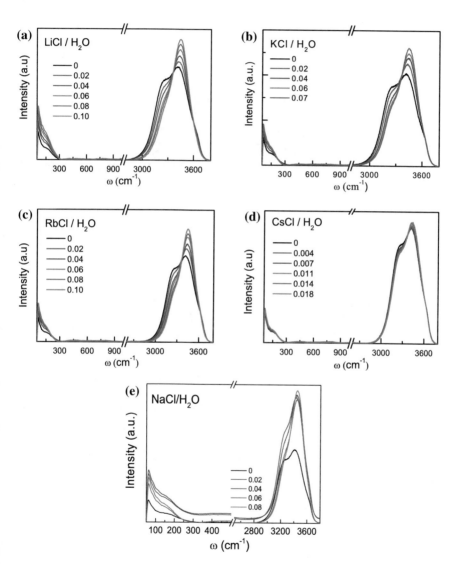

Fig. 6.9 Full-scan room-temperature Raman spectra for the concentrated **a** LiCl/H₂O, **b** KCl/H₂O, **c** RbCl/H₂O, **d** CsCl/H₂O and **e** NaCl/H₂O solutions compared with the spectrum of deionized water

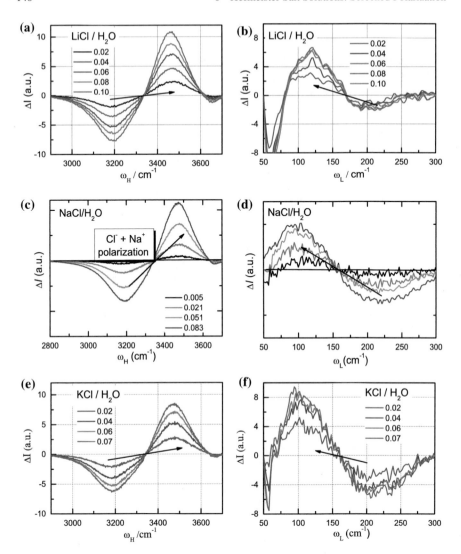

Fig. 6.10 Concentration dependent ω_x DPS for **a, b** LiCl/H$_2$O, **c, d** NaCl/H$_2$O, **e, f** KCl/H$_2$O and solutions. Ionic screened polarization shifts the ω_H from the vibration mode of bulk water to the first hydration subshell at ~3450 cm^{-1}. The second small valley fingerprints the anion preferential skin occupation that stiffens the H–O dangling bond vibration frequency from 3610 to 3620 cm^{-1} and screen the signal from being detected. Ionic screened polarization shifts the ω_L from 200 to 100–125 cm^{-1}, respectively. Greater frequency shift means a stronger electric field

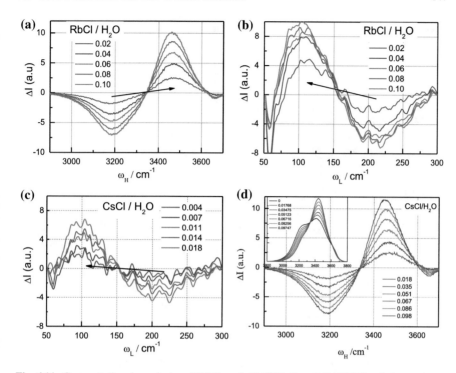

Fig. 6.11 Concentration dependent ω_x DPS for **a, b** RbCl/H$_2$O, **c, d** CsCl/H$_2$O solutions

6.2.2.3 Fraction of Bond Transition and Volume of Hydration

Figure 6.12 shows the $f_{YX}(C)$ for Y(I, Cl) solutions. (a) Y$^+$ ions have compatible fraction coefficients for O:H–O bond transition. The slight $f_Y(C)$ deviation (c) for Cs$^+$ and K$^+$ from the linear dependence at higher concentration may suggest the involvement of Y$^+ \leftrightarrow$ Y$^+$ repulsive interactions for the larger Y$^+$, though the $f_Y(C) = f_{YI}(C) - f_{HI}(C)$, may be subject to precision.

At solvation, an n number of H$_2$O dipoles dissolves the salt crystal into a pair (or cluster) of counterions. An n_{HY} number of H$_2$O dipoles hydrates a cation and n_{HX} dipoles hydrates one anion in their first-vicinal hydration shells. The rest n_r number of molecules is free or in the high-order hydration shells of the binary and the trinary salt solutions:

$$YX + nH_2O \Rightarrow Y^+ + X^- + (n_{HY} + n_{HX} + n_r)H_2O$$

$$ZX_2 + nH_2O \Rightarrow Z^{2+} + 2X^- + (n_{HY} + 2n_{HX} + n_r)H_2O$$

The molecular concentration of a salt solution is C = 1/(1 + n). The inverse of C, 1/C = 1 + n_{HY} + n_{HX} + n_r. The n_{HY} and n_{HX} are dipoles located in the first vicinal subshells of the respective ions.

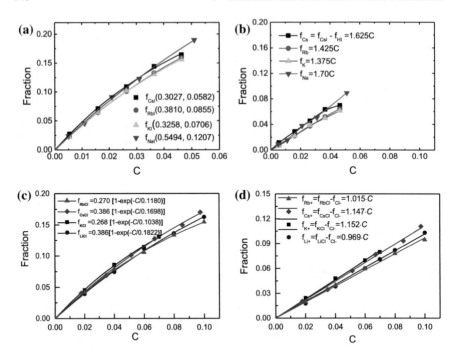

Fig. 6.12 The fraction coefficients for **a** YI/H$_2$O solutes and **b** Y$^+$ cations, **c** YCl/H$_2$O solutes and **d** Y$^+$ cations estimated from $f_Y(C) = f_{YI}(C) - f_{HI}(C)$ with the C scale larger than measured to show the trends. Reprinted with copyright permission from [54]

The division of the $f_Y(C)$ and $f_X(C)$ by the solute concentration C corresponds to the number of H–O bonds per ion (n_H/ion) in the first hydration subshell, which characterizes the ionic hydration volume and its local electric field. The derived n_H/ion may under-estimate the real value as the DPS distills only H–O bonds subjecting to strongest electric field. Those H–O bonds initially deformed by skin molecular undercoordination can hardly be further deformed by the polarization [39]. However, the derived n_H/ion values show us their concentration trends and contrast across their relative hydration volume sizes.

The quasilinear $f_Y(C)$ gives rises to a constant $n_H(C)$, which indicates that the Y$^+$ has a constant hydration volume whose size varies insignificantly with solute concentration and slightly with its counterions, as shown in Fig. 6.13c. However, the X$^-$ capability of H–O bonds transition follows the $f_X(C) = a[1-\exp(-C/C_0)]$ manner toward saturation, arising from the progressive involvement of anion-anion repulsion at higher concentrations. The $n_H(C) = f_X(C)/C$ per anion drops with the separation between adjacent anions. The insufficiency number of the ordered H$_2$O dipoles in the hydration volume only partly screens the anionic potential.

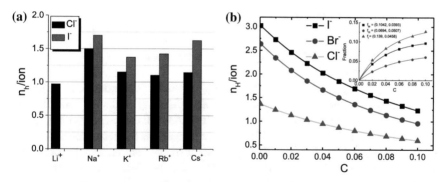

Fig. 6.13 **a** Effective number n_H of the first vicinal HBs per cation and **b** the concentration dependence of the $n_H(C)$ for X^- in HX/H$_2$O [87] solutions. Inset b shows the $f_{HX}(C)$ profiles

6.2.2.4 Surface Stress: Polarization Dominance

Figure 6.14 compares the contact angles between the glass substrate and Na(F, CL, Br, I) and YI/H$_2$O solutions at the ambient temperature, as an indicator of solute polarizability. At the same concentration, ions follow the ionic polarizability follows the Na$^+$ > K$^+$ > Rb$^+$ > Cs$^+$ and I$^-$ > Br$^-$ > Cl$^-$ > F$^-$ ~ 0. The cations show well-resolved capabilities of polarization.

The cation polarizability results from the surface charge density of the outermost orbitals of the tiny cations. The discriminative polarizability may infer the mechanism behind the activation or inactivation of the alkali ionic channels [91] and hypertension medication and prevention involving cations. For instance, NaCl solvation enhances hypertension more than KCl solvation. The cationic solutions show the similar capability of bond transition but well-resolved Na$^+$ > K$^+$ > Rb$^+$ > Cs$^+$ Hofmeister polarization order.

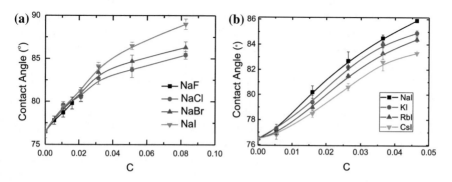

Fig. 6.14 Comparison of the contact angle for the concentrated NaX/H$_2$O and YI/H$_2$O solutions

6.2.3 Divalent (Mg, Ca, Sr)(Cl, Br)$_2$ Solvation

6.2.3.1 Interlock of ZX$_2$ and H$_2$O Bonding Networks

The ZX$_2$/H$_2$O solution bonding network is an interlock of the ordered, tetrahedrally-coordinated H$_2$O and the ZX$_2$ lattices. The ZX$_2$ dissolves into a Z^{2+} and 2X$^-$ ions and each of them serves as a center source of electric field interacting with hydrating H$_2$O molecules and other solutes. Figure 6.15a inset shows the 2H$_2$O solvent unit cell of four identical O:H–O bonds linking adjacent oxygen atoms. Figure 6.15b inset shows the proposed 2ZX$_2$ solute unit cell. Solutes do not form any ionic or covalent bonds to the solvent molecules but each of them clusters, polarizes, and stretches the neighboring O:H–O bonds. Ions are screened by their hydrating H$_2$O molecules to form the supersolid hydration volumes (inset of Fig. 6.15c)—the hydration volumes are highly polarized, gel-like, thermally more stable associated with H–O contraction and O:H elongation [38].

The dissolved cations and anions form a homogeneous tetrahedral unit, which is like the 2H$_2$O with four identical Z^{2+}~X$^-$~Z^{2+} electrostatic interactions. The smaller divalent Z^{2+} cations occupy the center and vertices of the tetrahedron. Four larger X$^-$ anions are located at the midpoint between two Z^{2+} cations. In the solute bonding network, the Z^{2+}~X$^-$ attraction and the Z^{2+}↔Z^{2+} and X$^-$↔X$^-$ repulsion coexist but the extent will be subject to the solute concentration and the extent of the hydrating H$_2$O dipoles screening in the hydration shells. Compared with the monovalent YX solution of the same concentration, there exist 2X$^-$ and one Z^{2+}. The folded number of X$^-$ and the folded valence of Z^{2+} stem the difference between the ZX$_2$/H$_2$O and the YX/H$_2$O solutions.

6.2.3.2 Raman Spectral Characteristics

The ionic size, charge sign, charge quantity, electronegativity, ionic separation, and the screen shielding by hydrating H$_2$O determine the strength of the solute ionic electric field. Ionic electrification shortens and stiffens the H–O bond associated always O:H nonbond elongation and softening [38] (see Fig. 6.15 inset d). The O:H–O segmental length, vibration frequency, and the phonon abundance transition in the DPS fingerprint directly the locally screened electronic field and the hydration volume size.

Figure 6.15 compares the full-frequency (50–3800 cm^{-1}) Raman spectra of ZX$_2$ solutions. One may focus on the cooperative relaxation of the O:H nonbond stretching vibration frequency ω_L at 50–300 cm^{-1} and the H–O bond stretching vibration ω_H at 3100–3800 cm^{-1}. Salt solvation shifts the O:H and the H–O phonon cooperatively without adding new features of bond formation. Solvation of the ZX$_2$ stiffens the H–O phonon and softens the O:H phonon simultaneously without being able to discriminate the charge sign difference between the 2X$^-$ and the Z^{2+} ions. The spectral features within 300–2000 cm^{-1} range due to bond angle bending and rotating

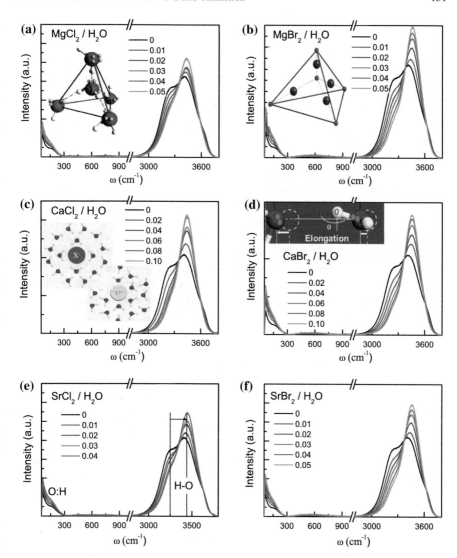

Fig. 6.15 Full-frequency Raman spectra of the divalent (Mg, Ca, Sr)(Cl, Br)₂/H₂O solutions as a function of sulte concentration (a-f). Inset **a** illustrates the 2H₂O structural unit cell consisting of four oriented O:H–O bonds with pairing dots standing for the lone pairs ":" of oxygen. Inset **b** illustrates the 2ZX₂ unit cell extending from the 2H₂O geometry by removing the ":". Inset **c** illustrates the effect of Z²⁺ and X⁻ charge injection on water molecules aligning, clustering, and H₂O dipolar shielding. Inset **d** shows electric stretching of the O:H–O bond. The ionic sizes are R_Z ~ 1.0 Å for Z²⁺ and R_X ~ 2.0 Å for X⁻. Inset **e** denotes frequency regimes of O:H and the H–O stretching vibrations and the shift $\Delta\omega_x$ as a function of the segmental length d_x and energy E_x. Reprinted from permission of [85]

vibration are out of immediate concern as these features fingerprint less the change of the O:H–O bond segmental length and energy. This approach is the advantage of the phonon spectroscopy based on the Fourier transition principle that sorts out the bonds vibrating in the same frequency disregarding the orientation, location or even the lattice geometry.

The DPS for the concentrated ZX_2/H_2O solutions shows the H–O phonon stiffness transition from $3200\,cm^{-1}$ (valley) to $\sim3450\,cm^{-1}$ (peak). The O:H nonbond responds to ionic polarization oppositely and shifts its phonon from $\sim200\,cm^{-1}$ to $\sim110\,cm^{-1}$ for ZCl_2 and to $\sim100\,cm^{-1}$ for ZBr_2 solutions, instead of shifting to $75\,cm^{-1}$ by YX monovalent salt solutions, which shows the weaker polarizability of the denser ZX_2 solutes than the YX. The second valley at $3620\,cm^{-1}$ results from the stiffening of the dangling H–O bond featured at $3610\,cm^{-1}$ by the stronger field of the X^- ions that occupy preferentially at the solution surface. However, the spectral signal of the dangling bond is annihilated by the anions during Raman detection. These features confirm the effect of divalent salt on polarizing O:H–O bonds. In contrast, the monovalent salt solutes transit the H–O phonon from 3200 to $\sim3500\,cm^{-1}$ and the O:H phonon from 200 to $\sim75\,cm^{-1}$. The local electric fields around the ions in the ZX_2 solutions is weaker because of the superposition of the repulsion between the like ions of shorter separations (Figs. 6.16 and 6.17).

6.2.3.3 Z^{2+} and X^- Capabilities of Bond Transition

Integration of the DPS H–O phonon peaks results in the fraction $f_{ZX2}(C)$, which is the number of bonds transiting from the mode of water to hydration. Given that the $f_H(C) \cong 0$ as the H^+ does not polarizes its surroundings but forms the H↔H anti-HB, one has $f_X(C) \cong f_{HX}(C)$ [87]. The X^- anion performs the same to polarize neighboring H_2O molecules to form the hydration shell. Therefore, $f_Z(C) \cong f_{ZX_2}(C) - f_{HX}(2C)$, where the $f_{HX}(2C) = f_{X2}(C)$ stands for the folded number of X^- anions in the ZX_2 solutions. The $f_Z(C)$ and $f_{X2}(C)$ can thus be separated from the $f_{ZX2}(C)$ obtained by integrating the ω_H DPS peak. Figure 6.18 compares the $f_{ZX2}(C)$, $f_Z(C)$, and $f_{X2}(C)$ for the concentrated ZX_2 solutions.

The six $f_{ZX2}(C)$ profiles growing in the exponential form toward saturation, as shown in Fig. 6.18a, are categorized in three groups according to the solute valences. Results show the solute capabilities of O:H–O bond polarization follows: $Sr^{2+} > Ca^{2+} > Mg^{2+}$ and $Br^- > Cl^-$ Hofmeister series order. The $f_{ZX2}(C)$ increases with the ion size and with the drop of electronegativity [ion(R/Å, η)] [89]: $Mg^{2+}(0.78, 1.30)$, Ca^{2+} (1.0, 1.06), $Sr^{2+}(1.27, 1.00)$; $Cl^-(1.81, 3.00)$, $Br^-(1.96, 2.80)$].

One can obtain the $f_{X2}(C) = f_X(2C)$ from Fig. 6.18a by converting the lateral axis C into C/2. From the $f_{X2}(C)$ slope analysis, $X^- \leftrightarrow X^-$ repulsion remains throughout the concentrated solutions because of the large X^- size [38]. Insufficient number of H_2O molecules in the highly-ordered hydration shells only partially screen the electric field of the X^- ions, being the same case of the monovalent salt solutions.

Figure 6.18b shows that at $C \le 0.05$, the $f_Z(C) \propto C$ and at higher C, the $f_Z(C)$ follows the same trend of $f_X(C) \propto 1 - \exp(-C/C_0)$ toward saturation. The subscript

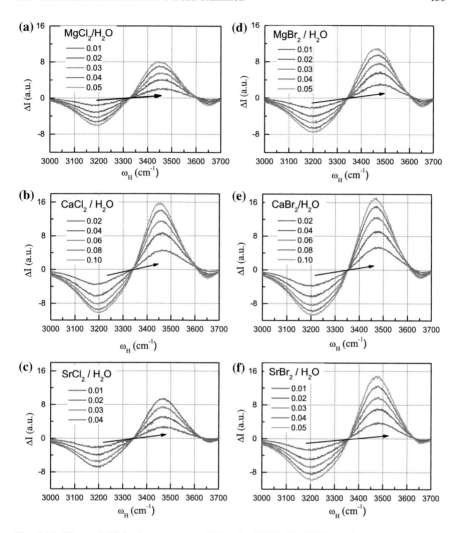

Fig. 6.16 The ω_H DPS for the concentrated (Mg, Ca, Sr)(Cl, Br)$_2$/H$_2$O solutions. Aqueous charge injection transits the H–O bonds from the mode of ordinary water centered at 3200 cm^{-1} to the hydration states at ~3450 cm^{-1}, as a result of H–O bond contraction and stiffening ($\omega_x^2 \propto E_x/d_x^2$). The valley at 3620 cm^{-1} features the skin preferential occupancy of anions that strengthen the H–O dangling bond and shield the signals of detection. Reprinted from permission of [85]

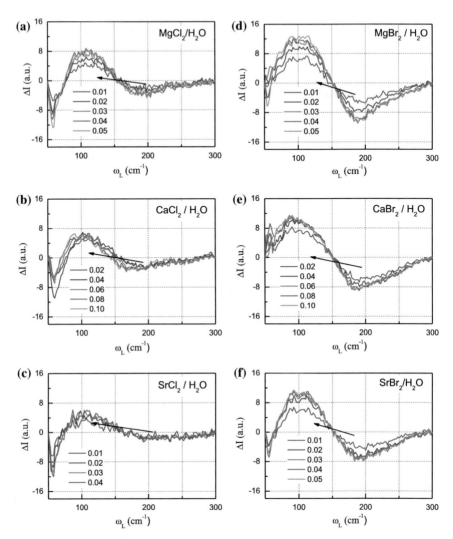

Fig. 6.17 The ω_L DPS for the concentrated (Mg, Ca, Sr)(Cl, Br)$_2$/H$_2$O solutions. Aqueous charge injection transits the O:H nonbonds from the mode of ordinary water centered at 200 cm^{-1} to the hydration states at ~100 cm^{-1}, resulting from the O:H nonbond elongation and softening by polarization. Reprinted from permission of [85]

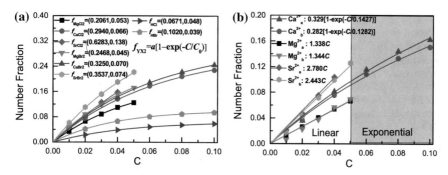

Fig. 6.18 Comparison of **a** the present $f_{ZX2}(C)$ and the $f_{HX}(C)$ for acid solutions [56] and **b** the $f_Z(C)$ for the divalent cations. The slopes and curvatures of the f(C) profiles provide information on the ionic local electric fields and the hydration volume sizes as a function of solute concentration. Reprinted from permission of [85]

a and b legend the Z^{2+} in the ZCl_2 solutions and the Z^{2+} in the ZBr_2 solutions, respectively. The critical molecular ratio $C = 0.05 = 1/20$ means that 20 H_2O molecules come surround one set of $Z^{2+} + 2X^-$ solutes. At such a concentration or below, the Z^{2+} is fully screened by its surrounding H_2O dipoles, and the $Z^{2+} \leftrightarrow Z^{2+}$ repulsion or the $Z^{2+} - X^-$ attraction is negligible. The $Z^{2+} - X^-$ attraction is annihilated by the $Z^{2+} \leftrightarrow Z^{2+}$ repulsion that becomes dominance. However, at higher concentration, the $Z^{2+} \leftrightarrow Z^{2+}$ repulsion occurs because of their shorter distance and high valences. However, the absence of $Y^+ \leftrightarrow Y^+$ repulsion from the monovalent salt solutions suggests that the Z^{2+} has a longer distance of its electric field. The charge quantity difference between the Z^{2+} and the Y^+ and the folded number of $2X^-$ discriminate the monovalent and divalent salt solutions of the same concentration.

For the trinary counterions in the divalent salt solutions, we need one more condition to discriminate the hydration volume of Z^{2+} from that of the $2X^-$. The $2X^-$ may not perform the same to Y^+ in the monovalent solution and the Z^{2+} is different from the Y^+ because of the solute density and charge quantity. However, we can estimate the $n_H(C)$ per set of $Z^{2+} + 2X^-$ for reference. Results in Fig. 6.19 show the $Sr^{2+} > Ca^{2+} > Mg^{2+}$ and $Br^- > Cl^-$ Hofmeister hydration volume size order.

6.2.4   Na(ClO₄, NO₃, HSO₄, SCN) Solvation

6.2.4.1   Complex Giant Dipolar Anions

Solvation dissolves an YX alkali halide into a monovalent Y^+ alkali cations and an X^- halogenic anion. Both Y^+ and X^- have then electrons filled closely their outermost shells. The spherical charge distribution surrounding the atomically sized ions. These ions do not form dipoles but create radial electric fields pointing from the positive to the negative, which serve each as a charge center polarizing neighboring

Fig. 6.19 Concentration resolved numbers of the first-vicinal H–O bonds hydrating per trinary ZX$_2$ solute counter ions (Z^{2+}+2X$^-$). Solutions saturates at 0.05 molar concentration except for the CaX$^-$

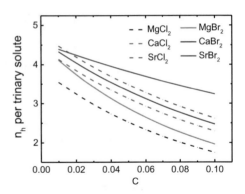

HBs. Conversely, the NaT (T = ClO$_4$, NO$_3$, HSO$_4$, SCN) complex salts dissolve into the Na$^+$ cations and T$^-$ complex anions whose volumes are much bigger than the halogenic anions. The Na$^+$ cation performed the same as it does in the NaX solutions. The less electronegative H(2.2) and Na(0.9) tends to bond to atoms of the more electronegative elements in the complex salts.

The giant T$^-$ dipolar anions contain electronegative O(3.5), S(2.5), N(3.0), P(2.1), Cl(3.0), and C(2.5) ions that undergo the electronic sp^3 orbital hybridization. The sp^3 orbital hybridization shows two key features: one is the lone pair occupation of the hybridized orbitals and the other is the oriented single bond between these ions. The hybridized orbitals are occupied by, from zero to four, electron lone pairs. The lone pairs tend to point outwardly to disperse the molecular anions one from another in solutions. These atoms form only single bond between them or other bonding parties [52].

Table 6.2 and Fig. 6.21 insets illustrate the revised configurations of the molecular solutes in the complex salt solutions in contrast to the conventional scheme without considering the orbital hybridization [52]. Upon solvation, the anionic intramolecular configuration and interaction follow the given regulations of lone pair occupation and bond number saturation and bond orientation, exemplified in oxygen chemisorption [92]. For Fig. 6.21a inset a instance, the conventional NaClO$_4$ configuration gives rise to: Cl^{7+} + 3O^{2-} + NaO$^-$ without involvement of the electron lone pairs. In contrast, the revised configuration takes the bonding dynamics with lone pair production into account. The rules of lone pair and bond number saturation dictate the solute-solvent molecular hydration bonding and nonbonding dynamics profoundly [52]. Firstly, the Na bonds to the Cl to form a NaCl and then the Cl$^-$ undergoes the sp^3-orbital hybridization with creation of three pairs of electron lone pairs surrounding the Cl$^-$. Upon solvation, NaCl dissolves into a Cl$^-$ and a Na$^+$. The Na$^+$ leaves its valence electron behind the Cl$^-$ that has then four pairs of lone pairs to fully occupy its sp^3 orbits.

The sp^3-orbital hybridization of both Cl(3.0) and O(3.5) and the electronegativity difference determines that the Cl$^-$ tends to donate an electron to each of the four O atoms and turns itself to be Cl^{3+}. The O$^-$ anions then paired up covalently and

Table 6.2 The conventional and revised configuration of the complex salts with involvement of the electron lone pairs and tetrahedral geometric requirements [52][a]

Complex salt (number f lone pairs)		Configuration	Notion (Fig. 6.21 inset)
$Na^+[ClO_4]^-$ (8)	Convention	$Cl^{7+} + 3O^{2-} + NaO^-$	**a** The central Cl^- gains an electron from Na^+ and then hybridizes its sp^3 orbits; the Cl^- bonds tetrahedrally to $4O$ and turns its self to be Cl^{3+}. The $O{-}O^-$ then paired up covalently with creation of each two ":", and a total of eight ":" surrounding symmetrically the anion cluster
	Revision	$Na^+ + Cl^{3+} + 2(O^-{-}O^-)$	
$Na^+[NO_3]^-$ (8)	Convention	$N^{5+} + 2O^{2-} + NaO^-$	**b** The central N bonds tetrahedrally to three O atoms and hybridizes its sp^3 orbits with creation of a ":"; two O pairs up covalently with creation of four ":". The third O^- gains an electron from Na with creation of three ":", being the same to HO^-. The molecular anion is surrounded asymmetrically by eight ":"
	Revision	$N^{3+} + (O^-{-}O^-) + [Na^+ + O^-]$	
$Na^+[HSO_3]^-$ (7)	Convention	$S^{4+} + O^{2-} + NaO^- + HO^-$	**c** The central HS^{3+} bonds covalently to three O and hybridizes its sp orbitals. One O^- gains an electron from Na with creation of three ":". The rest two O^- paired up covalently with creation of two ":" each. Seven ":" surround the anionic cluster
	Revision	$HS^{3+} + (O^-{-}O^-) + [Na^+ + O^-]$	
$Na^+[HSO_4]^-$ (10)	Convention	$S^{6+} + 2O^{2-} + NaO^- + HO^-$	**d** S^{4+} bonds to four O tetrahedrally. Two O paired up covalently and rest two O bond to the H and Na with creation of each two ":"; NaO^- dissociates leaving its electron behind the O to create a ":". The anionic cluster is surround by nine ":"
	Revision	$S^{4+} + (O^-{-}O^-) + [Na^+ + O^-] + HO^-$	

(continued)

Table 6.2 (continued)

Complex salt (number f lone pairs)		Configuration	Notion (Fig. 6.21 inset)
4[Na⁺(SCN)⁻] (12)	Convention	$N^{3-} + C^{4+} + NaS^-$	**e** Each C bonds to its two C neighbours and a N and a S. Two S atoms then paired up covalently with each creating two ":". Each N gains an electron from Na and bonds to two N neighbors and one C separately with creation of one ":". The anionic cluster is surrounded by twelf ":".
	Revision	$4(Na^+ + N^{3-} + C^{4+} + S^{2-})$ Only a collection of four molecules meets the bonding criteria.	

[a]Each O bonds to the central atom of the cluster and then the $O^- - O^-$ paired up covalently with each creating two pairs ":".

hybridize their sp orbits with creation of two pairs of lone pairs each, pointing outwardly of the $(ClO_4)^-$ cluster. This process meets the aforementioned criteria for sp^3-orbital hybridization. Furthermore, transiting the Cl^- into a Cl^{3+} is energetically much more favorable than transiting the Cl directly into the Cl^{7+} in convention.

Likewise, the $NaNO_3$ can be revised from the $N^{5+} + 2O^{2-} + NaO^-$ convention to the $N^{3+} + (O^--O^-) + [Na^+ + O^-]$ with inclusion of sp^3-orbital hybridization and creation of $1(N^{3+}) + 2\times2(O^--O^-) + 3(O^-) = 8$ lone pairs. As illustrated in Fig. 6.21b, the N bonds tetrahedrally to three O atoms and hybridizes its sp^3 orbits with creation of one pair of electron lone pair; the two O^- ions pair up covalently and then hybridize their sp^3 orbits with creation of each two pairs of electron lone pairs. The third O^- gains an electron from the Na with creation of three pairs of electron lone pairs—this NO^- is the same to the HO^- in the basic solutions. The anionic cluster is thus surrounded by 8 lone pairs and the sp^3 orbits of both N and O are hybridized.

The NaSCN is more complicated to meet the orbital hybridization criteria, which needs four molecules to form a cluster, as Fig. 6.21d shows. The convention of $NaSCN \rightarrow N^{3-} + C^{4+} + NaS^-$ is translated into $4NaSCN \rightarrow 4(NaN^{3-} + C^{4+} + S^{2-})$. All the S, N, C atoms undergo the sp^3-orbital hybridization in this anionic cluster. Each C bonds to its two C neighbours and a N and a S atom. Two S atoms then pair up covalently with each creating two pairs of lone pairs. Each N bonds to a C and a N neighbour and the Na. The N gains an electron from the Na upon solvation and hybridize its sp orbital and then bonds to one more N neighbors with creation of one lone pair. The anionic cluster is thus surrounded by $4 \times [2(S) + 1(N)] = 12$ lone pairs and thus this cluster meets the aforementioned sp^3-orbital hybridization criteria.

One needs to note that the larger anionic size has the same effect of higher solute concentration. The multiple lone pair distribution at cluster surface increases the effective charge of the anionic clusters to values greater than that of the halogenic ions. These features shall discriminate the complex Na-based salts from the alkali halides in the capabilities of polarization and transforming the solution HB network and the solution properties.

The extended anionic clusters with sp-orbital hybridization for the complex salts are more meaningful than the convention of a simple valence balance without counting the lone pairs and bond number saturation. Nevertheless, serving as the charge carriers of greater volume than halogenic ions, the complex salts are naturally the same to the YX salts despite greater anion size. The anionic clusters interact with their surrounding H_2O molecules through polarization, X:H–O bond formation, and possible H↔H and O:⇔:O repulsion. This sp^3-obital hybridization and lone pair production premise [82, 92, 93] can be extended to other complex molecules consisting atoms of electronegative elements with creation of different numbers of H^+ and ":" surrounding the molecular solute.

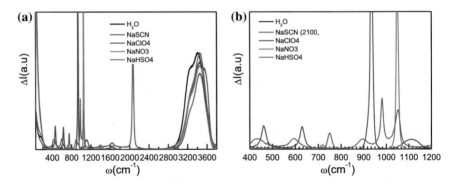

Fig. 6.20 Full-frequency Raman spectra for the **a** $NaClO_4/H_2O$, $NaNO_3/H_2O$, $NaHSO_4/H_2O$ ($NaHSO_3/H_2O$), and $NaSCN/H_2O$ complex salt solutions and the **b** magnified region for the intra-anion bonding vibration features. Reprinted with copyright permission from [52]

6.2.4.2   DPS Derivatives

To examine the solute capability of transiting the solution HB network and polariza-tion, one can repeat the DPS iteration and the contact angle detection with focuses on the HB segmental stiffness and phonon abundance transition and electric polar-ization. Figure 6.20 compares the full-frequency Raman spectra for the NaT/H_2O solutions of different 0.016 molar concentration. Except for the intra-anion bonding features, no solute-solvent bond vibration is resolved. The solvation shifts the H–O and O:H vibration frequencies in the similar manner of YI and NaX solutions.

The DPS derived information on the segmented O:H–O bond relaxation is shown in Figs. 6.21 and 6.22. Salt solvation transits a fraction of the O:H–O bonds from the mode of ordinary water into the hydration, which stiffens the H–O phonon from 3200 to ~3500 cm^{-1} and softens the O:H segment cooperatively from 180 to ~75 cm^{-1} by polarization, which is naturally the same to the alkali halide solvation. The DPS also resolves the preferential skin occupancy of the anions [19, 52]. The stronger ionic field stiffens the H–O dangling bond and the surface anions attenuate the H–O dangling bond vibration signal. SFG measurements resolve the ClO_4^- dipolar anisotropy of the concentrated anions at the air/solution interface with the polarizability tensor that is twice that of bulk water [94].

The frequency shifts change when the molar concentration increases from 0.010 to 0.082: $NaClO_4$ ($3600 \rightarrow 3550$ cm^{-1}), $NaHSO_4$ ($3500 \rightarrow 3520$ cm^{-1}), $NaNO_3$ (retains 3520 cm^{-1}), $NaSCN$ ($3500 \rightarrow 3480$ cm^{-1}). The H–O phonon softening with concen-tration increase suggests the presence of the pronounced anion-anion repulsion that weakens the electric field within the hydration shells, while the H–O stiffening means the higher saturation constant for the faction coefficient $f(C) \propto 1 - \exp(-C/C_0)$. The larger anionic volume and the higher surface charge density stem the high sensitivity of the Na-based complex salts to the variation of solute concentration compared to the alkali halides.

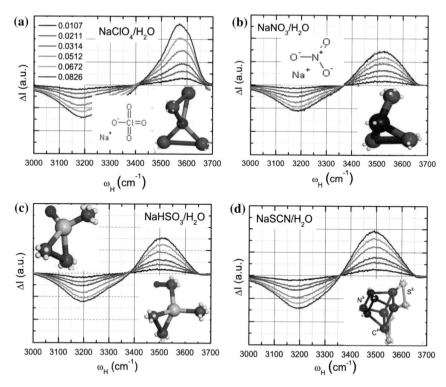

Fig. 6.21 Concentration dependence of the ω_H DPS for **a** NaClO$_4$/H$_2$O, **b** NaNO$_3$/H$_2$O, **c** NaHSO$_3$/H$_2$O (NaHSO$_4$/H$_2$O), and **d** NaSCN/H$_2$O complex solutions. Solute polarization stiffens the H–O phonon and raises the structure order (FWHM for the phonon abundance of the hydration shells and the reference water). Insets compare the solute structures from the perspective of valence balance and the revised molecular configuration with sp^3 orbits hybridization and electron lone pairs production. Reprinted with copyright permission from [52]

The DPS peak above the x-axis is the phonon abundance transition into the mode of hydration from the mode of ordinary water as the spectral loss below the x-axis. Being the same to alkali halides, complex salt solvation stiffens the ω_H phonons from 3200 to 3500 cm^{-1} and above and softens the O:H stretching phonons from 180 to ~75 cm^{-1}. Figure 6.22 compares the concentration dependence of the ω_L DPS for NaT/H$_2$O solutions. The O:H phonon frequency is more sensitive to the solute polarization, which varies from 70 to 80 cm^{-1} with higher order of molecular structures.

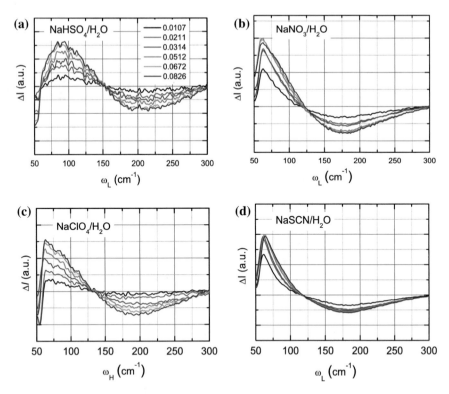

Fig. 6.22 The ω_L DPS for the concentrated **a** NaHSO$_4$/H$_2$O, **b** NaNO$_3$/H$_2$O, **c** NaClO$_4$/H$_2$O, and **d** NaSCN/H$_2$O solutions, showing discriminative polarizability for the O:H frequency shift and structure ordering. Reprinted with copyright permission from [52]

6.2.4.3 Bond Transition and Polarization

Figure 6.23 compares the solute capability of bond transition, polarization, and the effective number of hydrating dipoles of the concentrated NaT/H$_2$O solutions. Observations suggest the following:

(1) The $f(C)$ not only varies with the solute type but also increases with solute concentration in an exponential way towards saturation with larger decay constants. The complex anions decay their electric fields in length faster than the monatomic X$^-$ ions because of their dipolar nature in the anisotropic charge distribution surrounding the complex anions, as illustrated in Fig. 6.20 insets.

(2) The effective number of the H–O bonds per complex solute drops linearly with solute concentration slightly, which indicates the stronger yet shorter-range dipolar electric filed that is less influenced by the repulsion between the adjacent solutes.

(3) Compared with NaX ($C_0 = 0.11$–0.25), YI ($C_0 = 0.06$–0.12), the saturation constant C_0 for the Na-based complex salt is 0.4–0.8 (Fig. 6.23a). The $f(C)$ for the Na-based complex salt solutions approaches a linear dependence. A

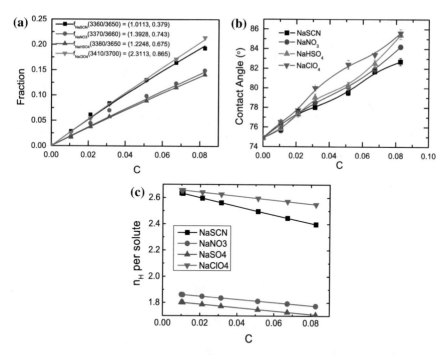

Fig. 6.23 The fraction coefficients **a** f(C), **b** the contact angle, **c** the effective number of HBs per solute, and **d** for the concentrated NaT/H$_2$O solutions. Reprinted with copyright permission from [52]

slower decay means the gradual increase of the anion-anion repulsion with the inter-anion separation. The greater saturation constant means a non-agressive and weaker anion↔anion repulsion.

(4) However, the greater H–O frequency shift shows the stronger local electric field in the hydration shells. Molecular anions in the Na-based complex salt solution have short-range yet stronger local electric fields, being identities of the typical dipolar anisotropic field.

(5) Solvation of all the considered complex salts raises the surface stress by polarization. The polarizability of the Na-based complex salts is slightly weaker than those of the NaX and the YI because of the greater anion sizes and the dipolar shorter interaction lengths of the complex salt anions.

(6) The concentration dependent H–O phonon frequency shift in the hydration shells is more sensitive to the type of the complex salts than the alkali halides. However, the concentration dependent fraction coefficient performs contrastingly, which means that the invariant hydration shell size of the anionic cluster though the local electric field changes more rapidly because of the larger dipolar volume. However, for the alkali halides, the anionic shell size and the local field change in the opposite trends of the complex salts.

Therefore, when examining the solvation, it would be more comprehensive and revealing to consider the solute capability of O:H–O bond transition from the mode of ordinary water to its hydrating in terms of the O:H–O bond stiffness, number fraction, structure order, and electron polarization. Consideration of the intermolecular nonbond interaction and intramolecular H–O bond cooperativity is necessary. Charge injection by monatomic ions or molecular dipoles mediate the solution network and polarize electrons in different ways, which dictates the solution properties such as the surface stress, solution temperature, critical energies for phase transition through the intra- and intermolecular cooperative bond relaxation as pursued presently.

6.3 Supersolid Hydration Shells

6.3.1 O:H–O Bond Segmental Length, Energy and Stiffness

Salt solvation results in $\Delta d_H < 0$, $\Delta \omega_H > \Delta E_H > 0$, and $\Delta dL > 0$, $\Delta \omega_L < \Delta E_L < 0$ for a fraction of O:H–O bonds by ionic polarization with respect to the ordinary bulk water. The ω_x peak shifts give rise immediate information of the segmental length and energy when calibrated against the known bulk and skin features at different temperatures [2]. One can estimate the segmental length and energy with a certain tolerance for the HBs in hydration shells of the aqueous solutions. Quantum computations revealed that the X·H–O bond in the closest hydration shells relaxes more than the subsequent O:H–O shell and the extent of bond relaxation also follows the Hofmeister series order for the ionic solutes in the protonated acidic solutions [56].

Table 6.3 estimates the H–O and O:H length in salt solutions. The relaxation trend of segmental length, phonon frequency, cohesive energy, and polarization would suffice to accounting for the functional activities of the solutions, including the solubility, surface stress, viscosity, and the critical temperatures for phase transition. For instance, one can elucidate that drinking salt water could raise blood pressure because O:H–O bond polarization turning the hydration molecules into the supersolid or the semi-rigid states. Salting not only promotes snow melting on the road but also accelerates sol-gel transition, called gelation, of colloid solution. Under these guidelines, one can find means to weaken the H–O bond for hydrogen generation and storage for green chemistry and energy management.

6.3.2 Thermal XAS and DPS: H–O Bond Stability

Oxygen K-edge XAS measurements [95] revealed that Li^+, Na^+, and K^+ cations shift the XAS pre-edge component energy more than the Cl^-, Br^-, and I^- anions. Figure 6.24 compares the temperature dependence of the XAS spectra for bulk water and 5 M LiCl solutions. Thermal heating shifts the pre-edge peak for the liquid

Table 6.3 O:H–O bond identities in the neat water and in the YX salt solutions[a,b]

	H_2O (277 K)	H_2O (skin)	Salt hydration shell
ω_L $(cm^{-1})^c$	200	75	65–90
ω_H (0.022)	3150	3450	≥ 3500
ω_H (0.059)	–	–	≥ 3500
d_L (Å)	1.70	2.0757	≥ 2.0
d_H (Å)	1.00	0.8893	≤ 0.89
E_L (meV)	95	59	≤ 59
E_H (eV)	3.98	4.97	≥ 4.43
O–O (Å)	2.695 [2]	2.965	≥ 2.965

[a] Change with respect to the ordinary bulk water
[b] All examined cases show the electrification effect on the O:H–O elongation with slight difference of the extent
[c] Subject to polarization and fluctuation. Polarization raises the ω_L slightly upward and fluctuation widens the peak

water negatively faster than that for the LiCl solutions, which is suggested as the O:H undergoes thermal elongation at different rates for the deionized water and LiCl solutions [95].

The NEXFAS data in Fig. 6.25 shows that the pre-edge component associated with water in the first hydration shell of Li^+ ion is thermally more stable than those beyond hydration. At 25 °C, the cation effect on the pre-edge shift from the value of 534.67 eV to deeper in alkali chlorides is remarkable: Li^+ (0.27 eV), Na^+ (0.09 eV), and K^+ (0.00 eV). The energy shift of Li^+ in the 5 M LiCl solution (0.30 eV) is close to that happened to the 3 M $LiCl/H_2O$ solution. On the other hand, in the NaCl solution, the anion effect is small: Cl^- (0.09 eV), Br^- (0.04 eV), and I^- (0.02 eV). The energy trend of the pre-edge shifts is the same as the DPS ω_H shifting positively for the solutions and the skin of water, see Fig. 6.26 [96].

The contributions from the Li:O polarization or the O:H binding energy are negligibly small. At polarization, the H–O bond becomes shorter and stiffer, the stiffer H–O bond is thermally more stable than those in the bulk of ordinary water, as it does in the water skin [96]. The identical O_{fw}–O_{bw} and O_{bw}–O_{bw} thermal expansion in both pure water and the 5 M LiCl solution indicates the invariance of the Li^+ hydration shell size, which does not interfere the O–O thermal behavior between the hydrated and non-hydrated oxygen anions. The O_{fw} and the O_{bw} represent oxygen in the first hydration shell and those beyond.

The full-frequency Raman spectra of the deionized water heated from 5 to 95 °C, shown in Fig. 6.26, revealed that heating shifts the ω_H upwardly and the ω_L downwardly because heating stiffens the H–O bond and softens the O:H nonbond due to the O:H–O cooperativity [96]. Inset a shows the ω_H decomposition into the bulk (3200 cm^{-1}), skin (3450 cm^{-1}) and the H – O free radical (3600 cm^{-1}) components. Panel b shows that the H–O bond in the bulk and in the skin undergoes

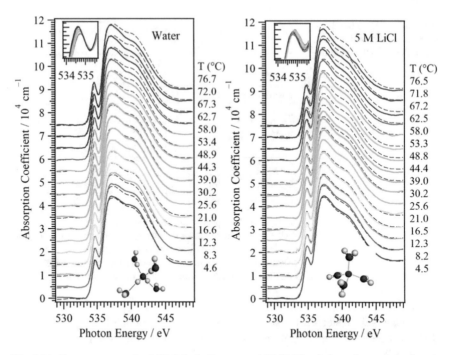

Fig. 6.24 Temperature resolved XAS for bulk water and 5 M LiCl solutions. Insets show the pre-edge negative shift and the structures of pure water and the hydrating molecules. Reprinted with permission from [95]

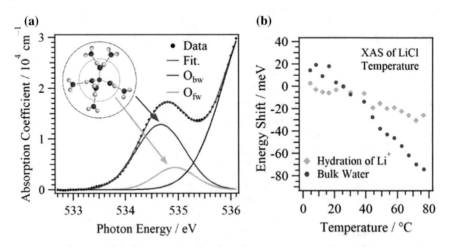

Fig. 6.25 Site and temperature resolved **a** NEXFAS pre–edge peak shift for LiCl solutions. **b** The E_{edge} for the first hydration shell (O_{fw}) shifts more than those beyond (O_{bw}) yet the O_{fw} energy is thermally more stable. The ΔE_{edge} for the O_{fw} shifts slower than that of pure water. Reprinted with permission from [95]

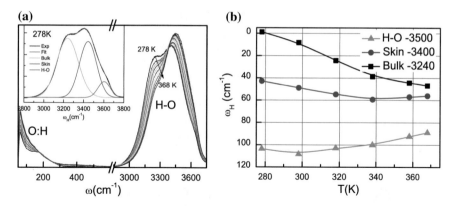

Fig. 6.26 **a** Full-frequency Raman spectra of deionized water heated from 5 to 95 °C revealed that heating shifts the ω_H and the ω_L phonon frequencies in opposite direction. The ω_H is decomposed into the bulk (3200 cm^{-1}), skin (3450 cm^{-1}) and the H–O free radical at 3610 cm^{-1} (inset a). **b** The DPS skin $\omega_H(T)$ components is thermally more stable than the bulk showing both thermal H–O contraction but the dangling H–O bond undergoes thermal expansion because the lacking of the O–O coupling at the open end of water surface. Reprinted with permission from [96]

thermal contraction and phonon stiffening, but the dangling H–O bond is subject to thermal expansion and softening because the dangling H–O bond lacks the O–O coupling at the open end of water surface. However, the skin $\omega_H(T)$ component shifts slower than the bulk component as the skin is thermally more stable than the bulk [96].

Figure 6.27 shows the temperature resolved Raman $\Delta\omega_H$ of the concentrated KCl/H$_2$O solutions encapsulated in a silica capillary tube [97]. Peak 1 at 3200 cm^{-1} corresponds to the ω_H for ordinary bulk water and peak 2 at 3450 cm^{-1} to the skin ω_H mode. At the ambient temperature, ionic polarization shifts the 3200 cm^{-1} up to overlap the 3450 cm^{-1}, as shown previously. Heating and polarization enhance each other to shorten and stiffen the H–O bond at different rates, depending on the solute concentration. In the liquid phase below 373 K, the ω_H shifts with temperature and more with the solute concentration. The ω_H shifts further as the temperature increases, but the increasing rate reverts with solute concentration. At 373 K and above, the liquid becomes partially vapor to the saturation pressure of 100 MPa. The thermal reversion of the ω_H shift shows that the hydrating H–O bonds are thermally more stable than that of the ordinary water in liquid and vapor phases. The polarization deformed H–O bonds are hardly further deformed, as observed from the XAS in Fig. 6.25.

Table 6.4 lists the MD estimation and neutron diffraction resolved the first O:Y$^+$ and O$\leftrightarrow$X$^-$ hydration shell sizes of the solutes. MD calculations agree with the DFT derived segmental strains of the O:H–O bonds surrounding X$^-$ solutes [56].

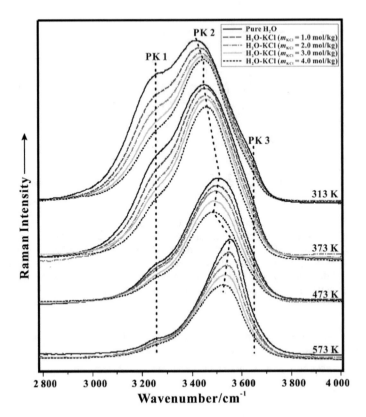

Fig. 6.27 Temperature resolved ω_H shift for the concentrated KCl/H$_2$O solutions. Reprinted with permission from [97]

6.3.3 Solution XAS: Polarization Dominance

Figures 6.28 and 6.29 compared the O–K edge XAS for the concentrated LiCl, 3 M YCl and NaX solutions. Salt solvation shifts the pre-edge peak to the positive direction in contrasting to the effect of heating that shifts the peak in the negative direction. It seemed conflicting that heating and ionic polarization share the same effect of H–O bond contraction. Bond contraction dictates the energy shift of all the core level and valence band [100] as the O:H binding energy contribution is negligible.

To clarify the inconsistency of pre-edge peak shift induced by heating and ionic polarization, one must consider the competition between H–O bond contraction and electrostatic polarization. Firstly, the pre-edge peak shift of the valence band (the occupied 4a$_1$ orbital) and the O 1 s level contribute simultaneously to the pre-edge shift, $\Delta E_{edge} = \Delta E_{1s} - \Delta E_{vb}$. Secondly, the electron in the 1 s level transiting to the upper edge of the valence band absorbs energy equal to the separation between the involved valance band and the O 1 s core level. Thirdly, the involved energy levels

Table 6.4 The O:Y$^+$ and O↔X$^-$ distances (Å) in the YX solutions [72, 95, 98][a]

	Li$^+$:O (5 M LiCl)	Na$^+$:O (Å)	K$^+$:O (Å)
Y$^+$:O distance MD [95]	2.00 (5 °C)	–	–
	1.99 (25 °C)	2.37	2.69
	1.98 (80 °C)	–	–
Neutron diffraction [72, 95, 98]	1.90	2.34	2.65
MD [95] X$^-$↔O distance	Cl$^-$·H–O	Br$^-$·H–O	I$^-$·H–O
	3.26	3.30	3.58
DFT(acid) [56][c]	Cl$^-$·(H–O:H)	Br$^-$·(H–O:H)	I$^-$·(H–O:H)
1st (ε_H; ε_L)%	−0.96; +26.1	−1.06; +30.8	−1.10; + 41.6
2nd (ε_H; ε_L)%	−0.73; +19.8	−0.78; +22.8	−0.83; +28.6
XAS (5 M LiCl) [95]	Li$^+$:(first O:H–O)	Li$^+$: (next O:H–O)	O:H–O (H$_2$O)
5 °C	$d_{O-O} = 2.71$ (at 4 °C, $d_{O-O} = d_H + d_L = 1.0004 + 1.6946$ = 2.695 [99])		
80 °C	2.76 ($\Delta d_L > -\Delta d_H(<0)$)		

[a](ε_H; ε_L)% is the DFT derived segmental strain for the first and the second O:H–O bonds radially away from X$^-$ anions in acid solutions. The strain is referred to the standard values of $d_H = 1.0004$ and $d_L = 1.6946$ Å for 4 °C water [99]. X$^-$·H represents the anion and H$^+$ Coulomb interaction

Fig. 6.28 O K-edge XAS spectra of concentrated LiCl solutions at 25 °C. Inset shows magnification of the pre-edge positive shift arising from polarization. Reprinted with permission from [102]

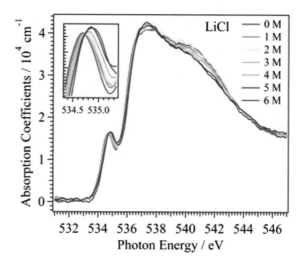

undergo positive shift because of the addition of the local crystal potential to the intra-atomic potential. The addition of the crystal potential shifts the energy level positively, according to the tight-binding approximation [101]. A contraction of the neighboring bonds shifts the energy levels further—causing the quantum entrapment of electron binding energy [100]. The amount of energy shift varies from level to level because of the screening of the potential by electrons in the outer orbitals. Therefore, the energy of the valence band composed of electrons in the outermost orbital shifts

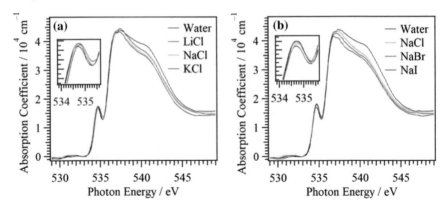

Fig. 6.29 O K-edge XAS spectra of 3 M **a** YCl and NaX solutions at 25 °C. Insets show magnification of the pre-edge positive shift due to polarization dominance. Reprinted with permission from [95]

more than an inner energy level. One has thus, $\Delta E_{vb} > \Delta E_{1s}$, and $\Delta E_{edge} = \Delta E_{1s} - \Delta E_{vb} < 0$, negative shift due to bond contraction.

On the other hand, electrostatic polarization changes the situation contrastingly. Charge polarization screens and splits the local potential and then shifts a proportion of electrons up in energy causing the negative shift, then, $-\Delta E_{vb} > -\Delta E_{1s}$, and thus $\Delta E_{edge} = \Delta E_{1s} - \Delta E_{vb} > 0$, positive shift takes place. In fact, bond contraction by molecular undercoordination or salt solvation is associated with polarization. H–O bond thermal contraction without polarization results in the pre-edge energy negative shift. For the salt solution, ionic polarization becomes dominance, which overweight the effect of H–O bond contraction on the binding energy shift. The polarization shifts the energy levels up and thus, $\Delta E_{edge} = -(\Delta E_{vb} - \Delta E_{1s}) > 0\,(-\Delta E_{vb} > -\Delta E_{1s} > 0)$. Therefore, the XAS pre-edge shift is sensitive to the local energetic environment, being capable of discriminating the effect of entrapment by bond contraction and charge polarization. If the entrapment and polarization is compatible, no shift will occur to the pre-edge peak.

6.3.4 XAS Capability: Entrapment and Polarization

The consistency between DPS and XAS observations evidences the thermal stability of the supersolid hydration volume and clarifies the origins of the pre-edge energy shift due to competition between thermal H–O bond contraction and charge polarization. Observations verified the following:

(1) The XAS pre-edge shift features the energy difference between the O 1 s core level shift ΔE_{1s} and its valence band shift ΔE_{vb} from their energy levels of an isolated O atom: $\Delta E_{edge} = \Delta E_{1s} - \Delta E_{vb} < 0$ [86] when the H–O contraction,

$\Delta E_H > 0$, is dominant [100]. In contrasting, $\Delta E_{edge} = \Delta E_{1s} - \Delta E_{vb} > 0$ when polarization becomes dominant, as polarization shifts all energy levels upwardly. Competition between H–O bond contraction and polarization dictates the pre-edge energy shift that is in a contrasting manner of the XPS O 1 s level shift.

(2) The H–O bond undergoes energy gain in liquid water heating [55], skin molecular undercoordination [86], and polarization [39]. The shortened H–O bonds are thermally and mechanically more stable because the stiffened bonds are less sensitive to perturbation. It is harder to further deform an already deformed H–O bond by stimulations such as heating in the present case: $(d|E_H|/dT)_{supersolid}/(d|E_H|/dT)_{regular} < 1$ and $(d\omega_H/dT)_{supersolid}/(d\omega_H/dT)_{regular} < 1$.

(3) The QS phase upper boundary is at 4 °C for regular water and it seems at 25 °C for the supersolid states according to the slopes of the XAS profile in Fig. 6.25b. The supersolidity disperses outwardly the QS phase. Outside the QS phase, O:H–O bond follows the regular thermodynamics: $dd_L/dT > 0$, $dd_H/dT < 0$.

(4) The local electric field of a small cation is stronger than that of a larger anion because of the $X^- \leftrightarrow X^-$ repulsion due to the insufficient number of the hydrating dipoles with the ordered molecular structure of the supersolid hydration shells [38].

6.4 Ionic Solutes Capabilities

There are two characters featuring the inter- and intra-molecular cooperative interactions that determine the hydration shell size and the HB stiffness in the hydration shells. One is the $\Delta\omega_x(C)$ shift and the other is the slope of the $f(C)$ in addition to polarization. The $\Delta\omega_x(C)$ and the $f(C)$ vary with the solute type and ionic size. As an integration of the residual phonon abundance in the DPS, the $f(C)$ is the relative number of bonds transiting from the mode of ordinary water to the hydrating. The ratio of the $f(C)/C$ is proportional to the hydration shell size or to the number of bonds associated with the solute. The $f(C)$ may be zero, linear, or nonlinear dependent towards saturation, which features how the local electric filed changes with the local bonding environment. The local bonding environment is subject to the solute-solvent, solute-solute interactions and water dipole shielding.

The following further explains the respective fraction coefficient, $f_X(C)$ and $f_Y(C)$, from the DPS profiles for the HX and the YX solutions. It has been intensively justified that $f_H(C) \cong 0$ [87]. One can thus assume $f_{HX}(C) - f_H(C) = f_X(C)$, which means that only the X^- polarizes its neighboring O:H–O bonds and transforms the H–O phonon from 3200 to 3500 cm^{-1} upon HX acid solvation. As the X^- anion in the (H, Y)X solutions is assumed to perform the same. The $f_X(C)$ and $f_Y(C)$ vary with the ionic radius and electronegativity of the same kind of ions. The best fitting of the integrals, see Fig. 6.30, results in: $f_H(C) \cong 0$, $f_{Y,X}(C) = a[1 - \exp(-C/C_0)]$ towards saturation, which would suffice to resolve the solute capabilities of bond transition upon solvation with involvement of solute-solute interactions. Results in Fig. 6.30 suggest the following:

(1) The high $f_{YX}(C)$ value at each concentration means that the resultant polarization capability of the Y^+ and the X^- is greater than either of them alone, and thus the relationships, $f_X(C) \cong f_{HX}(C)$ and $f_Y(C) \cong f_{YX}(C) - f_{HX}(C)$, hold true.

(2) The $f_Y(C)$ increases quasi-linearly and surpasses the $f_X(C)$ at a certain concentration, which suggests that the hydration shell size of the small Y^+ keeps constant and it is less sensitive to interference by its soundings. The constant slope means that the number of bonds per solute conserves in the hydration shell. The H_2O dipoles in its hydration shells fully-screen the electric field of a small Y^+ cation. Thus, a cation does not interact with any kind of solutes in the YX solutions.

(3) The $f_X(C) \propto 1 - \exp(-C/C_0)$ towards saturation means the number of H_2O molecules in the hydration volume is insufficient to fully screen the large X^- (radius ~2 Å) solute local electric field because of the geometric limitation to molecules packed in the hydration shells. On the other hand, the interstatial occupany of the large anion distorts the local structure of hydration volume. This number inadequacy further proves the highly ordered crystal-like solvent. The anion solute can thus interact with their alike—only anion-anion repulsion exists in the X^--based solutions to weaken the local electric field of the X^-. Therefore, the $f_X(C)$ increases gradually approaching to saturation, the hydration volume turns to be smaller, which limits the X^- solute capability of bond transition.

(4) Ions ($Z^{2+} + 2X^-$) in divalent salt solutions perform differently from YX monovalent because of the Z^{2+} folded charge quantity and the folded number of the $2X^-$. The $2X^-$ shows stronger repulsion because of its high density. The Z^{2+} also shows slight repulsion at high concentrations.

(5) The ability of X^- anions transiting the network O:H–O bonds follows the $I^- > Br^- > Cl^- > Na + > K^+ > Rb^+ > Cs^+$ Hofmeister series orders because of their exposure ability to the X^- interference.

(6) In contrast, the hydration shells of the complex molecular anions retain their sizes. The local electric field of the molecular anions is stronger, but the length of interaction is shorter, showing the nature of dipolar electric field—shorter yet anisotropic.

Figure 6.31 compares the ZX_2 solution conductivity $\sigma_{YX2}(C)$ and surface stress (contact angle $\theta_{YX2}(C)$). The $\sigma_{YX2}(C)$ and $\theta_{YX2}(C)$ share the similar manner of $f_{YX2}(C)$ exponential saturation toward high solute concentration. These identities follow the same $Sr^{2+} > Ca^{2+} > Mg^{2+}$ and $Br^- > Cl^-$ Hofmeister order. These concentration trend consistency evidences their common origin of solute polarization. The $\sigma_{YX2}(C)$ drops at higher concentration for the thicker solutions may arise from the lower carrier mobility because of the strong localization in the supersolid phase.

Table 6.5 summarizes the capabilities of the binary solutes in solutions of the proton dominated HX, lone-pair dominated H_2O_2 and YOH, and ionic dominated YX and ZX_2 solutions and their solvation consequence on the performance of the solutions.

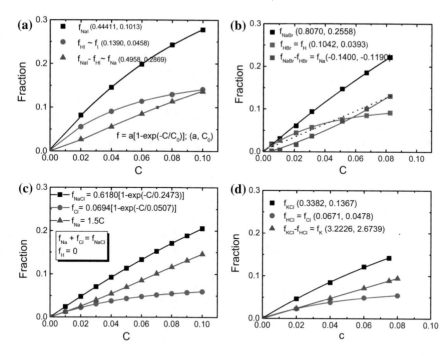

Fig. 6.30 The concentration dependence of the $f_{YX}(C)$ for **a–c** Na(I, Br, Cl) and **d** KCl and their decompositions into and nonlinear $f_X(C)$ and quasi-linear $f_Y(C)$. The constant slopes of the $f_Y(C)$ suggest the invariant hydration shell sizes because the hydrating H_2O molecules fully-screen the small Y^+ solutes. $X^- \leftrightarrow X^-$ repulsion weakens the X^- fields

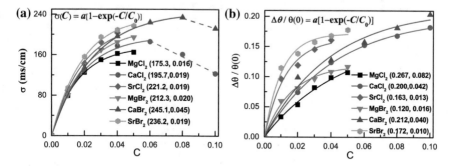

Fig. 6.31 Solute concentration and solute type resolved **a** solution conductivity and **b** surface stress (contact angle) follow the same trend of exponential decay of the fraction coefficient toward saturation and the: $Sr^{2+} > Ca^{2+} > Mg^{2+}$ and $Br^- > Cl^-$ Hofmeister capability order. The fact of conductivity drops at higher concentration suggests that thicker supersolid solution depresses the carrier mobility. Reprinted from permission of [85]

Table 6.5 Concentration C dependence of the solute capabilities of transforming the fraction (phonon abundance) of H–O bonds in the HX, YX, YOH and H₂O₂ aqueous solutions by transiting the H–O phonon frequency from 3200 cm^{-1} to the indicated wavenumbers in cm^{-1}

Solution		Fraction coefficient	Formulation	Functionality
HX [56, 57]	Excessive H$^+$ and Y$^+$ polarizer	f_H (3500)	0	H$_3$O$^+$ formation and H↔H repulsion
		f_{OH}(<3100) $f_{HX} = f_X$ (3500)	>≈0	H↔H repulsion elongated solvent H–O bond (mechanical compression [11, 103])
YX [19, 54, 55]	Ionic polarizers	$f_{YX} - f_X = f_Y$ (3500)	a[1−exp(−C/C_0)] shorter C_0 for X stable ω_H	X$^-$ shell size reduction and constant ω_H shift due to solute field shielding
				Constant Y$^+$ shell size Resolvable surface stress
NaX [52]		$f_Y \propto C$ (~3500)	Longer C_0 ω_H reverse shift	Larger Z$^-$ molecular anions
(Mg, Ca, Sr)(Cl, Br)$_2$ [85]		$f_Y = C$ (<0.05 and then exponential)	(3450; 100 cm^{-1})	Sr > Ca > Mg; Br > Cl
YOH [11]	Excessive lone pairs	f_Y annihilation f_{OH} (3610)	a[1−exp(−C/C_0)] longer C_0	HO$^-$ soluteH–O bond (H–O free radical [2, 86])
		f_{OH} (<3100)		O:⇔:O compression elongates the solvent H–O bond (mechanical compression [11, 63])
H$_2$O$_2$ [104]		f_{OH} (3550)		H$_2$O$_2$ solute H–O bond (bond order deficiency [86, 105, 106])

C_0 is the decay constant that features the effect of solute-solute shielding

6.5 Hofmeister Solution Properties

6.5.1 Diffusivity, Viscosity, Surface Stress, HB Transition

SFG measurements [68, 69] revealed that the SCN^- and CO_2 solution viscosity increases with solute concentration or the cooling rate of solution. The H–O phonon relaxation time increases with the viscosity and the molecular diffusivity according to convention. In aqueous solutions, solute and molecules drift randomly under thermal fluctuation by collision, which varies with the viscosity of the solution. Stokes-Einstein diffusivity $D(\eta, R, T)$ [64], Jones–Dale viscosity $\eta(C)$ [65], and the fraction $f(C)$ for O:H–O bonds transiting from the mode of water to hydrating are correlated as follows.

$$\begin{cases} \frac{D(\eta,R,T)}{D_0} = \frac{k_B T}{6\pi \eta R} & (Stokes - Einstein) \\ \frac{\Delta \eta(C)}{\eta(0)} = A\sqrt{C} + BC & (Jones - Dale) \\ f(C) = a\left[1 - \exp(-C/C_0)\right] & (Fraction - HB - transition) \end{cases}$$

One can adjust the factors A and B of the viscosity and formulate the surface stress to match the $f(C)$ for bond transition in salt solutions to clarify their origins, as illustrated in Fig. 6.32.

Therefore, ionic polarization reconciles quantitatively the number of bond transition, surface stress, solution viscosity, hydration-shell supersolidity, H–O phonon frequency, H–O phonon lifetime, and molecular drift diffusivity, consistently by shortening the H–O bond and lengthening the O:H nonbond. Salts solute overdosing endows both the relative viscosity and the $f_{YX}(C)$ with a negative curvature towards saturation because of the contribution of anion-anion repulsion. It is true that the viscosity and surface stress of the monovalent salt solutions arise from ionic polarization represented by $f_{YX}(C)$. The $f_Y(C) \propto C$ and the $f_X(C) \propto 1 - \exp(-C/C_0)$ contribute to the linear and nonlinear part of the Jones-Dale notion. The $f_X(C) \propto 1 - \exp(-C/C_0)$ is a superposition of the solute-solvent and anion-anion interactions.

However, the viscosity of (H, Li)Cl solutions shows the same increasing trend of $f_{XY}(C)$ for O:H–O polarization [107] or Jones-Dale expression but acidic solvation depresses the surface stress of an acidic solution, indicating different origins. The H↔H fragilization disrupts the surface stress and the Cl^- polarization and the H↔H and $Cl^- \leftrightarrow Cl^-$ repulsion raises the viscosity of an acidic solution.

The concentration resolved surface stress of acidic, basic, and salt solutions, show in Fig. 6.33, suggests the ionic or the paring of lone pairs polarization dominance unless the H↔H point disruption become dominant in acidic solutions. The H↔H fragilization overweighs the anion polarization, which makes the anionic hydration volume into individual fragments.

In contrast, the viscosities of the LiOH, complex $NaClO_4$ and $LiClO_4$, and divalent $ZX_2(Z = Mg, Ca, Sr, X = Cl, Br)$ solutions do not follow the negatively curvature saturation manner of the $f_{YX}(C)$ [52] nor the Jones-Dale description, as shown

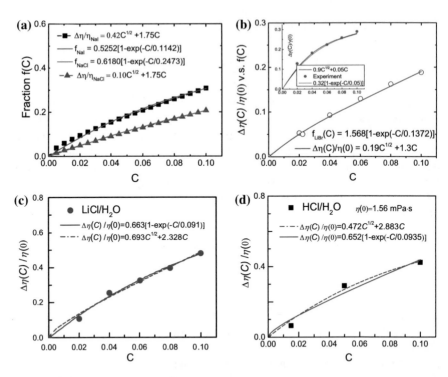

Fig. 6.32 Correlation between the Jones–Dale relative viscosity and the fraction coefficient $f_{YX}(C)$ for **a** Na(I, Cl)/H$_2$O [19], **b** LiBr/H$_2$O [38], **c** LiCl/H$_2$O [107], and **d** HCl/H$_2$O [107] solutions as a function of molar concentration. Inset b shows that the LiBr/H$_2$O surface stress follows the same exponential trend of the $f_{LiBr}(C)$ and the relative viscosity with different coefficients because of the molecular undercoordination effect that enhances the ionic polarization

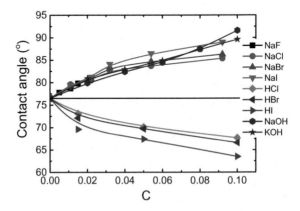

Fig. 6.33 The concentration dependent contact angles for acidic, basic and salt solutions showing evidence for the dominance or polarization in basic and salt solutions and $f_H \cong 0$ disruption in acidic solutions. Reprinted with permission from [90]

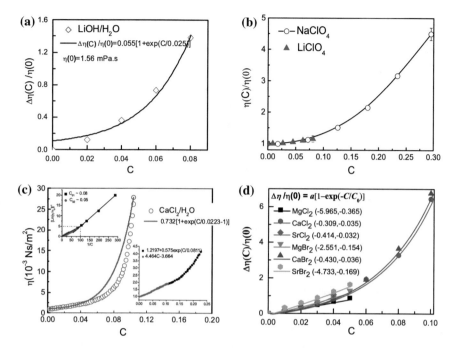

Fig. 6.34 Viscosity of the concentrated **a** LiOH/H_2O [107], **b** complex (Li, Na)ClO_4/H_2O [70], **c**, **d** divalent (Mg, Ca, Sr)(Cl, Br)$_2$/H_2O [85, 108] solutions do not follow the $f_{YX}(C)$ trend nor Jones-Dale description

in Fig. 6.34. The O:⇔:O strong repulsion derives the much greater viscosity with $d^2\eta/dC^2 > 0$ for LiOH solutions [107]. This unexpected concentration trend of viscosity elevation is consistent to those of the divalent (Ma, Ca, Cs)(Cl, Br)$_2$ solutions [85, 108] and the NaClO$_4$ and LiCiO$_4$ [70] solutions.

Likewise, (Mg, Ca, Sr)(Cl, Br)$_2$ dissolves into the Z^{2+} and $2X^-$ ions at the same concentration. The folded ionic density and the divalent Y^{2+} electric field are much stronger than the monovalent cations. The viscosity $[Ln(\eta/\eta_0)]^{-1-1}/C$ profiles for the (Mg, Ca, Sr) (Cl, Br)$_2$ solution shows two C_0 values at 0.08 and 0.05 at a molar concentration around 0.01. The viscosity of LiClO$_4$ and NaClO$_4$ solutions follows closely the trend of the CaCl$_2$ solutions. The folded $2X^-$ concentration and the large amount Y^{2+} charge lead to the $X^-\leftrightarrow X^-$ and $Y^{2+}\leftrightarrow Y^{2+}$ repulsions that are absent in monovalent salt solutions. These shorter and stronger interactions could be the origin of the viscosities disobeying Jones–Dale notion or the $f_{ZX2}(C)$.

Therefore, the mechanisms of solution viscosity and surface stress are different. Ionic polarization and H⇔H fragilization govern the surface stress yet the ionic polarization and anion-anion repulsion dictate the solution viscosity.

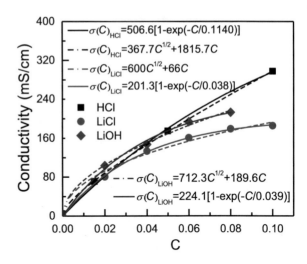

Fig. 6.35 Electric conductivity of the concentrated (H, Li)Cl and LiOH aqueous solutions. Reprinted with permission from [107]

6.5.2 Solution Electrical Conductivity

Figure 6.35 compares the electric conductivity for the concentrated (H, Li)Cl and LiOH solutions [107]. The conductivity follows the respective f(C) trends except for the LiOH solution whose f(C) depends linearly on the solute concentration [87] but the $\sigma(C)$ follows the Jones-Dale description. The viscosity of (H, Li)Cl solutions is proportional to the respective f(C). The solution conductivity is in the $\sigma_{HCl} > \sigma_{LiOH} > \sigma_{LiCl}$ order at higher concentrations. Ions and their hydration volume may serve as carriers for the current flow that is subject to scattering as an additional mechanism of transporting. H↔H and O:⇔:O fragmentation of the hydration volumes shall ease the transportation. Further investigation under variation of the measuring electric filed would derive the carrier mobility in solutions.

6.6 Quasisolid Phase Boundary Dispersivity: Anti-icing

6.6.1 QS Boundary Dispersion

One of the important phenomena of salt solvation is that solvation melts snow to improve the traffic condition in winter time, or lowers the freezing temperature T_N of the snow, as Fig. 6.36a inset shows the scenario. Figure 6.36a shows that the ΔT_N drops with the increase of NaCl or CaCl$_2$ concentration in their solutions The effect of CaCl$_2$ ⇒ Ca^{2+} + 2Cl$^-$ three-body solutes is more pronounced than the NaCl ⇒ Na$^+$ + Cl$^-$ two-body solvation on depressing the T_N that follows the $1 - \exp(C/C_0)$ manner. The C_0 of NaCl approaches three times that of the CaCl$_2$ because of the anions number and cation's charge quantity difference. MD investigations [109]

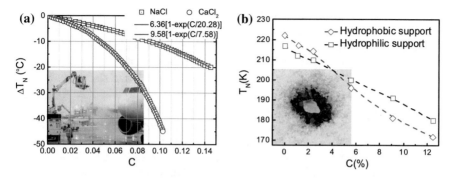

Fig. 6.36 a Concentration-resolved T_N depression by NaCl and CaCl$_2$ solvation [108] with inset a showing the airport salt anti-icing (from public domain). **b** The T_N of the concentrated LiCl solution on hydrophobic/hydrophilic graphite surfaces with inset showing snow melting by a piece of salt [109]. Reprinted with permission from [58, 108, 109]

show that the T_N of LiCl aqueous solutions is even lower on graphitic surface, see Fig. 6.36b. The concentration trend of the T_N transits at 5% for the hydrophobic and hydrophilic surfaces. The T_N transits from molecular undercoordination dominance to ionic polarization at the critical concentration.

How does the salt solvation depress the T_N of its solution?

Figure 3.1c in Chap. 3 illustrated the specific heat superposition of the O:H and the H–O over the full range of temperature [58]. The characteristics of a specific heat curve is the Debye temperature Θ_{Dx} and its thermal integral over its full temperature range [2]. The $\Theta_{Dx} \propto \omega_x \propto (E/d^2)_x^{1/2}$ and the integral is proportional to the segmental cohesive energy E_x. The superposition of the two specific heat curves derives two cross intersections that correspond to the temperatures of extreme densities. For the ordinary water, they are valued at 4 and -15 °C, which matches closely the critical temperatures for melting T_m and homogeneous ice nucleation, T_N, called freezing. The critical temperatures remain no constant but change with external stimulations such as molecular undercoordination [105, 106, 110], compression [111], salting [11, 112], electric polarization [18], etc.

The specific heat superposition divides the full-temperature range into five regimes with different specific heat η_H/η_L ratios: Ice XI ($\eta_H \approx \eta_L \approx 0$), solid I$_{h+c}$ ($\eta_H/\eta_L > 1$), quasisolid (QS, $\eta_H/\eta_L < 1$), liquid ($\eta_H/\eta_L > 1$), and Vapor ($\eta_H > 0$, $\eta_L \approx 0$). The segment of smaller η_x value follows the regular rule of thermal expansion, but the other part relaxes oppositely when the specimen is thermally excited. The O:H always relaxes more than the H–O does. In the QS phase, H–O contracts less than O:H expansion so ice floating takes place [58]. Little O:H–O frequency change of ice XI [113, 114] has been resolved below 100 K as both specific heat curves approach zero—the O:H and the H–O respond not to thermal excitation [58].

The intersecting temperatures change with the Θ_{Dx} that is proportional to the ω_x of the characteristic phonon. Therefore, supercooling or T_N depression and superheating or T_m elevation happen under a perturbation. From the dimensionality wise, a certain kind of energy determines the critical temperature for phase transition. As demonstrated, the T_m is proportional to the H–O energy and the T_N to the O:H energy through the relationship, $\Theta_{Dx} \propto \omega_x \propto (E/d^2)_x^{1/2}$ [63]. One may tune T_m and T_N by adjusting the O:H–O segmental length and energy through perturbation.

As salt solvation shifts the ω_H from 3200 to 3500 cm^{-1} and shifts the ω_L from 200 to 75 cm^{-1}, the Θ_{DL} for shifts from 192 K to some 80 K and the Θ_{DH} shifts from some 3000 K to its above. On the other hand, salt solvation shortens the H–O bond from 1.0 Å to some 0.95 Å and raises the E_H from 4.0 eV to its above, and meanwhile, lengthens the O:H from 1.70 to 1.95 Å and lowers the E_L from 0.095 eV to its below. These happenings depress the T_N and raises the T_m, and thus the salted snow at some -20 °C within the QS phase—a gel like state.

It is now clear why salt spreading melts snow—the Solid-QS phase transition results from its phase boundary outward dispersion by charge injection through salt solvation. Precisely, salt anti-icing is the process of T_N depression and T_m elevation rather than T_m depression that happens under compression—called regelation—ice melts by compression and freezes again when releasing the compression [63]. The anti-icing also happens to the molecular undercoordination at the water skin or water at the hydrophobic interface [110]. This mechanism is general to other solutions and water-protein hydrophobic interfaces. One can easily infer such information directly from Raman H–O phonon frequency blue or red shift.

Backscattering neutron spectroscopy [115] probing the dynamics of water molecules in LiCl and $CaCl_2$ aqueous solutions confined in 2.7, 1.9, and 1.4 nm diameter pores of silica matrices revealed that the pore size of 2.7 nm is sufficiently large for the confined liquids to exhibit characteristic traits of bulk behavior, such as a freezing–melting transition and a phase separation. On the other hand, none of the fluids in the 1.4 nm pores shows a clear freezing–melting transition; instead, their dynamics at low temperatures gradually became too slow for the nanosecond resolution of the experiment. The greatest suppression of water mobility in the $CaCl_2$ solutions suggests that cation charge and the cation hydration environment have a profound influence on the dynamics of the water molecules because of the short-range and stronger ionic fields. Quasi-elastic neutron scattering measurements of pure H_2O and 1 M $LiCl/H_2O$ solution confined in 1.9 nm pores revealed a dynamic transition in both liquids at the same temperature of 225–226 K, even though the dynamics of the solution at room temperature is one order of magnitude slower compared to the pure water.

6.6.2   Numbers of Hydrating H_2O Dipoles

According to the present understanding, ions in electrolyte solutions polarize their surrounding H_2O molecules to form each a supersolid hydration shell whose volume depends on its ionic radius and charge quantity. The DPS strategy distills information of the closest O:H–O bonds in the hydration shell of a solute in terms of abundance, stiffness, and fluctuation order by filtering out those molecules in the intermediate or high-order hydration shells.

The hydration volume may have multiple shells and leave some molecules free from being polarized. H_2O molecules in a solution can thus be categorized into groups of hydrating and free in a water-rich solution: the hydrating ones within the hydration shells freeze at a lower temperature than those free [18]. Wang et al. [116] developed a method to discriminate the free molecules from the hydrating ones by defining the number of H_2O molecules in the aqueous (n) solutions and the ones hydrating (n_h) the solute of W_m molar mass as follows:

Mass concentration of aqueous solute: $W_m/(W_m + n \times 18)$

Mass concentration of hydrated solute: $(W_m + n_h \times 18)/(W_m + n \times 18)$.

The aqueous solution is expressed as $M \cdot nH_2O$ with M and n being the respective number of solutes and solvent molecules. One can tune the n_h by matching the universal $T_H - C$ curve for a referential specimen of known n_H value. One can determine the n_H from measurement in the phase precipitation of the free from the hydrating water by cooling. The free water freezes first and leave the rest hydrating water to crystalize to form amorphous ice at even lower temperatures in a dilute electrolyte solutions. One can thus separate the remaining solution of purely hydrating water from the ice of free water and then to elucidate the n_H value. By doing so, $n_H = 10$ was derived for the $CaCl_2$ ($Ca^{2+} + 2Cl^-$) solutes, which means that one $CaCl_2$ molecule has ten H_2O molecules surrounding to hydrate the $CaCl_2$ at saturation. With the known $n_h = 10$ reference, one can then calibrate $n_h = 5$ from the T_H-C curves for a NaCl ($Na^+ + Cl^-$) solute instance [116]. Table 6.6 lists the hydrating H_2O molecular numbers per solute derived from Fig. 6.37.

Figure 6.37 shows the T_H variation of the aqueous solutions and the hydrated solute concentration of a collection of electrolyte solutions with a derivative of the n_H values. The T_H were obtained by measuring emulsified samples of 1–10 μm in diameter. The observed T_H depression match the expected QS boundary dispersion by salt solvation. According to the present O:H–O cooperativity notion, dissolved ions polarize water molecules to form each a hydration volume of supersolidity associated with T_m elevation and T_H depression [39] because of the QS boundaries dispersion by electrostatic polarization. The T_H variation reflects the local electric field that determines the O:H–O segmental energy and vibration frequencies. The ω_H blueshift raises the T_m and the cooperative ω_L redshift depresses the T_H.

6.7 Summary

Spectroscopic examination of the hydration bonding dynamics for the YX alkali halides, the ZX_2 divalent salts, and the Na-based complex salts revealed the following:

(1) Salt dissolves into the smaller Y^+ cations and X^- anions and each of them serves as a point polarizer that aligns, clusters, and stretches the surrounding O:H–O bonds to form the supersolid hydration volume.

Table 6.6 Hydrating H_2O molecular numbers per electrolyte solute (reprinted with permission from [116] supplementary information)

Solute	Hydration number, n_h
LiCl	6
$CaCl_2$	10
$Ca(NO_3)_2$	7
$MgCl_2$	12
$ZnCl_2$	7
$(Mg + Zn)Cl_2$	16.4
$MnCl_2$	9
$Mn(NO_3)_2$	11.7
$CrCl_3$	17
$Cr(NO_3)_3$	13.8
$AlCl_3$	19
$FeCl_3$	15
$Fe(NO_3)_3$	15.8
Glycerol	1.5
Dimethyl sulfoxide	3.0
Ethylene glycol	1.0
1,2,4-butanetriol	1.9
PEG300	6.8
H_2SO_4	6.9
HNO_3	4.1
H_2O_2	2.0

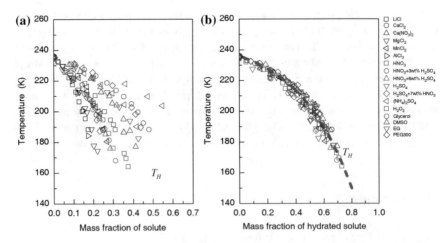

Fig. 6.37 The homogeneous ice nucleation temperature T_H as a function of **a** mass concentration of solute, $W_m/(W_m + n \times 18)$ and **b** mass fraction of hydrated solute, $(W_m + n_h \times 18)/(W_m + n \times 18)$, for various aqueous solutions [117–129]. For the mixed solutions, the composition is specified with the weight percentage of the minor solute. Reprinted with permission from [116]

(2) Salt solvation has the same effect of molecular undercoordination that shortens the H–O bond and stiffens its phonon but relaxes the O:H nonbond oppositely, which disperses the QS phase boundary to raise the T_m and depress the T_N.

(3) The fraction coefficient for HBs transition follows the $f_Y(C) \propto C$ and $f_X(C) \propto 1 - \exp(-C/C_0)$ form towards saturation, which features the fraction of bonds transiting from the mode of ordinary water to hydrating. Monovalent Y^+ cations effect quite the same in the quasi-linear $f_Y(C)$ manner but raises the surface stress in the order of $Na^+ > K^+ > Rb^+ > Cs^+$. The linear form of the $f_Y(C)$ indicates that the capability of the smaller Y^+ responds insignificantly to the solute-solute interactions and the invariant hydration volume. The linear $f_Y(C)$ contributes to the linear part of the solution viscosity and surface stress and the exponential $f_Y(C)$ form to the $C^{1/2}$ term of Jones-Dale viscosity.

(4) The exponential form of $f_X(C) \propto 1 - \exp(-C/C_0)$ arises from the anion-water and anion-anion interactions, which contributes nonlinearly to the viscosity and surface stress of the monovalent alkali halide solutions. The X^- capability of bond transition follows the order of $I^- > Br^- > Cl^- > F^- \sim 0$, which arises from the extent of screening by the hydrating H_2O dipoles. The shorter C_0 and the constant ω_H shift suggest the involvement of the anion-anion repulsion that reduces the hydration shell size but changes insignificantly the local electric field. The number inadequacy of the X^- hydrating H_2O evidences the high-order of the supersolid or the semi-rigid solvent water.

(5) The capability of the divalent Y^{2+} of polarization follows the order: $Sr^{2+} > Ca^{2+} > Mg^{2+}$. The $Y^{2+} \sim 2X^-$ is less capable of transiting O:H–O bond stiffness than the monovalent salt solutions as the superposition of the $Y^{2+} \sim 2X^-$ ionic fields. Divalent salt has a stronger effect than the monovalent salt on depressing the T_N and elevating the viscosity of the solution because of the number and charge-quantity of ionic solutes.

(6) The complex molecular anions show dipolar electric field feature having shorter and stronger local electric fields in the hydration shells, which shows sensitive $\Delta\omega_H$ but insensitive shell size to solute concentration. Higher solute concentration does not affect the shell size but weakens the electric field in the hydration shells.

(7) The concentration-resolved surface stress and solution viscosity follow the same manner of the $f_{YX}(C)$ for the YX salt but different for acidic, basic, and high valence salt solutions. The $\partial^2\eta/\partial C^2$ value offers criteria for solute-solute interactions: $\partial^2\eta/\partial C^2 = 0$, solute isolation (alkali cation); $\partial^2\eta/\partial C^2 > 0$, repulsive interaction (halide anions; O:⇔:O, H↔H, divalent and complex salt solvation).

(8) The concentration-trend consistency of $f_{YX2}(C)$, solution conductivity, and surface stress for YX_2 solutions clarifies their common origin of the creened ionic polarization.

References

1. F. Hofmeister, Zur Lehre von der Wirkung der Salze. Archi. Exp. Pathol. Pharmakol. **25**(1), 1–30 (1888)
2. Y.L. Huang, X. Zhang, Z.S. Ma, Y.C. Zhou, W.T. Zheng, J. Zhou, C.Q. Sun, Hydrogen-bond relaxation dynamics: resolving mysteries of water ice. Coord. Chem. Rev. **285**, 109–165 (2015)
3. P. Lo Nostro, B.W. Ninham, Hofmeister phenomena: an update on ion specificity in biology. Chem. Rev. **112**(4), 2286–2322 (2012)
4. C.M. Johnson, S. Baldelli, Vibrational sum frequency spectroscopy studies of the influence of solutes and phospholipids at vapor/water interfaces relevant to biological and environmental systems. Chem. Rev. **114**(17), 8416–8446 (2014)
5. Y. Liu, A. Kumar, S. Depauw, R. Nhili, M.H. David-Cordonnier, M.P. Lee, M.A. Ismail, A.A. Farahat, M. Say, S. Chackal-Catoen, A. Batista-Parra, S. Neidle, D.W. Boykin, W.D. Wilson, Water-mediated binding of agents that target the DNA minor groove. J. Am. Chem. Soc. **133**(26), 10171–10183 (2011)
6. E.K. Wilson, Hofmeister still mystifies. C&EN Arch. **90**(29), 42–43 (2012)
7. B. Wang, W. Jiang, Y. Gao, Z. Zhang, C. Sun, F. Liu, Z. Wang, Energetics competition in centrally four-coordinated water clusters and Raman spectroscopic signature for hydrogen bonding. RSC Adv. **7**(19), 11680–11683 (2017)
8. K.D. Collins, Why continuum electrostatics theories cannot explain biological structure, polyelectrolytes or ionic strength effects in ion-protein interactions. Biophys. Chem. **167**, 43–59 (2012)
9. L. Li, J.H. Ryu, S. Thayumanavan, Effect of Hofmeister ions on the size and encapsulation stability of polymer nanogels. Langmuir **29**(1), 50–55 (2013)
10. R. Đuričković, P. Claverie, M. Bourson, J.-M. Marchetti, M.D.Fontana Chassot, Water-ice phase transition probed by Raman spectroscopy. J. Raman Spectrosc. **42**(6), 1408–1412 (2011)
11. Q. Zeng, T. Yan, K. Wang, Y. Gong, Y. Zhou, Y. Huang, C.Q. Sun, B. Zou, Compression icing of room-temperature NaX solutions (X = F, Cl, Br, I). Phys. Chem. Chem. Phys. **18**(20), 14046–14054 (2016)
12. J. Li, C. Zhang, J. Luo, Superlubricity behavior with phosphoric acid-water network induced by rubbing. Langmuir **27**(15), 9413–9417 (2011)
13. J. Li, C. Zhang, J. Luo, Superlubricity achieved with mixtures of polyhydroxy alcohols and acids. Langmuir **29**(17), 5239–5245 (2013)
14. B.C. Donose, I.U. Vakarelski, K. Higashitani, Silica surfaces lubrication by hydrated cations adsorption from electrolyte solutions. Langmuir **21**(5), 1834–1839 (2005)
15. J. Abraham, K.S. Vasu, C.D. Williams, K. Gopinadhan, Y. Su, C.T. Cherian, J. Dix, E. Prestat, S.J. Haigh, I.V. Grigorieva, P. Carbone, A.K. Geim, R.R. Nair, Tunable sieving of ions using graphene oxide membranes. Nat. Nanotechnol. **12**(6), 546–550 (2017)
16. L. Chen, G. Shi, J. Shen, B. Peng, B. Zhang, Y. Wang, F. Bian, J. Wang, D. Li, Z. Qian, G. Xu, G. Liu, J. Zeng, L. Zhang, Y. Yang, G. Zhou, M. Wu, W. Jin, J. Li, H. Fang, Ion sieving in graphene oxide membranes via cationic control of interlayer spacing. Nature **550**(7676), 380–383 (2017)
17. Y. Zhang, P.S. Cremer, Chemistry of Hofmeister anions and osmolytes. Annu. Rev. Phys. Chem. **61**, 63–83 (2010)
18. C.Q. Sun, Y. Sun, in *The Attribute of Water: Single Notion, Multiple Myths*. Springer Series Chemical Physics, vol. 113 (Springer, Heidelberg, 2016), 494p
19. Y. Zhou, Y. Huang, Z. Ma, Y. Gong, X. Zhang, Y. Sun, C.Q. Sun, Water molecular structureorder in the NaX hydration shells (X = F, Cl, Br, I). J. Mol. Liq. **221**, 788–797 (2016)
20. X.P. Li, K. Huang, J.Y. Lin, Y.Z. Xu, H.Z. Liu, Hofmeister ion series and its mechanism of action on affecting the behavior of macromolecular solutes in aqueous solution. Prog. Chem. **26**(8), 1285–1291 (2014)

21. F. Hofmeister, Concerning regularities in the protein-precipitating effects of salts and the relationship of these effects to the physiological behaviour of salts. Arch. Exp. Pathol. Pharmacol. **24**, 247–260 (1888)

22. P. Jungwirth, P.S. Cremer, Beyond Hofmeister. Nat. Chem. **6**(4), 261–263 (2014)

23. W.M. Cox, J.H. Wolfenden, The viscosity of strong electrolytes measured by a differential method. Proc. R. Soc. Lond. A **145**(855), 475–488 (1934)

24. P. Ball, J.E. Hallsworth, Water structure and chaotropicity: their uses, abuses and biological implications. PCCP **17**(13), 8297–8305 (2015)

25. K.D. Collins, M.W. Washabaugh, The Hofmeister effect and the behaviour of water at interfaces. Q. Rev. Biophys. **18**(04), 323–422 (1985)

26. K.D. Collins, Charge density-dependent strength of hydration and biological structure. Biophys. J. **72**(1), 65–76 (1997)

27. T.T. Duignan, D.F. Parsons, B.W. Ninham, Collins's rule, Hofmeister effects and ionic dispersion interactions. Chem. Phys. Lett. **608**, 55–59 (2014)

28. X. Liu, H. Li, R. Li, D. Xie, J. Ni, L. Wu, Strong non-classical induction forces in ion-surface interactions: general origin of Hofmeister effects. Sci. Rep. **4**, 5047 (2014). http://www.naturecom/srep/2013/131021/srep03005/metrics

29. W.J. Xie, Y.Q. Gao, A simple theory for the Hofmeister series. J. Phys. Chem. Lett. **4**, 4247–4252 (2013)

30. H. Zhao, D. Huang, Hydrogen bonding penalty upon ligand binding. PLoS ONE **6**(6), e19923 (2011)

31. W.B. O'Dell, D.C. Baker, S.E. McLain, Structural evidence for inter-residue hydrogen bonding observed for cellobiose in aqueous solution. PLoS ONE **7**(10), e45311 (2012)

32. M. Cacace, E. Landau, J. Ramsden, The Hofmeister series: salt and solvent effects on interfacial phenomena. Q. Rev. Biophys. **30**(3), 241–277 (1997)

33. I.S. Perelygin, G.P. Mikhailov, S.V. Tuchkov, Vibrational and orientational relaxation of polyatomic anions and ion-molecular hydrogen bond in aqueous solutions. J. Mol. Struct. **381**(1–3), 189–192 (1996)

34. F. Bruni, S. Imberti, R. Mancinelli, M.A. Ricci, Aqueous solutions of divalent chlorides: ions hydration shell and water structure. J. Chem. Phys. **136**(6), 137–148 (2012)

35. M. Andreev, A. Chremos, J. de Pablo, J.F. Douglas, Coarse-grained model of the dynamics of electrolyte solutions. J. Phys. Chem. B **121**(34), 8195–8202 (2017)

36. M. Andreev, J.J. de Pablo, A. Chremos, J.F. Douglas, Influence of ion solvation on the properties of electrolyte solutions. J. Phys. Chem. B **122**(14), 4029–4034 (2018)

37. P.H.K.D. Jong, G.W. Neilson, M.C. Bellissent-Funel, Hydration of Ni^{2+} and Cl^- in a concentrated nickel chloride solution at 100 °C and 300 °C. J. Chem. Phys. **105**(12), 5155–5159 (1996)

38. C.Q. Sun, J. Chen, Y. Gong, X. Zhang, Y. Huang, (H, Li)Br and LiOH solvation bonding dynamics: molecular nonbond interactions and solute extraordinary capabilities. J. Phys. Chem. B **122**(3), 1228–1238 (2018)

39. C.Q. Sun, Perspective: supersolidity of undercoordinated and hydrating water. Phys. Chem. Chem. Phys. **20**, 30104–30119 (2018)

40. B. Hess, N.F.A. van der Vegt, Cation specific binding with protein surface charges. Proc. Natl. Acad. Sci. **106**(32), 13296–13300 (2009)

41. J.S. Uejio, C.P. Schwartz, A.M. Duffin, W.S. Drisdell, R.C. Cohen, R.J. Saykally, Characterization of selective binding of alkali cations with carboxylate by x-ray absorption spectroscopy of liquid microjets. Proc. Natl. Acad. Sci. **105**(19), 6809–6812 (2008)

42. L. Vrbka, J. Vondrášek, B. Jagoda-Cwiklik, R. Vácha, P. Jungwirth, Quantification and rationalization of the higher affinity of sodium over potassium to protein surfaces. Proc. Natl. Acad. Sci. **103**(42), 15440–15444 (2006)

43. J. Paterová, K.B. Rembert, J. Heyda, Y. Kurra, H.I. Okur, W.R. Liu, C. Hilty, P.S. Cremer, P. Jungwirth, Reversal of the hofmeister series: specific ion effects on peptides. J. Phys. Chem. B **117**(27), 8150–8158 (2013)

44. J. Heyda, T. Hrobárik, P. Jungwirth, Ion-specific interactions between halides and basic amino acids in water†. J. Phys. Chem. A **113**(10), 1969–1975 (2009)

45. J.D. Smith, R.J. Saykally, P.L. Geissler, The effects of dissolved halide anions on hydrogen bonding in liquid water. J. Am. Chem. Soc. **129**(45), 13847–13856 (2007)

46. S. Park, M.D. Fayer, Hydrogen bond dynamics in aqueous NaBr solutions. Proc. Natl. Acad. Sci. U.S.A. **104**(43), 16731–16738 (2007)

47. Q. Sun, Raman spectroscopic study of the effects of dissolved NaCl on water structure. Vib. Spectrosc. **62**, 110–114 (2012)

48. F. Aliotta, M. Pochylski, R. Ponterio, F. Saija, G. Salvato, C. Vasi, Structure of bulk water from Raman measurements of supercooled pure liquid and LiCl solutions. Phys. Rev. B **86**(13), 134301 (2012)

49. S. Park, M.B. Ji, K.J. Gaffney, Ligand exchange dynamics in aqueous solution studied with 2DIR spectroscopy. J. Phys. Chem. B **114**(19), 6693–6702 (2010)

50. S. Park, M. Odelius, K.J. Gaffney, Ultrafast dynamics of hydrogen bond exchange in aqueous ionic solutions. J. Phys. Chem. B **113**(22), 7825–7835 (2009)

51. K.J. Gaffney, M. Ji, M. Odelius, S. Park, Z. Sun, H-bond switching and ligand exchange dynamics in aqueous ionic solution. Chem. Phys. Lett. **504**(1–3), 1–6 (2011)

52. Y. Zhou, Y. Zhong, X. Liu, Y. Huang, X. Zhang, C.Q. Sun, NaX solvation bonding dynamics: hydrogen bond and surface stress transition (X = HSO$_4$, NO$_3$, ClO$_4$, SCN). J. Mol. Liq. **248**, 432–438 (2017)

53. Y. Chen, H.I.I. Okur, C. Liang, S. Roke, Orientational ordering of water in extended hydration shells of cations is ion-specific and correlates directly with viscosity and hydration free energy. Phys. Chem. Chem. Phys. **19**(36), 24678–24688 (2017)

54. Y. Gong, Y. Zhou, H. Wu, D. Wu, Y. Huang, C.Q. Sun, Raman spectroscopy of alkali halide hydration: hydrogen bond relaxation and polarization. J. Raman Spectrosc. **47**(11), 1351–1359 (2016)

55. X. Zhang, T. Yan, Y. Huang, Z. Ma, X. Liu, B. Zou, C.Q. Sun, Mediating relaxation and polarization of hydrogen-bonds in water by NaCl salting and heating. Phys. Chem. Chem. Phys. **16**(45), 24666–24671 (2014)

56. X. Zhang, Y. Zhou, Y. Gong, Y. Huang, C. Sun, Resolving H(Cl, Br, I) capabilities of transforming solution hydrogen-bond and surface-stress. Chem. Phys. Lett. **678**, 233–240 (2017)

57. X. Zhang, Y. Xu, Y. Zhou, Y. Gong, Y. Huang, C.Q. Sun, HCl, KCl and KOH solvation resolved solute-solvent interactions and solution surface stress. Appl. Surf. Sci. **422**, 475–481 (2017)

58. C.Q. Sun, X. Zhang, X. Fu, W. Zheng, J.-L. Kuo, Y. Zhou, Z. Shen, J. Zhou, Density and phonon-stiffness anomalies of water and ice in the full temperature range. J. Phys. Chem. Lett. **4**, 3238–3244 (2013)

59. L. Wang, Y. Guo, P. Li, Y. Song, Anion-specific effects on the assembly of collagen layers mediated by magnesium ion on mica surface. J. Phys. Chem. B **118**(2), 511–518 (2014)

60. Y. Gong, Y. Xu, Y. Zhou, C. Li, X. Liu, L. Niu, Y. Huang, X. Zhang, C.Q. Sun, Hydrogen bond network relaxation resolved by alcohol hydration (methanol, ethanol, and glycerol). J. Raman Spectrosc. **48**(3), 393–398 (2017)

61. C. Yan, Z. Xue, W. Zhao, J. Wang, T. Mu, Surprising Hofmeister effects on the bending vibration of water. ChemPhysChem **17**(20), 3309–3314 (2016)

62. Z. Yin, L. Inhester, S. Thekku Veedu, W. Quevedo, A. Pietzsch, P. Wernet, G. Groenhof, A. Foehlisch, H. Grubmüller, S.A. Techert, Cationic and anionic impact on the electronic structure of liquid water. J. Phys. Chem. Lett. **8**(16), 3759–3764 (2017)

63. X. Zhang, P. Sun, Y. Huang, T. Yan, Z. Ma, X. Liu, B. Zou, J. Zhou, W. Zheng, C.Q. Sun, Water's phase diagram: from the notion of thermodynamics to hydrogen-bond cooperativity. Prog. Solid State Chem. **43**, 71–81 (2015)

64. J.C. Araque, S.K. Yadav, M. Shadeck, M. Maroncelli, C.J. Margulis, How is diffusion of neutral and charged tracers related to the structure and dynamics of a room-temperature ionic liquid? Large deviations from Stokes-Einstein behavior explained. J. Phys. Chem. B **119**(23), 7015–7029 (2015)

65. G. Jones, M. Dole, The viscosity of aqueous solutions of strong electrolytes with special reference to barium chloride. J. Am. Chem. Soc. **51**(10), 2950–2964 (1929)

66. K. Wynne, The mayonnaise effect. J. Phys. Chem. Lett. **8**(24), 6189–6192 (2017)

67. S.T. van der Post, C.S. Hsieh, M. Okuno, Y. Nagata, H.J. Bakker, M. Bonn, J. Hunger, Strong frequency dependence of vibrational relaxation in bulk and surface water reveals sub-picosecond structural heterogeneity. Nat. Commun. **6**, 8384 (2015)

68. T. Brinzer, E.J. Berquist, Z. Ren (任哲), S. Dutta, C.A. Johnson, C.S. Krisher, D.S. Lambrecht, S. Garrett-Roe, Ultrafast vibrational spectroscopy (2D-IR) of CO_2 in ionic liquids: carbon capture from carbon dioxide's point of view. J. Chem. Phys. **142**(21), 212425 (2015)

69. Z. Ren, A.S. Ivanova, D. Couchot-Vore, S. Garrett-Roe, Ultrafast structure and dynamics in ionic liquids: 2D-IR spectroscopy probes the molecular origin of viscosity. J. Phys. Chem. Lett. **5**(9), 1541–1546 (2014)

70. Q. Wei, D. Zhou, H. Bian, Negligible cation effect on the vibrational relaxation dynamics of water molecules in $NaClO_4$ and $LiClO_4$ aqueous electrolyte solutions. RSC Adv. **7**(82), 52111–52117 (2017)

71. A.W. Omta, M.F. Kropman, S. Woutersen, H.J. Bakker, Negligible effect of ions on the hydrogen-bond structure in liquid water. Science **301**(5631), 347–349 (2003)

72. R. Mancinelli, A. Botti, F. Bruni, M.A. Ricci, A.K. Soper, Hydration of sodium, potassium, and chloride ions in solution and the concept of structure maker/breaker. J. Phys. Chem. B **111**, 13570–13577 (2007)

73. K.D. Collins, Ions from the Hofmeister series and osmolytes: effects on proteins in solution and in the crystallization process. Methods **34**(3), 300–311 (2004)

74. K. Tielrooij, N. Garcia-Araez, M. Bonn, H. Bakker, Cooperativity in ion hydration. Science **328**(5981), 1006–1009 (2010)

75. Z.S. Nickolov, J. Miller, Water structure in aqueous solutions of alkali halide salts: FTIR spectroscopy of the OD stretching band. J. Colloid Interface Sci. **287**(2), 572–580 (2005)

76. X. Zhang, Y. Huang, Z. Ma, Y. Zhou, W. Zheng, J. Zhou, C.Q. Sun, A common supersolid skin covering both water and ice. Phys. Chem. Chem. Phys. **16**(42), 22987–22994 (2014)

77. X. Zhang, Y. Huang, Z. Ma, Y. Zhou, J. Zhou, W. Zheng, Q. Jiang, C.Q. Sun, Hydrogen-bond memory and water-skin supersolidity resolving the Mpemba paradox. Phys. Chem. Chem. Phys. **16**(42), 22995–23002 (2014)

78. R. Zangi, B. Berne, Aggregation and dispersion of small hydrophobic particles in aqueous electrolyte solutions. J. Phys. Chem. B **110**(45), 22736–22741 (2006)

79. Y. Levin, Polarizable ions at interfaces. Phys. Rev. Lett. **102**(14), 147803 (2009)

80. H.I. Okur, J. Hladílková, K.B. Rembert, Y. Cho, J. Heyda, J. Dzubiella, P.S. Cremer, P. Jungwirth, Beyond the Hofmeister series: ion-specific effects on proteins and their biological functions. J. Phys. Chem. B **121**(9), 1997–2014 (2017)

81. L. Pauling, *The Nature of the Chemical Bond*, 3rd edn. (Cornell University Press, Ithaca, NY, 1960)

82. C.Q. Sun, *Relaxation of the Chemical Bond*. Springer Series Chemical Physics, vol. 108 (Springer, Heidelberg, 2014), 807p

83. C.Q. Sun, Y. Sun, Y.G. Ni, X. Zhang, J.S. Pan, X.H. Wang, J. Zhou, L.T. Li, W.T. Zheng, S.S. Yu, L.K. Pan, Z. Sun, Coulomb repulsion at the nanometer-sized contact: a force driving superhydrophobicity, superfluidity, superlubricity, and supersolidity. J. Phys. Chem. C **113**(46), 20009–20019 (2009)

84. X. Zhang, Y. Huang, Z. Ma, L. Niu, C.Q. Sun, From ice superlubricity to quantum friction: electronic repulsivity and phononic elasticity. Friction **3**(4), 294–319 (2015)

85. H. Fang, Z. Tang, X. Liu, Y. Huang, C.Q. Sun, Capabilities of anion and cation on hydrogen-bond transition from the mode of ordinary water to (Mg, Ca, Sr)(Cl, Br)$_2$ hydration. J. Mol. Liq. **279**, 485–491 (2019)

86. C.Q. Sun, X. Zhang, J. Zhou, Y. Huang, Y. Zhou, W. Zheng, Density, elasticity, and stability anomalies of water molecules with fewer than four neighbors. J. Phys. Chem. Lett. **4**, 2565–2570 (2013)

87. C.Q. Sun, Perspective: unprecedented O:⇔:O compression and H↔H fragilization in Lewis solutions. Phys. Chem. Chem. Phys. **21**, 2234–2250 (2019)

88. L.M. Levering, M.R. Sierra-Hernández, H.C. Allen, Observation of hydronium ions at the air–aqueous acid interface: vibrational spectroscopic studies of aqueous HCl, HBr, and HI. J. Phys. Chem. C **111**(25), 8814–8826 (2007)

89. C.Q. Sun, Size dependence of nanostructures: impact of bond order deficiency. Prog. Solid State Chem. **35**(1), 1–159 (2007)

90. C.Q. Sun, Aqueous charge injection: solvation bonding dynamics, molecular nonbond interactions, and extraordinary solute capabilities. Int. Rev. Phys. Chem. **37**(3–4), 363–558 (2018)

91. J. Ostmeyer, S. Chakrapani, A.C. Pan, E. Perozo, B. Roux, Recovery from slow inactivation in K channels is controlled by water molecules. Nature **501**(7465), 121–124 (2013)

92. C.Q. Sun, Oxidation electronics: bond-band-barrier correlation and its applications. Prog. Mater Sci. **48**(6), 521–685 (2003)

93. W.T. Zheng, C.Q. Sun, Electronic process of nitriding: mechanism and applications. Prog. Solid State Chem. **34**(1), 1–20 (2006)

94. Y. Tong, I.Y. Zhang, R.K. Campen, Experimentally quantifying anion polarizability at the air/water interface. Nat. Commun. **9**, 1313 (2018)

95. M. Nagasaka, H. Yuzawa, N. Kosugi, Interaction between water and alkali metal ions and its temperature dependence revealed by oxygen K-edge X-ray absorption spectroscopy. J. Phys. Chem. B **121**(48), 10957–10964 (2017)

96. Y. Zhou, Y. Zhong, Y. Gong, X. Zhang, Z. Ma, Y. Huang, C.Q. Sun, Unprecedented thermal stability of water supersolid skin. J. Mol. Liq. **220**, 865–869 (2016)

97. Q. Hu, H. Zhao, Understanding the effects of chlorine ion on water structure from a Raman spectroscopic investigation up to 573 K. J. Mol. Struct. **1182**, 191–196 (2019)

98. N. Ohtomo, K. Arakawa, Neutron diffraction study of aqueous ionic solutions. I. Aqueous solutions of lithium chloride and cesium chloride. Bull. Chem. Soc. Jpn **52**, 2755–2759 (1979)

99. Y. Huang, X. Zhang, Z. Ma, Y. Zhou, J. Zhou, W. Zheng, C.Q. Sun, Size, separation, structure order, and mass density of molecules packing in water and ice. Sci. Rep. **3**, 3005 (2013)

100. X.J. Liu, M.L. Bo, X. Zhang, L. Li, Y.G. Nie, H. TIan, Y. Sun, S. Xu, Y. Wang, W. Zheng, C.Q. Sun, Coordination-resolved electron spectrometrics. Chem. Rev. **115**(14), 6746–6810 (2015)

101. M.A. Omar, *Elementary Solid State Physics: Principles and Applications* (Addison-Wesley, New York, 1993)

102. M. Nagasaka, H. Yuzawa, N. Kosugi, Development and application of in situ/operando soft X-ray transmission cells to aqueous solutions and catalytic and electrochemical reactions. J. Electron Spectrosc. Relat. Phenom. **200**, 293–310 (2015)

103. C.Q. Sun, X. Zhang, W.T. Zheng, Hidden force opposing ice compression. Chem. Sci. **3**, 1455–1460 (2012)

104. J. Chen, C. Yao, X. Liu, X. Zhang, C.Q. Sun, Y. Huang, H_2O_2 and HO^- solvation dynamics: solute capabilities and solute-solvent molecular interactions. Chem. Sel. **2**(27), 8517–8523 (2017)

105. X. Zhang, P. Sun, Y. Huang, Z. Ma, X. Liu, J. Zhou, W. Zheng, C.Q. Sun, Water nanodroplet thermodynamics: quasi-solid phase-boundary dispersivity. J. Phys. Chem. B **119**(16), 5265–5269 (2015)

106. X. Zhang, X. Liu, Y. Zhong, Z. Zhou, Y. Huang, C.Q. Sun, Nanobubble skin supersolidity. Langmuir **32**(43), 11321–11327 (2016)

107. C.Q. Sun, C. Yao, Y. Sun, X. Liu, H. Fang, Y. Huang, (H, Li)Cl and LiOH hydration: surface tension, solution conductivity and viscosity, and exothermic dynamics. J. Mol. Liq. (2019). https://doi.org/10.1016/j.molliq.2019.03.077

108. D.R. Lide, *CRC Handbook of Chemistry and Physics*, 80th edn. (CRC Press, Boca Raton, 1999)

109. A.K. Metya, J.K. Singh, Nucleation of aqueous salt solutions on solid surfaces. J. Phys. Chem. C **122**(15), 8277–8287 (2018)

110. M.A. Sánchez, T. Kling, T. Ishiyama, M.-J. van Zadel, P.J. Bisson, M. Mezger, M.N. Jochum, J.D. Cyran, W.J. Smit, H.J. Bakker, Experimental and theoretical evidence for bilayer-by-bilayer surface melting of crystalline ice. Proc. Natl. Acad. Sci. **114**(2), 227–232 (2017)

111. X. Zhang, Y. Huang, P. Sun, X. Liu, Z. Ma, Y. Zhou, J. Zhou, W. Zheng, C.Q. Sun, Ice regelation: hydrogen-bond extraordinary recoverability and water quasisolid-phase-boundary dispersivity. Sci. Rep. **5**, 13655 (2015)

112. Q. Zeng, C. Yao, K. Wang, C.Q. Sun, B. Zou, Room-temperature NaI/H_2O compression icing: solute–solute interactions. PCCP **19**, 26645–26650 (2017)

113. L. Wong, R. Shi, D. Auchettl, D.R. McNaughton, E.G.Robertson Appadoo, Heavy snow: IR spectroscopy of isotope mixed crystalline water ice. Phys. Chem. Chem. Phys. **18**(6), 4978–4993 (2016)

114. C. Medcraft, D. McNaughton, C.D. Thompson, D. Appadoo, S. Bauerecker, E.G. Robertson, Size and temperature dependence in the far-ir spectra of water ice particles. Astrophys. J. **758**(1), 17 (2012)

115. E. Mamontov, D.R. Cole, S. Dai, M.D. Pawel, C. Liang, T. Jenkins, G. Gasparovic, E. Kintzel, Dynamics of water in LiCl and $CaCl_2$ aqueous solutions confined in silica matrices: a backscattering neutron spectroscopy study. Chem. Phys. **352**(1), 117–124 (2008)

116. Q. Wang, L. Zhao, C. Li, Z. Cao, The decisive role of free water in determining homogenous ice nucleation behavior of aqueous solutions. Sci. Rep. **6**, 26831 (2016). http://www.naturecom/srep/2013/131021/srep03005/metrics

117. C.A. Angell, E.J. Sare, J. Donnella, D.R. Macfarlane, Homogeneous nucleation and glass-transition temperatures in solutions of Li salts in D_2O and H_2O—doubly unstable glass regions. J. Phys. Chem. **85**(11), 1461–1464 (1981)

118. A. Kumar, Homogeneous nucleation temperatures in aqueous mixed salt solutions. J. Phys. Chem. B **111**(37), 10985–10991 (2007)

119. B. Zobrist, C. Marcolli, T. Peter, T. Koop, Heterogeneous ice nucleation in aqueous solutions: the role of water activity. J. Phys. Chem. A **112**(17), 3965–3975 (2008)

120. K. Miyata, H. Kanno, T. Tomizawa, Y. Yoshimura, Supercooling of aqueous solutions of alkali chlorides and acetates. Bull. Chem. Soc. Jpn. **74**(9), 1629–1633 (2001)

121. K. Miyata, H. Kanno, Supercooling behavior of aqueous solutions of alcohols and saccharides. J. Mol. Liq. **119**(1–3), 189–193 (2005)

122. B. Zobrist, U. Weers, T. Koop, Ice nucleation in aqueous solutions of poly[ethylene glycol] with different molar mass. J. Chem. Phys. **118**(22), 10254–10261 (2003)

123. M. Oguni, C.A. Angell, Heat capacities of $H_2O + H_2O_2$, and $H_2O + N_2H_4$, binary solutions: isolation of a singular component for Cp of supercooled water. J. Chem. Phys. **73**(4), 1948 (1980)

124. A. Bogdan, T. Loerting, Impact of substrate, aging, and size on the two freezing events of $(NH_4)_2SO_4/H_2O$ droplets. J. Phys. Chem. C **115**(21), 10682–10693 (2011)

125. A. Bogdan, M.J. Molina, H. Tenhu, E. Mayer, T. Loerting, Formation of mixed-phase particles during the freezing of polar stratospheric ice clouds. Nat. Chem. **2**(3), 197–201 (2010)

126. H.Y.A. Chang, T. Koop, L.T. Molina, M.J. Molina, Phase transitions in emulsified HNO_3/H_2O and $HNO_3/H_2SO_4/H_2O$ solutions. J. Phys. Chem. A **103**(15), 2673–2679 (1999)

127. K. Murata, H. Tanaka, Liquid-liquid transition without macroscopic phase separation in a water-glycerol mixture. Nat. Mater. **11**(5), 436–443 (2012)

128. J.M. Hey, D.R. MacFarlane, Crystallization of ice in aqueous solutions of glycerol and dimethyl sulfoxide. 1. A comparison of mechanisms. Cryobiology **33**(2), 205–216 (1996)

129. K.D. Beyer, A.R. Hansen, N. Raddatz, Experimental determination of the $H_2SO_4/HNO_3/H_2O$ phase diagram in regions of stratospheric importance. J. Phys. Chem. A **108**(5), 770–780 (2004)

Chapter 7
Organic Molecules: Dipolar Solutes

Contents

Abstract The excessive number of H^+ or ":" and their asymmetrical distribution determines the performance of their surrounding water molecules in a way different from that of ordinary water. The naked lone pairs and protons are equally capable of interacting with the solvent H_2O molecules to form O:H vdW bond, O:⇔:O super–HB or H↔H anti-HB without charge sharing or new bond forming. Solvation

© Springer Nature Singapore Pte Ltd. 2019
C. Q. Sun, *Solvation Dynamics*, Springer Series in Chemical
Physics 121, https://doi.org/10.1007/978-981-13-8441-7_7

examination of alcohols, aldehydes, formic acids, and sugars reveals that O:H–O formation enables the solubility and hydrophilicity of alcohol; the H↔H anti-HB formation and interface structure distortion disrupt the hydration network and surface stress. The O:H phonon redshift depresses the freezing point of sugar solution of anti-icing.

Highlight

- Solvation dissolves an organic crystal into molecular dipoles covered with ":" and H⁺.
- Solute-solvent interacts through O:H vdW, H↔H and O:⇔:O repulsion, and dipolar induction.
- Molecular nonbond interactions relax the intramolecular covalent bond cooperatively.
- Nonbond-bond cooperativity dictates the phonon frequency shift and performance of solutions.

7.1 Wonders of Organic Molecular Solvation

Organic molecular solvation makes cases even more complicated, which is ubiquitous to food and pharmaceutical industrial sectors. For instance, alcohols modify the HBs of Vodka, Whiskey, and thus make these distilled spirits taste differently; aldehydes and carboxylic acids could damage DNA, causing caners; glycine and its N-methylated derivatives denature proteins; salt (NaCl) and sodium glutamate ($NaC_5H_8NO_4$) could cause but ascorbic acids (vitamin C) lower the blood pressure. Sugar addition could lower the freezing temperature of water solvent for low-temperature storage of bio-species from being crystalized. However, understanding organic molecular solv8ation is quite infancy. This chapter shows the essentiality of considering the sp^2 and sp^3-orbital hybridization of C, N, O in organic solutes to produce the H⁺ and ":" covering the dipolar solutes. The H⁺ and ":" interact with their alike or unlike of the neighboring water molecules to create the O:H, H↔H, and O:⇔:O, which deforms the local bonding network of the solution. An systematic examination of solvation of typical organic molecules derived the conventionally-unexpected, preliminary knowledge.

7.2 Alcohols: Excessive Protons and Exothermic Solvation

7.2.1 Wonders of Alcohol Hydration

Alcohol solutions are widely used in biological, biomedical, chemical and industrial processes, and our daily life including beverages, solvents, protein hydration, as well as self-assembly of bio-molecular structures [1–4]. For example, the carboxylic acids modify the HBs of Vodka, Whiskey, and Japanese Shochu, and thus make these distilled spirits taste differently from the corresponding ethanol-water mixtures with the same ethanol/water volume ratio [1, 2]. One feels drunken went with dizziness, blush, chest stuffiness, and sensation of a lump or discomfort in the throat when drinking alcoholic beverages because of the presence of the methyl group $C-H^+$ and the electron lone pairs of the O^{2-}.

Alcohols are well-known agent for anti-icing and heat emission at solvation. The naked lone pairs and protons form the important functional groups for methanol, ethanol, and glycerol and many drugs as well, like the Artemisinin treating malaria [5]. Moreover, alcohols are the simplest amphiphilic molecules, and serve as a modeling system for investigating hydrophobic and hydrophilic interactions. The understanding of hydrophobic and hydrophilic phenomena is critical to know the properties of more complicated systems such as oil-water, and surfactant-water mixtures at the molecular level. Although the alcohol–water mixtures are ubiquitously important [6], it remains yet puzzling how the alcohols interact with H_2O molecules and how the alcohol molecules functionalize the HB network and properties of the solution.

Infrared absorption [7, 8], Raman scattering [6, 8–13], Brillouin scattering [14], neutron scattering [15], X-ray scattering [16, 17], SFG [8, 18–21], and NMR [22–25], as well as MD simulations [18, 26, 27] have been broadly used for investigating alcohol solutions. For instances, Li and coworkers [22] found using the terahertz and the PEG-NMR spectroscopies that in the diluted alcohol solutions, water molecules tend to form large hydration shell surrounding the hydrophobic alcohol molecules. Concentrating the solution could promote the formation of an extended HB network between alcohol and water molecules until the alcohol are aggregated.

Using a combination of the polarization-resolved femtosecond infrared (fs-IR), Raman spectroscopy, and MD simulations, Rankin and coworkers [28] studied the hydrophobic interactions of aqueous tertiary butyl alcohol (TBA) solutions. They found that the force potentials among hydration shells are repulsive, which drives the hydrocarbon groups apart rather than pulling them closer together. On the other hand, the results of femtosecond fluorescence up conversion measurements of aqueous TBA solutions revealed that although macroscopically homogeneous, the TBA molecules tend to aggregate micro-heterogeneously upon mixing with water and the aggregation becomes more pronounced at higher temperatures [29].

Microscopically, Davis and coworkers [11, 12] studied the water structure at the hydrophobic interfaces using Raman scattering. They found that within the hydration shells water molecules have greater tetrahedral order than they are in the bulk water at low temperatures; while at higher temperatures this structure becomes less ordered when the hydrophobic chain is longer than 1 nm. They also observed that the presence of the hydrophobic interfaces promotes the formation of the dangling OH^- groups with a probability depending nonlinearly on the size of the hydrophobic groups. Moreover, Juurinen et al. [9] studied the effect of the hydrophobic chain length on the water structure by combing X-ray Raman scattering and MD simulations. They suggested that alcohols neither break or make HBs in the mixture nor change the tetrahedrality of the H_2O motifs surrounding alcohol molecules [19]. The formation of the hydrophobic interface means the dominance of repulsion due to $H{\leftrightarrow}H$ or $O{:}{\Leftrightarrow}{:}O$ interactions.

In addition to hydrophobicity, interaction between the hydrophilic groups of alcohols and water also play a critical role in determining the anomalous properties of alcohol-water systems [7, 17, 30, 31]. For instance, Prémont-Schwarz et al. [7] investigated hydrogen bonding of the N–H group of N-methylaniline (NMA) with oxygen (N–O:H or N:H–O) from liquid DMSO and acetone. They found that the N–H stretching vibration mode of the NMA undergoes a redshift due to interaction with the hydroxyl of the DMSO and the acetone. The intramolecular N–H bond elongation occurs due to local $H{\leftrightarrow}H$ inter-proton or $X{:}{\Leftrightarrow}{:}Y$ inter-lone-pair repulsion, as they do in the acid and in the basic solutions [32–34].

Extended x-ray absorption fine structure spectroscopy (EXAFS) revealed that the O–O distance undergo a 4.6% surface contraction (2.76 Å for bulk) of liquid methanol in contrast to 5.9% expansion at the liquid water surface with respect to 2.85 Å for the bulk water [35]. At 4 °C, the O–O distance is 2.70 Å [36]. Despite the similar properties such as abnormal heats of vaporization, boiling points, dipole moments, etc., methanol and liquid water have dramatic differences in the surface HB structure, stemming surface tension of these liquids ($\gamma_{H_2O}/\gamma_{alcohol} = 75/22$).

However, accounting for the effect of both hydrophobic and hydrophilic interactions between alcohol and water molecules remains an open task. Much attention has paid to either the C–H stretching vibration mode of alcohol molecules or the H–O stretching vibration of water without caring about the intermolecular O:H stretching vibration in the wavenumber range of 70–300 cm^{-1} and their cooperativity.

This section shows that an incorporation of the O:H–O bond cooperativity and the DPS strategy [37–39] enables the resolution of the alcohol-water molecular interaction. Measurements revealed that alcohol hydration softens both the H–O bond and the O:H nonbond, which is different from solvation of acid [32], base [40], and salts [41, 42]. Observations suggested that each of the dangling hydroxide group or lone pairs and the dangling methyl group H^+ is equally capable of interacting with their alike or unlike of the solvent H_2O molecules. The O:H–O formation enables the solubility and hydrophilicity. The excessive number of H^+ forms the $H{\leftrightarrow}H$ anti-HB that not only disrupts the hydration network but also lengthens the solvent H–O bond. The $H{\leftrightarrow}H$ repulsion results in the hydrophobicity. The spatially asymmetrical distribution of the H^+ and the ":" makes an alcohol molecule a dipole, which polarizes its surrounding O:H–O bonds and distorts their geometries. H–O softening releases heat and depresses the solution melting temperature; the O:H nonbond softening lowers the critical temperatures for freezing.

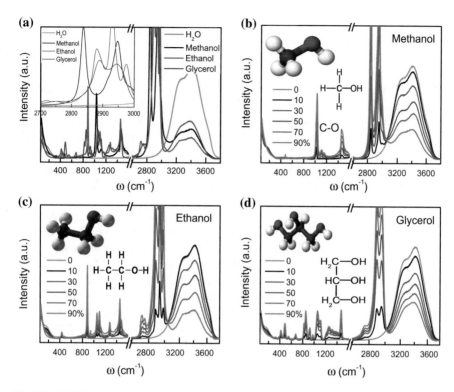

Fig. 7.1 Full-frequency Raman spectra for the **a** alcohol type and **b** concentrated methanol (CH$_3$OH), **c** ethanol (CH$_5$OH), and **d** glycerol (C$_3$H$_5$(OH)$_3$) solutions probed at 298 K. Inset a shows magnification of the 2900 ± 100 cm^{-1} sharp C–H$_3$ features. 800–1400 cm^{-1} correspond to C–O and C–C vibration. Insets **b–d** show the respective molecular structures composed of H$^+$(white), tetrahedrally-coordinated C (black) and O (red) with each O having two pairs of electron lone pairs. Peaks at $\geq$3000 cm^{-1} feature the solvent H–O stretching vibration, and $\leq$300 cm^{-1} the intermolecular O:H vibrations of the solvent water. Reprinted with copyright permission from [39]

7.2.2 H↔H Formation and Local HB Distortion

Methanol, ethanol, and glycerol molecules consist of multiple C–H$^+$ protons and O-H$^+$ groups with each O^{2-} having two pairs of electron lone pairs, see Fig. 7.1 insets b–d. Distributed asymmetrically, the lone pairs ":" and the H$^+$ protons are capable of interacting with their alike or unlike of the neighboring solvent O^{2-} by forming the O:H vdW bond, H↔H anti-HB, or O:⇔:O super–HB. Unfortunately, the presence of the ":" and its functionality is conventionally oversighted in dealing with molecular solvation.

Table 7.1 counts the number of the H$^+$, ":", and the nonbonds formed between an alcohol molecule and its surrounding tetrahedrally coordinated O^{2-} anions. The p is the number of protons and n the number of lone pairs. For all the three species, $n <$

p, so there will be $2n + (p-n)/2$ number of O:H vdW bonds and $(p-n)/2$ number of H↔H anti-HBs without O:⇔:O being formed between the solute and solvent.

On the other hand, the alcohol solvation reorders the local coordination environment and the alcohol-water interface, which distorts the surrounding O:H–O bonds. The solute dipole also polarizes the surrounding O:H–O bonds in the subsequent hydration shell. Therefore, the solute polarization, local O:H–O bond distortion, and the H↔H anti-HB formation dictate the performance of the solution. The Raman profiles for the H–O and the O:H stretching vibration modes will respond sensitively and cooperatively to the structural distortion and the H↔H anti-HB repulsion. Additional vibration features of the intra-solute C–H bond and the C–O bond will present to the spectrum, which is out of immediate concern about the solution HB network relaxation.

For alcohols, because $n < p$, there will be $2n + (p-n)/2$ number of O:H vdW bonds and $(p-n)/2$ number of H↔H anti-HBs between the solute and the H_2O solvent. For the pure alcohols, there will be $2n$ number of O:H vdW bonds and $p-n$ number of H↔H anti-HBs. The cooperation of the H↔H repulsion and the O:H–O tension not only balances the alcohol molecular interactions but also shortens the intramolecular bonds slightly though this expectation is subject to confirmation. One cannot ignore the presence of the ":" associated with O^{2-} for the nonbond counting and solute-solvent interaction. The high number of H↔H anti-HB repulsion may explain why the alcohol is easier to evaporate than it is in the solution.

7.2.3 DPS Derivatives: O:H–O Bonds Transition

Figure 7.1 plots the full-frequency Raman spectra for volumetrically concentrated alcohol solutions. Raman peaks at 2900 ± 100 cm^{-1} feature the intramolecular CH_3 symmetrical and asymmetrical stretching vibrations and the Fermi resonance modes of CH_3 bending overtone [10, 43–45], whose intensity increases with alcohol concentration. The 3000–3700 cm^{-1} band corresponds to the H–O stretching vibration (ω_H) and <300 cm^{-1} band to the O:H stretching vibration of the solvent water matrix. Features between 800 and 1400 cm^{-1} correspond to the C–C (1330, 1500 cm^{-1}) and C–O (1000, 1500 cm^{-1}) vibrations.

Table 7.1 Number counting of the H$^+$, ":", H↔H and the O:H between an alcohol molecule and its solvent H_2O neighbors and in their pure liquids

Alcohols	Molecules	p(H$^+$)	n(:)	O:H–O	H↔H	H↔H[a]	O:H–O[a]
Methanol	CH_3OH	4	2	5	1	2	4
Ethanol	C_2H_5OH	6	2	6	2	4	4
Glycerol	$C_3H_5(OH)_3$	8	6	13	1	2	12

[a]Numbers of the alcohol intermolecular H↔H anti-HBs and O:H–O (C) bonds in the alcohol liquids

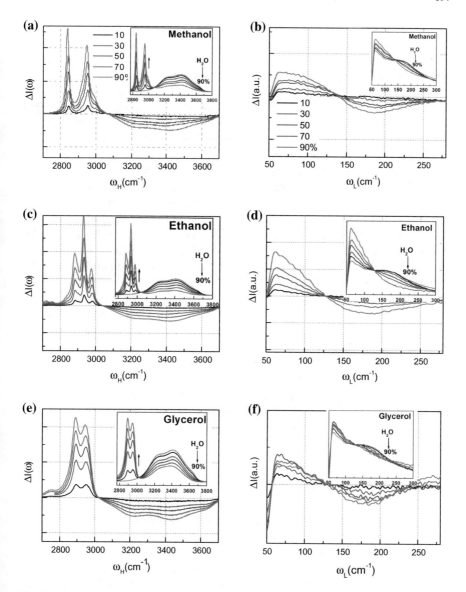

Fig. 7.2 Solvation of the concentrated **a**, **b** methanol, **c**, **d** ethanol, **e**, **f** and glycerol transits the H–O phonon from 3400 to 2900 cm^{-1} and O:H phonon from 180 to 70 cm^{-1}. The C–H phonon centered at 2900 $\pm$ 100 cm^{-1}. Reprinted with copyright permission from [39]

Alcohol solvation depresses the ω_H intensity and softens its frequency. The ω_{H-C} peak intensity increases with concentration without significant frequency shift. The stiffness relaxation of the ω_H is insensitive to the length of the hydrocarbon chain, which suggests that the local O:H–O bond distortion plays a more important role than the H↔H point repulsion. Walrafen [46] suggested that the HB stretching and bending, i.e., vibration of the neighboring H_2O molecules along with and perpendicular to the HB could yield Raman features with a maximum at 190 (ω_{L2}) and 60 cm^{-1} (ω_{L1}), respectively. Here, the Raman band at 60 cm^{-1} in Fig. 7.1 is not resolvable because of its low-intensity.

Figure 7.2 compares the volume concentration dependent ω_x DPS of all solutions in the 2800–3800 cm^{-1} frequency range. The ω_H redshifts proceeded by solvent H–O bond elongation, which is possible by inter-proton or inter-lone-pair repulsion. The asymmetrical distribution of the H$^+$ and the ":" of alcohol molecules may allow such O:⇔:O or H↔H formation between the alcohol solute and the solvent H_2O molecules. The O:⇔:O has strong capability of polarization but the H↔H is capable of disruption. The absence of the 3610 cm^{-1} features in the solution DPS suggests that the dangling H–O bonds of alcohol solutions stay as they are at the deionized water surface. Alcohol hydration transits neither the stiffness nor the abundance of the H–O dangling bonds. The alcohol solute H–O bond should be subject to contraction as the H_3O^+, the HO$^-$ and the H_2O_2 do as solvation [32, 40, 47], which could be shielded by other yet known effects.

The C–H_3 phonon abundance increases with the alcohol concentration, while the H–O bond vibration modes decrease, agreeing with observations by other researchers [6, 48]. Besides the spectral intensity, other properties of alcohol-water mixtures such as entropy, microstructure, sound absorption coefficient, viscosity, and volume contraction also depend on the concentrations [49–54].

7.2.4 Exothermic Solvation Dynamics

Alcohol solvation heats up the solution as it occurs for YOH and H_2O_2 solvation, which is why one feels burning when drinking alcohol. Repeating the same iteration of DPS peak area integration, one can obtain the fraction coefficient for the solvent O:H–O bonds transiting from the mode of the ordinary water to the hydration elongated H–O bonds, as shown in Fig. 7.3a. The invariant f(C) slopes for the methanol and the ethanol solutions suggest their constant hydration shell sizes but the slightly nonlinear f(C) suggests the presence of the Glycerol solute-solute repulsion at higher concentrations.

Unlike the YOH and the H_2O_2 having excessive lone pairs for the O:⇔:O compressive super–HB formation, alcohol-solvent interaction distorts the local O:H–O bonds to elongate both the O:H and the H–O by the H↔H anti-HB repulsion and dipolar interface distortion. Polarization elongates the O:H and H↔H anti-HB compression lengthens and weakens the H–O bond. The elongated H–O bond emits its energy heating up the solution. Therefore, exothermic solvation of alcohol, YOH

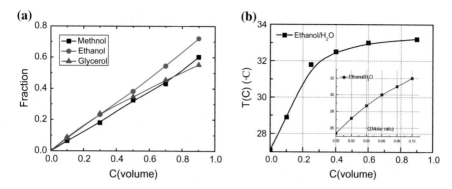

Fig. 7.3 Volume concentration dependence of **a** the alcohol fraction coefficients and **b** the ethanol solution temperature T(C) as a function of the volume ratio and molar ratio (inset) concentration. The invariant f(C) slopes for the methanol and ethanol solutions suggest their constant hydration shell sizes but the nonlinear f(C) suggests the presence of the Glycerol solute-solute interaction

base and H_2O_2 share the same mechanism of H–O bond elongation albeit different origins. Figure 7.3b shows in situ calorimetric detection of the solution temperature against volume and molar (inset) concentrations of ethanol solvation. The solution temperature T(C) increases from 25.5 up to 33 °C toward saturation, which means that the H-O elongates less by H↔H repulsion than it is by O:⇔:O compression in basic solutions.

The H–O bond softening explains why alcohol-water mixture is exothermic and why the solvation lowers the freezing and melting temperatures of solutions [6, 55]. The O:H softening depresses the freezing temperatures [56–58]. The liquid-quasisolid-solid phase transition for ordinary water at 277 and 258 K, which closes to the melting and freezing temperatures. The ω_L and ω_H correspond directly to the Debye temperature of the O:H–O segmental specific heats whose superposition derives two crossing temperatures to specify the quasisolid phase between liquid and ice, as shown in Fig. 3.1c [37, 59].

7.2.5 Summary

Unlike acid, base, and salt, alcohol solvation softens both the ω_H and ω_L phonons at the alcohol-water interface. The formation of the H↔H anti-HB and the asymmetrical H^+ and ":" distribution surrounding the solute distort and polarize the hydrating O:H–O bonds. The interfacial O:H–O bond relaxation and distortion determines the hydrophilicity while the H↔H repulsion triggers the hydrophobicity of the alcohol molecules. The exothermic pross of ethanol further evidences H–O bond elongation by H↔H repulsion and distortion upon alcohol solvation as the O:H bond elongation emit negligible energy. The relaxation of the intramolecular covalent bonds release energy.

7.3 Aldehydes and Carboxylic Acids: DNA Damage

7.3.1 DNA Damage by Aldehydes

Aldehydes are the most common indoor pollutants that are extremely harmful to our health by DNA fragmentation. With the characteristics of high-level toxicity and long duration, aldehydes have been recognized as the key elements causing cancers of indoor pollution. Aldehyde contamination damages the DNA-protein cross-linking and binding to form adducts in human peripheral blood lymphocytes [60–63]. Likewise, carboxylic acids are also important biomedical and cosmetic ingredients [64]. Oral contamination of concentrated acetic acid can corrode one's mouth and cause oral mucosa, esophageal and gastric mucosal damage, and even lead to gastric perforation [65]. Once human body absorbs the acetic acids, hemolysis, acute renal failure, acute liver failure, disseminated intravascular coagulation and circulatory shock may occur [66, 67]. There are numerous applications of propionic acid in the food and chemical industries, such as food preservation, components in plasticizers, perfumes and pharmaceuticals production [68, 69]. However, molecular scale understanding of DNA-aldehyde and aldehyde-aldehyde interactions and the functionalities of formic acids are still challenge.

From the perspective of solvent hydrogen bond transition from the mode of ordinary water to the hydrating states, Chen et al. [70] examined the solvation dynamics and the solute capabilities of aldehydes and carboxylic acids using DPS and contact angle detection. Results suggest that besides the intermolecular O:H, H↔H, and O:⇔:O interactions, solute dipolar interaction and the solute-solvent interface structural distortion play important roles in disrupting the solution network and surface stress.

7.3.2 Molecular Geometric Configuration

One needs to clarify the molecular solute structure of aldehydes and formic (carboxylic) acids with consideration of the sp^3-orbiotal hybridization for C and O atoms upon reaction. Solvation dissolves a molecular crystallite into individual molecular solute covered with the dangling H^+ and electronic lone pairs ":" for the group of methanoic, acetic, methyl-acetic acids and the group of form-, acet-, propion-aldehyde. The unequal numbers of the H^+ and the ":" and their asymmetrical distribution surrounding the solute discriminate these molecules one from another with different properties.

To meet the criteria of sp^3-robital hybridization, pairing up two aldehyde molecules could be necessary. This molecular pairing gives rise to the geometrical mirror symmetry along the CH_2 chains. What differs the carboxylic acids from the aldehydes is the single or the pairing CH_2 chains. Both the H↔H and the O:⇔:O form in the carboxylic acidic solutions but only the H↔H forms in the aldehydes

Table 7.2 Formulation, geometric configuration and the intermolecular interactions for the formic acids and aldehydes. C and O atoms are subject to sp^3-orbital hybridization with oxygen producing each two pairs of ":" (paired yellow dots on oxygen in red). The H^+ and ":" attached to the molecular solutes

Solute	Carboxylic			Aldehyde		
	Methanoic	Ethanoic	Methyl-acetic	Form-	Acetal-	Propion-
Formulae	CH_2O_2	$C_2H_4O_2$	$C_3H_6O_2$	$2[CH_2O]$	$2[C_2H_4O]$	$2[C_3H_6O]$
Geometry						
$p(H^+)$	2	4	6	4	8	12
$n(:)$	4	4	4	4	4	4
O:H–O	5	8	9	8	10	12
H↔H	0	0	1	0	2	4
O:◇:O	1	0	0	0	0	0

The grey spheres represent for C and the blue ones for H^+ (reprinted with permission from [70])

upon solvation, in addition to the O:H–O bonds formed between the solute and the solvent O^{2-} anions for all samples. The difference between carboxylic acid and aldehydes exists in their functional groups, which gives them very different chemical and physical properties.

Table 7.2 counts the numbers of the O:H vdW bonds, the H↔H anti-HBs and the O:⇔:O super–HBs formed between the organic molecules in their aqueous solutions. Each of the solute H^+ and the ":" has half probability to connect with its alike or unlike of the solvent H_2O molecules. If a solute has n(:) and p(H^+), there will be n + p nonbonds between the solute and its surrounding solvent molecules. Amongst these nonbonds, there will be |p-n|/2 number of H↔H (for p < n) or O:⇔:O (for p < n) and the rest O:H vdW bonds.

The allotropicity of the aldehydes and the carboxylic acids determines the geometry and the polarizability of these dipoles, which govern the solute capabilities of solvent bond transition and the disruption or construction of the network and surface stress.

7.3.3 DPS Derivatives

Figure 7.4 compares the Raman spectra for the organic solutions and ordinary water probed under the ambient conditions. Besides the intramolecular C–H and O–O vibration modes featured at 2900 ± 100 and 877 cm^{-1}, the spectra show different reflectivity of the solution compared with pure water without significant change of the peak shape. The solute having longer CH_2 chain shows weaker reflectivity at the same concentration, and the reflectivity drops with the increases of the solute concentration. All the spectra show the polarization effect that shifts the O:H frequencies <300 cm^{-1} to lower values, being the same to solutions of the inorganic salts. The O:H and H–O phonon cooperative relaxation is the main concern and the peak discriminative peak intensities will be minimized by the peak area normalization in the DPS processing.

The concentration dependent ω_x DPS in Figs. 7.5 and 7.6 shows clearly the polarization effect on the H–O stretching vibration phonon blueshift from 3200 to ~3500 cm^{-1}. The carboxylic acids shift the H–O phonon more to above 3500 than to aldehydes centered below 3500 cm^{-1}. The O:H phonon redshifts from 160 to 75 cm^{-1}. The capability of phonon abundance transition of acid is stronger than the aldehydes. The solute electric field will cluster, stretch and polarize the solvent H_2O molecules in the hydration shell. The stretching lengthens and softens the O:H and shortens and stiffens the H–O bond, resulting in the probed phonon frequency shifts [71], as Fig. 7.5a inset illustrates.

These molecules form each a giant dipole with the lone pairs ":" gathered at one end and the H^+ protons clustered at the other. The DPS confirms the solute dipole polarization effect on the hydrating O:H nonbond and H–O bond, which evidences the presence of the stronger electric field in the hydration shell.

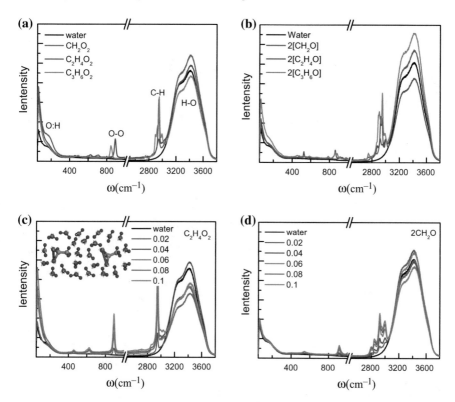

Fig. 7.4 Full-frequency Raman spectra for **a** the methanoic-, acetic-, and methylacetic-acidic and **b** form-, acet-, propion-aldehyde solutions at 0.04 molar concentration, and the concentration dependence of the **c** acetic acid and **d** formaldehyde solutions with reference to the spectrum of H_2O collected under the same ambient temperature. Inset **c** illustrates the hydration of the carboxylic acid solute that aligns and polarizes the neighboring water molecules. Reprinted with permission from [70]

7.3.4   Fraction Coefficients and Surface Stress

Figure 7.7 shows the surface stress and the fraction coefficient as a function of the solute concentration. The concentration dependence of contact angles between the solutions and the copper foil substrates reveals the manners of solution network and surface stress disruption, which indicates that these molecular solutes have the similar effect of HX acid solvation. Variation of the solution surface stress is related directly to the polarization or disruption. Ionic polarization [71] and O:⇔:O compression [71] raises the surface stress of salt and basic solutions, but thermal fluctuation [41] and H↔H fragilation [32] disrupt the surface stress. However, solvation of the formic acid and aldehyde has the same effect of salt solvation on the O:H–O phonon relaxation but has the same effect of acid on surface stress disruption. The concentration dependence of the solution surface stress shows that the slope of stress

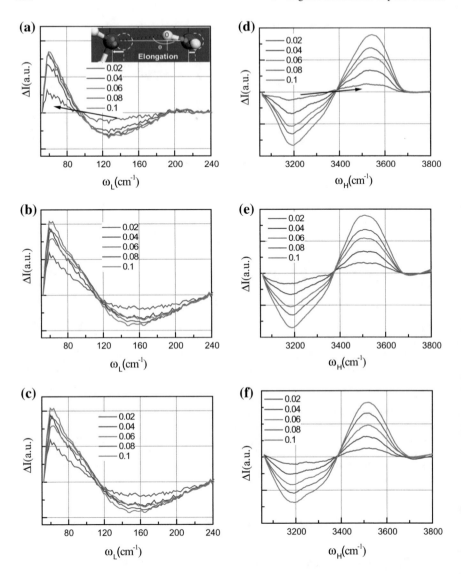

Fig. 7.5 Concentration dependence of the (**a**–**c**) ω_L and (**d**–**f**) ω_H DPS for (**a**) methanoic, (**b**) acetic and (**c**) methyl-acetic acids under the ambient conditions. Results show consistently the effect of polarization that stiffens the H–O phonon from 3200 to >3500 cm^{-1}, and meanwhile, softens the O:H stretching vibration from 160 to 75 cm^{-1}. Inset a shows the O:H–O bond segmental cooperative relaxation under an electric field [71]. Reprinted with permission from [70]

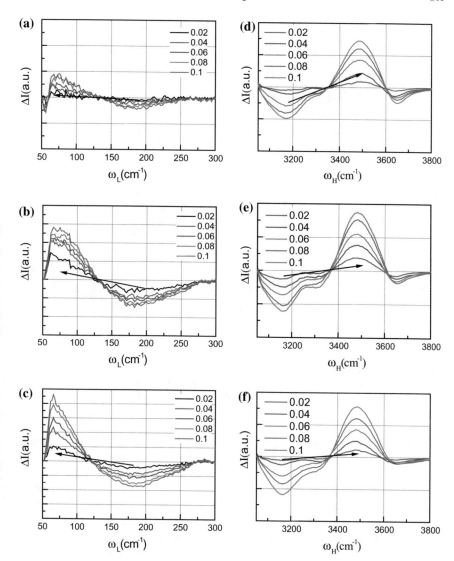

Fig. 7.6 Concentration resolved DPS (**a–c**) ω_L and (**d–f**) ω_H for the (**a, d**) formaldehyde, (**b, e**) acetaldehyde, and (**c, f**) propionaldehyde showing the stronger polarization and weaker transition of H–O phonon abundance to the acid solutions. Reprinted with permission from [70]

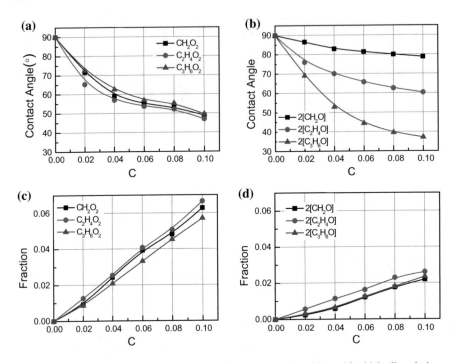

Fig. 7.7 Concentration resolved contact angles for **a** carboxylic acidic and **b** aldehydic solutions and the fraction coefficients for **c** carboxylic acidic and **b** aldehydic solutions. Copper foils serves as the substrate for contact angle detection. Reprinted with permission from [70]

depression is proportional to the CH_2 chain length, or the number of H^+, which discriminates the manner of stress disruption of the acids and the aldehydes.

The fraction coefficient features the solute capability of bond transition from the mode of ordinary water to the hydration shells. The linear dependence of the fraction coefficient for all solutes indicates the invariance of the local electric field and the constant hydration shell sizes of the solutes because of their dipolar nature. The electric field is anisotropic, locally stronger, and short-range order. The smaller $f(C)$ for aldehydes but stronger surface stress disruption may infer the aldehyde capability of DNA damaging. The linear concentration dependence of the fraction coefficients shows the invariance of the solute dipolar electric fields.

7.3.5   Summary

The asymmetrical distribution of the H^+ and the ":" makes the molecular solute a dipole with anisotropic, short-range, and strongly localized electric field. The solute dipolar polarization and local structure distortion has the similar effect on the solvent O:H–O bond relaxation and surface stress disruption of protonated acids. The linear concentration trend of the fraction coefficient for bond transition suggests the invariance of the local electric field and the constant hydration-shell size. The fact that a smaller fraction of aldehyde disrupts significantly the surface stress may infer its manner of DNA fragmentation, which might be of reference in understanding the cancer factor induced by aldehyde absorption - network disruption and interface local polarization.

7.4 Glycine and Its N-Methylated Derivatives: Protein Denaturation

7.4.1 Solvation of Glycine and Its N-Methylated Derivatives

As a kind of a simple organic compound, osmolytes commonly exist in the complicated and susceptible living organisms to counteract with the hostile environment [72], such as variations of temperature [73], pressure [73, 74], salinity [74], and so on so forth [75, 76]. Glycine is the simplest yet essential amino acid presenting in the formation and performance of living species. Besides, methylated glycine osmolytes including sarcosine (NMG); dimethylglycine (DMG); and glycine betaine (TMG), are widely used in many fields including biomedicine, food industry, agriculture, etc. [77–79]. In addition, the glycine and its N-methyl derivatives serve as a class of small organic osmolytes in the zwitterionic form.

In general, organic osmolytes mediate the structure and performance of proteins indirectly by solvent hydration or by directly interacting with the proteins. The urea-denaturation is most likely electrostatic in origin, which affect directly the linear hydrocarbon chain [80]. For the case of denaturation by urea and guanidine hydrochloride, researchers realized that the unfolding process is accompanied with the direct binding of the denaturant with protein molecules. However, the sucrose and glycerol prefer to form stable structures and influence the hydration of proteins through the participation of water molecules [81].

Intensive investigation has been carried out using IR transmission [82–84], Raman reflection [85, 86], dielectric relaxation [87], and theoretically by *ab initial* [88, 89] or by MD calculations [90, 91]. Experimental investigations suggested that the trimethylamine N-oxide (TMAO) stabilize proteins via water structure enhancement [87, 92, 93]. The femtosecond mid-IR pump-probe spectral measurements suggest that the TMAO locally enhances the orientational mobility of the solvent HB network [94]. Moreover, glycine and its N-methyl derivatives prefer to work via the indirect mode [95, 96]. In the indirect mode, the HBs (referred to O:H interaction in most cases) become slightly more stable because of the perturbation of the omsolyte [97], according to the computational datum in terms of the HB network structures and the mean energy. Although the interactions between organic osmolytes and proteins have been flourished, mechanisms behind are still controversial. The HB network relaxation dynamics in the solute hydration shells is still uncertain [82, 98–104].

Fang and co-workers [105] examined the O:H–O bond transition upon the solvation of glycine and its methylated derivatives solvation to clarity the occurrence at the solute-solvent molecular interface and the solute capabilities of transiting the O:H–O bonds from the mode of ordinary water to hydration in terms of phonon abundance, bond stiffness, and fluctuation order. They also examined the correlation between the bond transition and the surface stress (contact angles) and solution viscosities as a function of solute concentration. Observations confirmed that the solute dipolar electric field polarizes the O:H–O bonds by stiffening the H–O bond and softening the O:H nonbond.

Phonon spectrometric results suggest that the glycine and its N-methyl derivatives strongly affect the surrounding solvent molecules through H↔H repulsion and dipolar polarization. The H↔H interproton repulsion disrupts the surface stress and the polarization enhances the solution viscosity.

7.4.2 Osmolyte Solute-Solvent Interaction

From the perspective of nonbond counting regulation and the fact of N sp^3-orbital hybridization with a pair ":" creation [33, 106], there is no covalent bond formed or charge sharing occurs at the solute-solvent interface. On the other hand, the sp^3-orbital hybridization disallows the C=O double bond to form in an aqueous solution [107]. In addition, the covalent pairing of two O atoms is essential to ensure the sp^3-orbital hybridization of an O^{2-}. Given these constraints, the optimal molecular structures of the glycine and its methylated group in their aqueous solutions could be established, as illustrated in Table 7.3. The asymmetrical distribution of the H^+ and the ":" on the solute resolves the geometry and the polarity of these molecular dipoles upon solvation, which reins the performance of the solute-solvent interface and the solutions.

7.4.3 DPS Identities

Figure 7.8 shows the full-frequency (50–4000 cm^{-1}) Raman spectra of the concentrated glycine and its N-methylated aqueous solutions. The raw Raman spectra manifest the concentration-dependent information about diversiform chemical bonds in these aqueous solutions. The bending vibration modes are less sensitive to the perturbations; one can thus focus on the O:H nonbond stretching vibration at 50–300 cm^{-1} and the H–O bond stretching vibration at 3100–3800 cm^{-1} because they sufficiently feature the binding energy and length change. Furthermore, the O–O and C–C stretching vibrational modes are overlapped at 897 cm^{-1}. The C–H symmetric stretching vibration adds 2800–3100 cm^{-1} features to the H–O phonon band. Compared with the H–O reference spectrum of pure water, solvation narrows

Table 7.3 Molecular formulae, the geometry, the numbers and the asymmetrical distribution of H^+ and ":" in the solute molecules for the glycine and its methylated derivatives, which derives the $O:H$ nonbonds and $H{\leftrightarrow}H$ anti-HBs without presence of the $O:{\Leftrightarrow}:O$ super–HBs

Solute	Gly	NMG	DMG	TMG
Formula	$C_2H_5NO_2$	$C_3H_7NO_2$	$C_4H_9NO_2$	$C_5H_{11}NO_2$
Molecular geometry				
H^+	5	7	9	11
Lone pair (:)	5			
$O:{\Leftrightarrow}:O$	0			
$H{\leftrightarrow}H$	0	1	2	3
$O:H$	10	11	12	13

The gray and blue spheres represent for C and N atoms, respectively. The white ones represent for H^+, and the red spheres are O anions with paired yellow dots being electric lone pairs (reprinted with permission from [105])

the skin-like component centered at 3420 cm^{-1}, as a result of the supersolidity of the solute hydration shells due to polarization [108].

Figures 7.9 and 7.10 show the DPS for the segmental O:H–O bond. The frequency of the H–O bond shifts from ~3200 (valley) to ~ 3450 cm^{-1} (peak) and the O:H phonon shifts in the order according to polarization from 180 to 75 cm^{-1} for TMG, 91 cm^{-1} for NMG, 95 cm^{-1} for DMG, and 102 cm^{-1} for Gly. This order illustrates that the increasing number of H^+ of solute molecules enhances the local polarization. The phonon frequency shifts evidence the supersolidity of the hydration shells [108]. Besides, the valley at 3650 cm^{-1} arises from annihilation of the H–O dangling bonds initially at 3610 cm^{-1} by the polarization. The organic acids of glycine and sarcosine, and the organic bases of glycine betaine share the same effect on the O:H–O bond relaxation at the solute-solvent interfaces. This observation suggests that the gigantic solute molecular dipoles and their protons H^+ and lone pairs ":" are the essential factors relaxing the interfacial hydrating O:H–O bonds.

7.4.4 Volume Fraction of HB Polarization Versus Surface Stress and Viscosity

The fraction coefficient derived from the DPS, as shown in Fig. 7.11, exhibits the quasi-linear correlation, $f_x(C) \propto 1 - \exp(-C/C_0)$, whose slope corresponds to the number of the hydrating H_2O molecules per solute. The saturation constant C_0 describes the solute-solute repulsive interaction. If the C_0 is sufficiently large, the

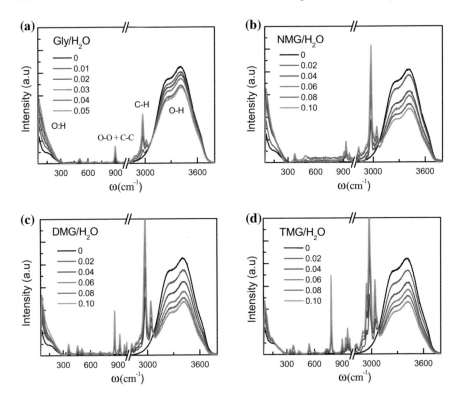

Fig. 7.8 Full-frequency Raman spectra for the concentrated **a** glycine/H$_2$O, **b** sarcosine/H$_2$O, **c** N,N-dimethylglycine/H$_2$O, and **d** glycine betaine/H$_2$O solutions with components indicated in panel (**a**). Reprinted with permission from [105]

solute-solute interaction is negligible because of the screening of the solute electric field by the hydrating H$_2$O dipoles. In the present case, the solute-solute interaction is weak because of the quasi-linear feature of the $f_x(C)$.

However, Fig. 7.12 shows opposite concentration trends between the relative viscosities and the relative contact angles (surface stress). The $\Delta\eta(C)/\eta(0) \propto 1 + \exp(C/C_0)$ and the $\Delta\theta(C)/\theta(0) \propto 1 - \exp(-C/C_0)$, which is different from those for other inorganic solutions where both the relative viscosities and the relative contact angles follow the same trend of depression or elevation [108]. The surface stress disruption is directly related to the H↔H fragilization while the viscosity to the solute molecular polarization. Besides, observations show the distinct discrepancy that the slopes $df_x(C)/dC$ follow the order: TMG > DMG > NMG > Gly, and the same order also appears on the concentrated variation of viscosities and contact angle. The fact suggests that the number rise of H$^+$ protons promotes the solute molecular polarization and the solution surface stress disruption.

The significant rise of the viscosity and the surface stress disruption imply the solute capabilities of glycine and its N-methyl derivatives of solution-protein inter-

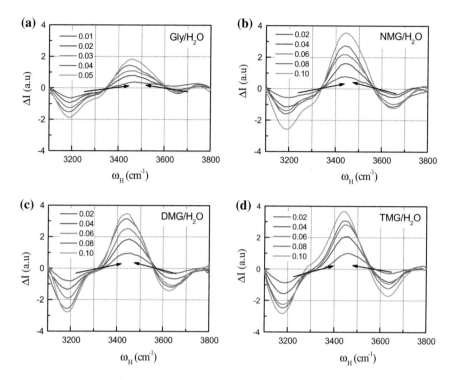

Fig. 7.9 The H–O bonds DPS for the concentrated **a** glycine/H$_2$O, **b** sarcosine/H$_2$O, **c** N,N-dimethylglycine/H$_2$O, and **d** glycine betaine/H$_2$O solutions showing O:H–O transition from the mode of water to solvation shells. Reprinted with permission from [105]

action that is different from other solutions. Hence, the glycine and its groups are labelled as a kind of protein stabilization osmolytes as previously reported by Bruździak and Panuszko [93, 95].

7.4.5 Summary

It is essential to consider the contribution of the nitrogen lone pair at solvation. The solute polarization creates hydration volume screened by hydrating water molecules which raises the solution viscosity, while the H↔H fragmentation disrupts surface stress. Besides, the increase of the H$^+$ number of a solute enhances the solute polarization and solution surface stress disruption. Within the hydration shell, the H–O bond is shorter and stiffer, while the O:H nonbond is longer and softer. From the nearly-linear trend of the fraction coefficient $f_x(C)$, one can hardly expect strong long-range solute-solute interactions in the organic solutions because of the dipolar interaction.

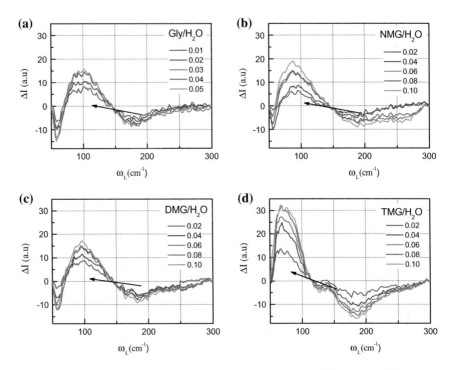

Fig. 7.10 The O:H nonbonds DPS for the concentrated **a** glycine/H$_2$O, **b** sarcosine/H$_2$O, **c** N,N-dimethylglycine/H$_2$O and **d** glycine betaine/H$_2$O solutions showing the extent order of polarization: **d** (75 cm^{-1}) > **b** (91 cm^{-1}) > **c** (95 cm^{-1}) > **a** (102 cm^{-1}). Reprinted with permission from [105]

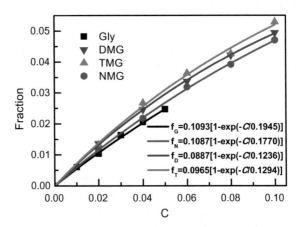

Fig. 7.11 The fraction coefficients for the concentrated solutions of glycine and its N-methyl derivatives. The quasi-linear $f_x(C)$ implies the stabilization of the osmotic molecular hydration shell size and weak inter-solute interaction. Reprinted with permission from [105]

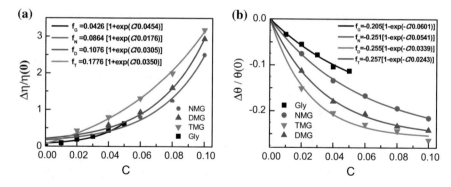

Fig. 7.12 Concentration dependence of **a** the relative viscosities and **b** relative contact angles between the copper flake substrate and the drops of the glycine groups solutions, and the referential water absolute viscosity $\eta(0)$ and contact angle $\theta(0)$ are 0.84 MPa·s and 88.4°, respectively. The relative viscosities and relative contact angles of these concentrated solutions follow the opposite exponential trends. Reprinted with permission from [105]

7.5 Hypertension Mediation by Salt and Acid Intake

7.5.1 Hypertension: Silent Killer

Having been recognized as increasing risk for coronary artery disease [109, 110], heart failure [111, 112], stroke [113], chronic kidney disease [114, 115], and cardiovascular diseases [116], hypertension or high blood-pressure (BP) has affected an increase number of people, estimated up to 1.5 billion by 2025, all around the world without receiving deserved attention [117]. Compared with many other diseases, hypertension has no apparent symptoms, so it is called the "silent killer" that threatens the human health gradually [118, 119]. The British Hypertension Society guidelines [120] and those outlined by the US National High-BP Education Program [121] suggested that one has to change lifestyle to prevent hypertension by reducing dietary aginomoto and salt intake and eating more fruits and vegetables of slightly acidic.

Salt (NaCl) and sodium glutamate ($NaC_5H_8NO_4$) are widely used in food industry to enhance flavor and to provide umami sodium glutamate [122]. However, overdosing both can raise the systolic and the diastolic blood pressure [123–127]. In 1968, Kwok [128] firstly realized that sodium glutamate is the main cause of the 'Chinese restaurant' syndrome—people like eating food with heavily-dosed aginomoto. The fact of Na-based ingredient on raising blood pressure was explained as the Na-based doses reset the hypothalamic neuronal activity [126].

Moderate intake fruits and vegetables that are rich in ascorbic acid (vitamin C) can lower the blood pressure, instead [129], and hence, ascorbic acid is a useful adjunctive therapy for effectively relieving hypertension [130–133]. Mullan et al [131] studied the hemodynamic effect of oral supplementation of ascorbic acid in a random double-blind trial and suggested that ascorbic acid lowers the blood pressure by enhancing endothelial nitric–oxide biological activity. What is more, acetic acid, which is abundant in vinegar, as another familiar acid can lower blood pressure by reducing renin activity and Angiotensin II according to Kondo et al [134]. However, the underlying mechanism for the acid and salt effect on mediating blood pressure stays unclear though focus has been on the pathology wise.

Blood is composed of its cells and 70% water that dissolves the Na-based salt into Na^+ cations and anions and the carboxylic acids into individual molecular dipoles surrounded asymmetrically by H^+ and electron lone pairs ":". These solutes react with the hydrogen-bonding network and blood cells in turn through ionic polarization [135, 136] and dipolar fragilization [70] as the H↔H anti-HB does in the H-based acid [32]. These facts may clarify how the NaC, $NaC_5H_8NO_4$ and $C_6H_8O_6$ ascorbic and the $C_2H_4O_2$ acetic acid mediate the blood pressure. Consistence between theory predictions and the spectrometric and surface-stress measurements reported here verifies that the ionic polarization enhances but the HB network fragilization relieves the hypertension. Understanding may offer insight into the mechanism behind hypertension towards effective medication and prevention.

7.5.2 Phonon Relaxation Detection

Figure 7.13a, c, e shows the full-frequency Raman (50–3800 cm^{-1}) spectra of all samples. Besides features for the O:H, H–O and C–H stretching vibrations, the Raman features between 500 and 3000 cm^{-1} in Fig. 7.13a–d feature the solute intramolecular bonding vibrations [137–139]. The facts that ω_L undergoes a red shift and ω_H a blue shift upon salt and acid solvation resulted from ionic or molecular polarization. Insets show the structures of the molecular solutes, of which the O, N and C follow the rule of sp^3 orbital hybridization. However, the full-frequency Raman spectra can hardly offer transition information on the stiffness, fraction and fluctuation of bonds in the hydration volume.

Figure 7.13b, d, f compares the ω_H DPS for the examined solutions. All the ω_H peaks undergo a blueshift from ~3200 to ~3500 cm^{-1}, because of O:H–O bond polarization that shortens and stiffens the H–O bonds, raises structure ordering, solution viscosity and surface stress and lowers molecular dynamics [42]. The peak area is proportional to the number of the HBs transition from the mode of ordinary water to the first shells of the hydration volume. The Na-based dose and acid has the same function of polarization but different abilities of amount transition. Acid solvation shifts the ω_L from 175 to <75 cm^{-1} but salt solvation shifts the ω_L to >75 cm^{-1} associated with different phonon abundances and structural orders.

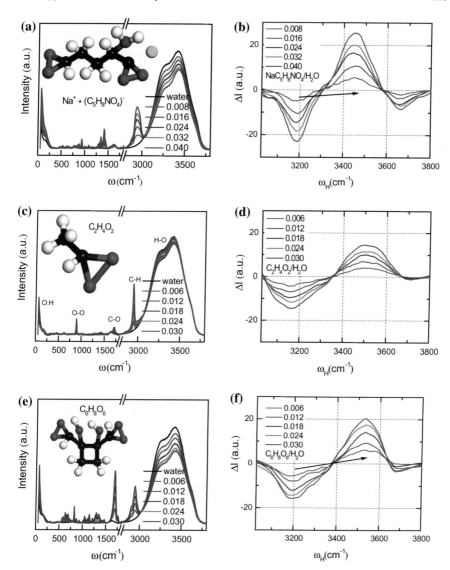

Fig. 7.13 Full-frequency and ω_H DPS for the concentrated **a** sodium glutamate, **c** acetic acid and **e** ascorbic acid and **b, d, f** the corresponding ω_H DPS. Insets show the corresponding molecular structures. Panel (**c**) denotes the solute intramolecular bond vibrations within 500 and 3000 cm^{-1}. Solvation shifts the DPS shift from ~3200 to ~3500 cm^{-1}. Reprinted with permission from [140]

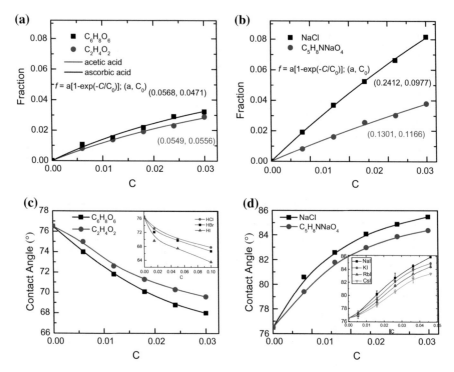

Fig. 7.14 a, b Fraction of O:H–O bonds transition from water to hydration and **c, d** contact angle (surface stress and solution viscosity [108]) of concentrated (**a, c**) acidic and (**b, d**) salt solutions. The fraction coefficient for the H-based is much smaller than the Na-Based salts. Inset **c** and **d** compare the contact angles of the concentrated H(Cl, Br, I) acidic and the (Na, K, Rb, Cs)I salt solutions [32, 42]. Reprinted with permission from [140]

7.5.3 Surface Stress and Bond Transition Coefficient

The integral of the specific O-H phonon DPS quantifies the solute capability of transiting the number of hydrogen bonds from the mode of ordinary water to hydration by polarization. Figure 7.14a, b show the concentration-resolved fraction coefficient for the salt and acid solutions, which follows the $f = a[1 − \exp(−C/C_0)$ towards saturation, where a and C_0 are constants and C is the molar concentration. The slope of the coefficient corresponds to the change of the hydration volume that is subject to the solute-solute interactions. The shorter C_0 indicates stronger solute-solute repulsion that weakens the local electric field and reduces the solute hydration volume. If the slope remains a constant, neither repulsion nor attraction takes place between solutes because of the screen shielding by surrounding water dipoles.

The fraction coefficient for the H-based is much smaller than the Na-based salts. Their difference is proportional to the non-polarizability of H⁺. Contact angle detection confirmed the dominance of polarization by salt solvation and network fragmentation by acid solvation, see Fig. 7.14c, d. The insets compare the contact angles between (Na, K, Rb, Cs)I salt solutions and the H(Cl, Br, I) acid. Trend consistence of the surface stress suggests that polarization raises the solution viscosity and surface stress and fragmentation dilutes the solution helping to the solution flow.

7.5.4 Inference to the Hypertension

Although the salt and acid solutions share the same tendency of phonon frequency shift due to polarization, the effect of altering blood pressure are complete opposite. Acetic acid and ascorbic acid are antihypertensive but excessive intake of sodium glutamate and sodium chloride raises hypertension. Observations offer a mechanism for the hypertension medication. Dominance of polarization of the Na-based dose solvation raises the blood pressure by increasing the viscosity of the blood, which impair the blood flow and raise the blood pressure. Fragmentation dominance of acidic solutions dilutes the flowing blood by decreasing structure ordering and viscosity to make blood running well, though both Na-based solutions and acidic solutions are subject to solute polarization. The Na^+ has much stronger capability of polarization than other alkali cations, K^+, Li^+, Rb^+ and Cs^+ because of their ionic radii and charge distribution [42].

7.5.5 Summary

From the perspective of O:H–O bond cooperative relaxation, we examined the solute capabilities of solvent O:H–O bond transition and the solution surface stress with inference of hypertension mediation. It is suggested that Na-based dose raises the blood pressure by ionic polarization that raises the viscosity and surface stress of the solution, impairing blood flow. However, acidic solvation depresses the blood pressure by disrupting the solution network. Findings should help in understanding the mechanism of hypertension and developing means for hypertension prevention and medication.

7.6 Sugar Solvation: The Attribute of Anti-frozen

7.6.1 Wonders of T_N Depression

Aqueous sugar solutions are important nutritional supplements [141, 142], catalysts [143], biological materials [144], and medicines [145]. Being excellent bioprotectant agents, sugars make the living organisms to survive in cold and snowy climates

[146]. For example, wood frog [147] and some insects [148, 149] can survive at extremely low temperatures due to their high levels of sugar content in their body fluids. Costanzo et al. [150] found that as temperature drops below the freezing point of pure water, the high level glucose by injection can reduce ice content in wood frogs' bodies and thus protect them from being injured during hibernation. Rapid cooling can injure wood frogs without intentional injection of glucose, which hinders the self-synthesis and delivery of glucose [151–153]. Kanno et al [154] examined the molality (mol/L) dependence of the T_N and T_m for aqueous solutions of sucrose, trehalose and maltose at higher concentration C and established the linear dependence of $\Delta T_N(C) \propto \Delta T_m(C)$. The T_N drops from -45 to -62 °C while the T_m drops from 0 to -11 °C as the solute concentration is increased from 0 to 3.8 mol/L according to their differential scanning calorimeter (DSC) measurements.

Various experimental techniques including FTIR [155], neutron scattering [156, 157], Rayleigh scattering [158], Brillouin ultraviolet light scattering [159], terahertz spectroscopy [160], nuclear magnetic resonance spectroscopy [161], density and viscosity analyzer [162], impedance/material analyzer [163] as well as numerical simulations [146, 156, 164, 165] have been applied to investigate how the sugar solvation influences the water HB network. For example, Thomas and coworkers [166] studied the O:H–C bonds geometry in sugar crystals by neutron diffraction and found that a fraction of protons interact with O atoms through the weaker O:H nonbonds.

Using Rayleigh scattering spectroscopy, Paolantoni and coworkers [158] observed that sugar solvation slows down water molecular motion dynamics and destroys the tetrahedral structure. Branca and coworkers [161] reported that trehalose solvation lowers the tetrahedral degree of water. Based on the terahertz (THz) spectra measurements, Heyden and coworkers [160] revealed that sugar concentration and the average number of HBs between the sugar molecules and the neighboring water molecules play an important role in depressing the T_N, and suggested that disaccharide is more effective than monosaccharide for bio-protection.

Kirschner et al. [165] suggested that water weakens sugar-sugar HB based on their results of MD simulations. Lerbret and coworkers [164] proposed that sugar solvation lowers remarkably the local molecular translational diffusion dynamics. Using maltose and trehalose as models, Magonet et al [146] suggested that there are two relaxation processes, the faster one behaving like bulk water while the other slower one as the hydration water molecules, being similar to the hydration shell of halogenic ions in salt solutions [71].

Although sugars play a significant role on biological living-body reservation, the underlying mechanism remains yet great challenge. Since the body fluids have a large amount of water as a solvent for sugar, it is critical to understand the sugar-water molecular interaction and the solvent O:H–O bond relaxation upon solvation. However, little is yet known about the microscopic mechanism for the anti-frozen attribute of sugar solutions. The lacking knowledge of the solute-solvent molecular interaction and the O:H–O cooperativity at relaxation [167] might have prevent the anti-frozen mechanism from being understood, in particular, the correlation between the O:H–O bond segmental stiffness relaxation and the T_N depression upon sugar solvation.

This section is focused on the sugar-water molecular interaction and the sugar solute capability of perturbing the HB network in terms of O:H–O cooperative relaxation and its consequence on the phonon frequency shift, Debye temperature depression, quasisolid phase boundary dispersion. Examined also the effect of sugar type and concentration on depressing the T_N of fructose, glucose and trehalose solutions. Results show that sugar solvation softens both the low-frequency O:H phonon and the high-frequency H–O phonon, being similar to the effect of alcohol solvation [39]. It is thus suggested that the sugar molecular dipoles form hydration shells by clustering and polarizing their adjacent water molecules. Spatial asymmetrical distribution of the H^+ and the ":" of the sugar solutes could distort the local solvent O:H–O bonds. The softened O:H phonon lowers the Debye temperature for the O:H specific heat, depressing the T_N.

7.6.2 Sugar-Water Molecular Interactions

Table 7.4 summarizes basic information of O:H–O bond counting between a sugar molecule and the surrounding solvent water molecules. The crystal geometry shows the dangling H^+ and O^{2-} anions that have each two pairs of ":". Because of the number harmonicity of the H^+ and the ":", neither the H↔H nor the O:⇔:O can form between sugar and water molecules so the sugar molecular dipolar polarization and solvent local distortion are the dominant factors. Figure 7.15a inset illustrates the hydration shell of the sugar solute.

Ideally, sugar molecules interact with the hydrating H_2O molecules through the dangling H^+ and ":", leading to the O:H–O bond, H↔H anti-HB or O:⇔:O super–HB. However, the equal number of H^+ and ":", prevents the H↔H anti-H or the O:⇔:O super–HB from formation. The sugar-water molecular interaction proceeds mainly through the O:H–O bond subjecting to local distortion. The spatially asymmetrical distribution of the H^+ and the ":" distorts the local hydrogen bonds in the hydration shells. In contrast, a number of H↔H anti-HBs may form between sugar and water because of the geometrical mismatch at the interface [39]. The O:H–O bonds distortion by the solute dipolar fields relax the O:H nonbond and the H–O bond cooperatively or not.

7.6.3 Low-Temperature Raman Spectroscopy

Figures 7.16 and 7.17 show representative Raman spectra of aqueous sugar solutions and pure water with temperature varying from 298 to 253 K. The extra Raman features in frequencies of 300–800 cm^{-1} and 2800–3000 cm^{-1} correspond to the intramolecular CCC, CCO and COC and CH_3 stretching and bending vibrations

Table 7.4 Molecular weight, molecular formula, spatial symmetry, structural formula and the number of the proton H^+ and the electron lone pair "$:$" of three sugars (reprinted with copyright permission from [168])

	Fructose	Glucose	Trehalose
Molecular weight	180.16	180.16	342.30
Molecular formula	$C_6H_{12}O_6$	$C_6H_{12}O_6$	$C_{12}H_{22}O_{11}$
Number of H^+	12	12	22
Number of "$:$"	12	12	22
Number of O:H–O bonds	24	24	44
Structural formula			

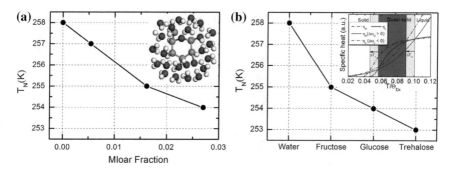

Fig. 7.15 **a** Trehalose molar concentration and **b** sugar type dependent T_N depression. Schematic illustration of inset (**a**) the hydration-shell surrounding a glucose molecular solute. The gray, red, and white dots stand for carbon oxygen, hydrogen ions; and inset (**b**) the superposition of the specific heat curves for the O:H nonbond and the H–O covalent bond. The superposition results in two interceptions as the quasisolid phase boundaries nearby the T_N and T_m of the solution. The redshift of both the O:H phonon and the H–O phonon in sugar solutions offset both Θ_{DL} and Θ_{DH} downward, resulting in depression of both T_N and T_m, being different from the situation of the ordinary water [169]. Reprinted with copyright permission from [168]

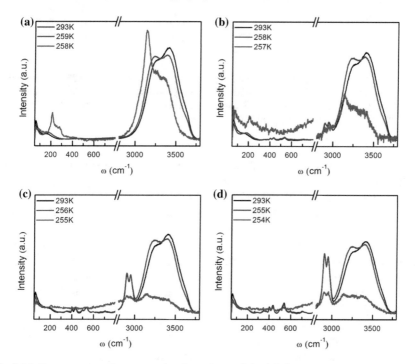

Fig. 7.16 Temperature and concentration dependence of the full-frequency Raman spectra for **a** pure water, **b** 0.0054, **c** 0.0162 and **d** 0.027 molar ratio trehalose solutions. The abrupt transition indicates the T_N for ice formation. The T_N for water, 0.0054, 0.0162 and 0.027 trehalose solutions are 258, 257, 255, and 254 K, respectively. Reprinted with copyright permission from [168]

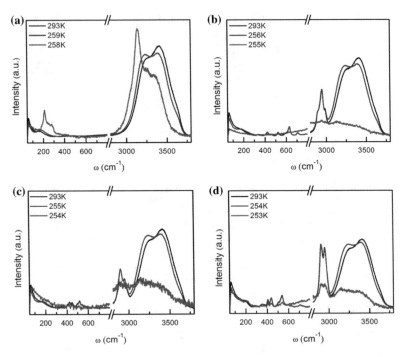

Fig. 7.17 Temperature dependence of the full-frequency Raman spectra for 0.0324 molar ratio **a** fructose **b** glucose, and **c** trehalose solutions. The abrupt transition indicates ice formation. The T_N for the 0.0324 mol ratio fructose, glucose and trehalose solutions are 255, 254, and 253 K, respectively. Reprinted with copyright permission from [168]

[170–172], which is out of immediate concern of the solvation dynamics. At frozen, the low-frequency O:H phonon undergoes a blueshift while the high-frequency H–O phonon undergoes a sudden redshift from 3200 to 3150 cm^{-1} [173] because of crystallization associated cooling expansion that weakens the O–O Coulomb repulsion [41].

The Raman peak abruption is an indicative of freezing [174], and thus the T_N varies with both sugar solute type and concentration, as summarized in Fig. 7.15. Under the subzero climate, the solution has ability to maintain the solution in the quasisolid (or quasiliquid) phase [169] till the T_N to reduce the formation of ice crystals destroying the organism cells. Sugars are thus efficient cryo-protecting agents preventing organisms from freezing. For one type of sugar, the increase of concentration can promote cryoprotective effectiveness. Under the same concentration, the cryoprotective effectiveness follows the series: trehalose > glucose > fructose. The extent of T_N depression may vary with the sample size as observed by Kanno et al [154] using DSC for large trunk of specimen.

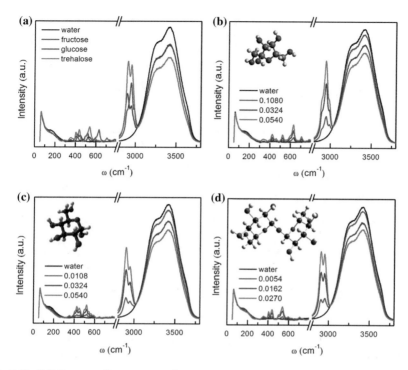

Fig. 7.18 Full-frequency Raman spectra for **a** water and fructose, glucose and trehalose solutions with a sugar/water molar ratio at 0.0324, and the concentrated **b** fructose, **c** glucose, **d** trehalose solutions. Insets of (**b**–**d**) plot the molecular structures of fructose, glucose and trehalose. Reprinted with copyright permission from [168]

7.6.4 Ambient DPS Verification

To correlate the O:H–O bond cooperative relaxation to the T_N and T_m depression, Raman spectra verification was conducted as a function of solute type and concentration. Figure 7.18a compares the spectra for the aqueous fructose, glucose and trehalose solutions at 0.0324 molar concentration with deionized water measured at 298 K. Figure 7.18b–d show that solvation of sugar molecules softens both the ω_L and the ω_H phonons. Moreover, fructose and glucose solutions show similar ω_H Raman bands whereas trehalose dissolution has a more pronounced effect on H–O bond relaxation. The Raman spectral features varied more pounced at higher concentrations.

As shown in Fig. 7.19, sugar solvation softens the O:H phonon from 190 to 75 cm^{-1}, which lowers the Θ_{DL} and depress the T_N of the deionized water. The $\Delta\omega_L$ depends both on the sugar type in the sequence of trehalose > glucose > fructose and concentrations. Observations confirmed the positive correlation between T_N and ω_L [169].

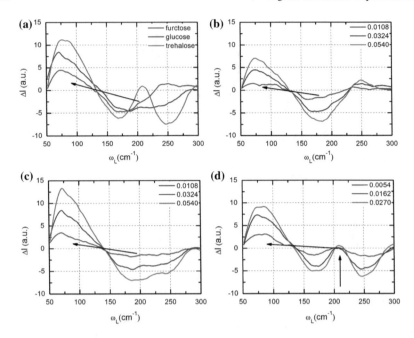

Fig. 7.19 The ω_L DPS redshift for **a** sugars solutions at 0.0324 molar ratio and the concentrated **b** fructose, **c** glucose, and **d** trehalose solutions. Reprinted with copyright permission from [168]

Figure 7.20 shows that sugar solvation softens the ω_H from 3650 to 3450 cm^1. The measured Raman spectra and DPS are insensitive to the size of the molecular ring. The feature at 3610 cm^{-1} and above arises from the skin dangling H–O bonds. Sugar solvation effects insignificantly to the bulk water but just the dipolar hydration shell where the electric field is anisotropic, local and strong. Sugar solvation removes the dangling bond feature by a certain mechanism subjecting to further clarification.

7.6.5 Summary

From the perspective of sugar-water molecular interaction, one may understand the anti-frozen mechanism of sugar solutions for cryoprotective effectiveness as the asymmetrical distribution of the protons and electron lone pairs. The sugar solute dipoles polarize and distort the local O:H–O bonds in the hydration shells. The structural distortion softens both the O:H and the H–O phonons and offsets the quasisolid phase boundaries and the T_N and T_m downward. The solute capability of T_N depression follows the trehalose > glucose > fructose order of the same concentration and the extent of T_N compression for a specific sugar is proportional to its concentration and varies with the solution sample size.

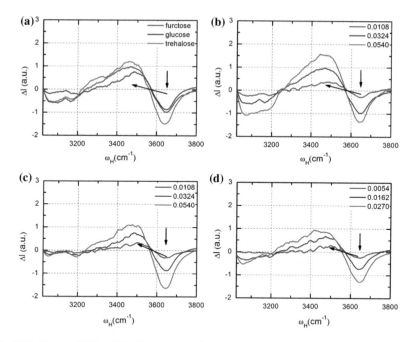

Fig. 7.20 The ω_H DPS redshift for **a** sugar solutions of 0.0324 molecular ratio and concentrated **b** fructose, **c** glucose and **d** trehalose solutions. Sugar solvation module more the interface O:H–O bonds than in the bulk. Reprinted with copyright permission from [168]

References

1. N. Hu, D. Wu, K. Cross, S. Burikov, T. Dolenko, S. Patsaeva, D.W. Schaefer, Structurability: a collective measure of the structural differences in vodkas. J. Agric. Food Chem. **58**, 7394–7401 (2010)
2. A. Nose, T. Hamasaki, M. Hojo, R. Kato, K. Uehara, and T. Ueda, Hydrogen bonding in alcoholic beverages (distilled spirits) and water-ethanol mixtures. J Agric Food Chem. **53**, 7074–7081 (2005)
3. H.-Y. Hsu, Y.-C. Tsai, C.-C. Fu, J.S.-B. Wu, Degradation of ascorbic acid inethanolic solutions. J. Agri. Food Chem. **60**(42), 10696–10701 (2012)
4. R. Zhang, Q. Wu, Y. Xu, Lichenysin, a cyclooctapeptide occurring in Chinese liquor Jiannanchun reduced the headspace concentration of phenolic off-flavors via hydrogen-bond interactions. J. Agric. Food Chem. **62**(33), 8302–8307 (2014)
5. The 2015 Nobel Prize in Physiology or Medicine—Press Release. 2016; Available from: http://www.nobelprize.org/nobel_prizes/medicine/laureates/2015/press.html
6. T.A. Dolenko, S.A. Burikov, S.A. Dolenko, A.O. Efitorov, I.V. Plastinin, V.I. Yuzhakov, S.V. Patsaeva, Raman spectroscopy of water-ethanol solutions: the estimation of hydrogen bonding energy and the appearance of clathrate-like structures in solutions. J. Phys. Chem. A **119**(44), 10806–10815 (2015)
7. M. Premont-Schwarz, S. Schreck, M. Iannuzzi, E.T.J. Nibbering, M. Odelius, P. Wernet, Correlating Infrared and x-ray absorption energies for molecular-level insight into hdyrogen bond makingand breaking in solution. J. Phys. Chem. B **119**, 8115–8124 (2015)
8. G. Ma, H.C. Allen, Surface studies of aqueous methanol solutions by vibrational broad bandwidth sum frequency generation spectroscopy. J. Phys. Chem. B **107**, 6343–6349 (2003)

9. L. Juurinen, T. Pylkkanen, C.J. Sahle, L. Simonelli, J. Hamalainen, S. Huotari, M. Hakala, Effect of the hydrophobic alcohol chain length on the hydrogen-bond network of water. J. Phys. Chem. B **118**, 8750–8755 (2014)

10. L. Xu, V. Molinero, Is there a liquid-liquid transition in confined water? J. Phys. Chem. B **115**(48), 14210–14216 (2011)

11. J.G. Davis, B.M. Rankin, K.P. Gierszal, D. Ben-Amotz, On the cooperative formation of non-hydrogen-bonded water at molecular hydrophobic interfaces. Nat. Chem. **5**, 796–802 (2013)

12. J.G. Davis, K.P. Gierszal, P. Wang, D. Ben-Amotz, Water strutural transformation at molecular hydrophobic interfaces. Nature **494**, 582–585 (2012)

13. M. Ahmed, A.K. Singh, J.A. Mondal, Hydrogen-bonidng and vibrational coupling of water in hydrophobic hydration shell as observed by Raman-MCR and isotopic dilution spectroscopy. Phys. Chem. Chem. Phys. **18**, 2767–2775 (2016)

14. L. Comez, L. Lupi, M. Paolantoni, F. Picchio, D. Fioretto, hydration properties of small hydrophobic molecules by brillouin light scattering. J. Chem. Phys. **137**(11), 114509 (2012)

15. D.T. Bowron, J.L. Finney, Anion bridges drive salting out of a simple amphiphile from aqueous solution. Phys. Rev. Lett. **89**, 215508 (2002)

16. K. NIshikawa, H. Hayashi, T. Iijima, Temperature-dependence of the concentration fluctuation, the kirkwood-buff parameters, and the correlation length of tert-butyl alcohol and water mixtures studied by small-angle X-Ray-scattering. J. Phys. Chem. B **93**, 6559–6595 (1989)

17. O. Gereben, L. Pusztai, Investigation of the structure of ethanol-water mixtures by molecular dynamics simulation I: analyses concerning the hydrogen-bonded pairs. J. Phys. Chem. B **119**, 3070–3084 (2015)

18. T. Ishihara, T. Ishiyama, A. Morita, Surface structure of methanol/water solutions via sum frequency orientational analysis and molecular dynamics simulation. J. Phys. Chem. C **119**, 9879–9889 (2015)

19. R.A. Livingstone, Y. Nagata, M. Bonn, E.H.G. Backus, Two types of water at the water-surfactant interface revealed by time-resolved vibrational spectroscopy. J. Am. Chem. Soc. **137**, 14912–14919 (2015)

20. S. Roy, S.M. Gruenbaum, J.L. Skinner, Theoretical vibrational sum-frequency generation spectroscopy of water near lipid and surfactant monolayer interfaces. J. Chem. Phys. **141**, 18C502 (2014)

21. H. Chen, W. Gan, B.-h. Wu, D. Wu, Y. Guo, H.-F. Wang, Determination of structure and energetics for Gibbs surface adsorption layers of binary liquid mixture 1. Acetone+ water. J. Phys. Chem. B **109**(16), 8053–8063 (2005)

22. R. Li, C. D'Agostino, J. McGregor, M.D. Mantle, J.A. Zeitler, L.F. Gladden, Mesoscopic structuring and dyanmics of alcohol/water solutions probed by terahertz time-domain spectroscopy and pulsed field gradient nuclear magnetic resonance. J. Phys. Chem. B **118**, 10156–10166 (2014)

23. C. Totland, R.T. Lewis, W. Nerdal, Long-range surface-induced water structures and the effect of 1-butanol studied by ^{1}H nuclear magnetic resonance. Langmuir **29**, 11055–11061 (2013)

24. O. Carrier, E.H.G. Backus, N. Shahidzadeh, J. Franz, M. Wagner, Y. Nagata, M. Bonn, D. Bonn, *Oppositely charged ions at water-air and water-oil interfaces: contrasting the molecular picture with thermodynamics*. J. Phys. Chem. Lett. **7**, 825–830

25. C. Carmelo, J. Spooren, C. Branca, N. Leone, M. Broccio, C. Kim, S.-H. Chen, H.E. Stanley, F. Mallamace, Clustering dynamics in water/methanol mixtures: a nuclear magnetic resonance study at 205 K < T < 295 K. J. Phys. Chem. B **112**, 10449–10454 (2008)

26. C.M. Phan, C.V. Nguyen, T.T. Pham, Molecular arrangement and surface tension of alcohol solutions. J. Phys. Chem. B **120**, 3914–3919 (2016)

27. Y. Nagata, S. Mukamel, Vibrational sum-frequency generation spectroscopy at the water/lipid interface: molecular dynamics simulation study. J. Am. Chem. Soc. **132**, 6434–6442 (2010)

28. B.M. Rankin, D. Ben-Amotz, S.T.V.D. Post, H.J. Bakker, Contacts between alcohols in water are random rather than hydrophobic. J. Phys. Chem. Lett. **6**, 688–692 (2015)

29. D. Banik, A. Roy, N. Kundu, N. Sarkar, Picosecond solvation and rotational dynamics: an attempt to reinvestigate the mystery of alcohol-water binary mixtures. J. Phys. Chem. B **119**, 9905–9919 (2016)
30. D. Gonzalez-Salgado, I. Nezbeda, Excess properties of aqueous mixtures of methanol: simulation versus experiment. Fluid Phase Equilib. **240**(2), 161–166 (2006)
31. P. Petong, R. Pottel, U. Kaatze, Water-ethanol mixtures at different compositions and temperatures. A dielectric relaxation study. J. Phys. Chem. A **104**(32), 7420–7428 (2000)
32. X. Zhang, Y. Zhou, Y. Gong, Y. Huang, C. Sun, Resolving H(Cl, Br, I) capabilities of transforming solution hydrogen-bond and surface-stress. Chem. Phys. Lett. **678**, 233–240 (2017)
33. X. Zhang, Y. Xu, Y. Zhou, Y. Gong, Y. Huang, and C.Q. Sun, HCl, KCl and koh solvation resolved solute-solvent interactions and solution surface stress. Appl. Surf. Sci. **422**, 475–481 (2017)
34. Y. Zhou, Y. Gong, Y. Huang, Z. Ma, X. Zhang, C.Q. Sun, Fraction and stiffness transition from the H-O vibrational mode of ordinary water to the HI, NaI, and NaOH hydration states. J. Mol. Liquids **244**, 415–421 (2017)
35. K.R. Wilson, R.D. Schaller, D.T. Co, R.J. Saykally, B.S. Rude, T. Catalano, J.D. Bozek, Surface relaxation in liquid water and methanol studied by X-ray absorption spectroscopy. J. Chem. Phys. **117**(16), 7738–7744 (2002)
36. Y. Huang, X. Zhang, Z. Ma, Y. Zhou, J. Zhou, W. Zheng, C.Q. Sun, Size, separation, structure order, and mass density of molecules packing in water and ice. Sci. Rep. **3**, 3005 (2013)
37. Y.L. Huang, X. Zhang, Z.S. Ma, Y.C. Zhou, W.T. Zheng, J. Zhou, C.Q. Sun, Hydrogen-bond relaxation dynamics: resolving mysteries of water ice. Coord. Chem. Rev. **285**, 109–165 (2015)
38. Y. Zhou, Y. Huang, Z. Ma, Y. Gong, X. Zhang, Y. Sun, C.Q. Sun, Water molecular structure-order in the NaX hydration shells (X=F, Cl, Br, I). J. Mol. Liq. **221**, 788–797 (2016)
39. Y. Gong, Y. Xu, Y. Zhou, C. Li, X. Liu, L. Niu, Y. Huang, X. Zhang, C.Q. Sun, Hydrogen bond network relaxation resolved by alcohol hydration (methanol, ethanol, and glycerol). J. Raman Spectrosc. **48**(3), 393–398 (2017)
40. Q. Zeng, T. Yan, K. Wang, Y. Gong, Y. Zhou, Y. Huang, C.Q. Sun, B. Zou, Compression icing of room-temperature NaX solutions (X= F, Cl, Br, I). Phys. Chem. Chem. Physics **18**(20), 14046–14054 (2016)
41. X. Zhang, T. Yan, Y. Huang, Z. Ma, X. Liu, B. Zou, C.Q. Sun, Mediating relaxation and polarization of hydrogen-bonds in water by NaCl salting and heating. Phys. Chem. Chem. Phys. **16**(45), 24666–24671 (2014)
42. Y. Gong, Y. Zhou, H. Wu, D. Wu, Y. Huang, C.Q. Sun, Raman spectroscopy of alkali halide hydration: hydrogen bond relaxation and polarization. J. Raman Spectrosc. **47**(11), 1351–1359 (2016)
43. N. Abe and M.I. to, Effects of hydrogen bonding on the Raman intensities of methanol, ethanol and water. J. Raman Spectrosc. **7**(3), 161–167 (1978)
44. L. Chen, W. Zhu, K. Lin, N. Hu, Y. Yu, X. Zhou, L.-F. Yuan, S.-M. Hu, Y. Luo, Identification of alcohol conformers by Raman spectra in the C-H stretching region. J. Phys. Chem. A **119**(13), 3209–3217 (2015)
45. Y. Yu, Y. Wang, N. Hu, K. Lin, X. Zhou, S. Liu, Overlapping spectral features and new assignment of 2-propanol in the C-H stretching region. J. Raman Spectrosc. **45**, 259–265 (2014)
46. G.E. Walrafen, Raman spectral studies of water structure. J. Chem. Phys **40**, 3249–3256 (1964)
47. J. Chen, C. Yao, X. Liu, X. Zhang, C.Q. Sun, Y. Huang, H_2O_2 and HO- solvation dynamics: solute capabilities and solute-solvent molecular interactions. Chem. Select **2**(27), 8517–8523 (2017)
48. B. Milorey, S. Farrell, S.T. Toal, R. Schweitzer-Stenner, Demixing of water and ethanol causes conformational redistribution and gelation of the cationic GAG tripeptide. Chem. Commun. **51**, 16498–16501 (2015)

49. S.K. Allison, J.P. Fox, R. Hargreaves, S. Bates, Clustering and microimmiscibility in alcohol-water mixtures: evidence from molecular-dynamics simulations. Phys. Rev. B **71**, 024201 (2005)

50. S. Banerjee, R. Ghosh, B. Bagchi, Structural transformations, compositioin anomalies and a dramatic collapse of linear polymer chains in dilute ethanol-water mixtures. J. Phys. Chem. B **116**, 3713–3722 (2012)

51. F. Franks, D.J.G. Ives, The structural properties of alcohol-water mixtures. Quart. Rev. Chem. Soc. **20**, 1–44 (1966)

52. J.-H. Guo, Y. Luo, A. Augustsson, S. Kashtanov, J.-E. Rubensson, D.K. Shuh, H. Agren, J. Nordgren, Molecular structure of alcohol-water mixtures. Phys. Rev. Lett. **91**, 157401 (2003)

53. I. Lee, K. Park, J. Lee, Precision density and volume contraction measurements of ethanol-water binary mixtures using suspended microchannel resonators. Sens. Actuators A **194**, 62–66 (2013)

54. K. Mizuno, Y. Miyashita, Y. Shindo, H. Ogawa, NMR and FT-IR studies of hydrogen bonds in ethanol-water mixtures. J. Phys. Chem. **99**, 3225–3228 (1995)

55. M.J. Costigan, L.J. Hodges, K.N. Marsh, R.H. Stokes, C.W. Tuxford, The isothermal displacement calorimeter: design modifications for measuring exothermic enthalpies of mixing. Austr. J. Chem. **33**(10), 2103–2119 (1980)

56. J.E. Hallsworth, Y. Nomura, A simple method to determine the water activity of ethanol-containing samples. Biotechnol. Bioeng. **62**, 242–245 (1999)

57. K. Koga, H. Yoshizumi, Differential scanning calorimetry (DSC) studies on the freezing processes of water-ethanol mixtures and distilled spirits. J. Food Sci. **44**, 1386–1389 (1979)

58. C.R. Lerici, M. Piva, M.D. Rosa, Water activity and freezing point depression of aqueous solutions and liquid foods. Food Sci. **48**, 1667–1669 (2006)

59. C.Q. Sun, X. Zhang, X. Fu, W. Zheng, J.-L. Kuo, Y. Zhou, Z. Shen, J. Zhou, Density and phonon-stiffness anomalies of water and ice in the full temperature range. J. Phys. Chem. Lett. **4**, 3238–3244 (2013)

60. K. Dong, X. Rao, X. Yang, J. Lin, P. Zhang, Raman spectroscopy of aldehyde molecules. Opt. Spectrosc. Spectrosc. Anal. **31**(12), 3277–3280 (2011). (Chinese)

61. X. Xi, S. Dai, Y. Sun, DNA-aldehyde molecular interaction. Envron. Sci. **22**(1), 19–22 (2001)

62. Z. Xi, F. Tao, D. Yang, Y. Sun, G. Li, H. Zhang, W. Zhang, Y. Yang, H. Liu, DNA damaged by aldehyde. J. Environ. Sci. **24**(4), 719–722 (2004). (Chinese)

63. R. Li, Z. Lu, Y. Qiao, H. Yao, F. Yu, X. Yang, DNA damage by aldehyde adsorption. Bull. Experim. Biol. **37**(4), 262–268 (2004). (in Chinese)

64. Y.N. Jo, I.C. Um, Effects of solvent on the solution properties, structural characteristics and properties of silk sericin. Int. J. Biol. Macromol. **78**, 287 (2015)

65. F. Greif, O. Kaplan, Acid ingestion: another cause of disseminated intravascular coagulopathy. Crit. Care Med. **14**(11), 990–1 (1986)

66. K. Yoshitomi, Y. Matayoshi, H. Tamura, S. Shibasaki, M. Uchida, Y. Haranishi, K. Nakamura, H. Oka, A case of acetic acid poisoning. J. Jap. Soc. Intensive Care Med. **11**, 217–221 (2009)

67. G.M. Tong, S.K. Mak, P.N. Wong, L.O. Kin-Yee, S.O. Sheung-On, C.L. Watt, A.K. Wong, Successful treatment of oral acetic acid poisoning with plasmapheresis. Hong Kong J. Nephrol. **2**(2), 110–112 (2000)

68. S. Kumar, B. Babu. A brief review on propionic acid: a renewal energy source, in *Proceedings of the National Conference on environmental conservation (NCEC-2006)* (2006)

69. S. Suwannakham, S.T. Yang, Enhanced propionic acid fermentation by *Propionibacterium acidipropionici* mutant obtained by adaptation in a fibrous-bed bioreactor. Biotechnol. Bioeng. **91**(3), 325 (2005)

70. J. Chen, C. Yao, X. Zhang, C.Q. Sun, Y. Huang, Hydrogen bond and surface stress relaxation by aldehydic and formic acidic molecular solvation. J. Mol. Liq. **249**, 494–500 (2018)

71. Y. Zhou, D. Wu, Y. Gong, Z. Ma, Y. Huang, X. Zhang, C.Q. Sun, Base-hydration-resolved hydrogen-bond networking dynamics: quantum point compression. J. Mol. Liq. **223**, 1277–1283 (2016)

72. P.H. Yancey, M.E. Clark, S.C. Hand, R.D. Bowlus, G.N. Somero, Living with water stress: evolution of osmolyte systems. Science 217(4566), 1214–22 (1982)
73. P.H. Yancey, Compatible and counteracting solutes: protecting cells from the Dead Sea to the deep sea. Sci. Prog. 87(1), 1–24 (2004)
74. B. Kempf, E. Bremer, Uptake and synthesis of compatible solutes as microbial stress responses to high-osmolality environments. Arch. Microbiol. 170(5), 319 (1998)
75. G.N. Somero, Protons, osmolytes, and fitness of internal milieu for protein function. Am. J. Physiol. 251(2), 197–213 (1986)
76. J.A. Raymond, A.L. Devries, Elevated concentrations and synthetic pathways of trimethylamine oxide and urea in some teleost fishes of McMurdo Sound, Antarctica. Fish Physiol. Biochem. 18(4), 387–398 (1998)
77. N.P. Davies, M. Wilson, K. Natarajan, Y. Sun, L. Macpherson, M.A. Brundler, T.N. Arvanitis, R.G. Grundy, A.C. Peet, non-invasive detection of glycine as a biomarker of malignancy in childhood brain tumours using in-vivo 1H MRS at 1.5 tesla confirmed by ex-vivo high-resolution magic-angle spinning NMR. NMR in Biomed. 23(1), 80–87 (2010)
78. T. Bessaire, A. Tarres, R.H. Stadler, T. Delatour, Role of choline and glycine betaine in the formation of N, N-dimethylpiperidinium (mepiquat) under Maillard reaction conditions. Food Additives Contam. Part A Chem. Anal. Control Exposure Risk Assess. 31(12), 1949–1958 (2014)
79. S. Chaum, C. Kirdmanee, Effect of glycinebetaine on proline, water use, and photosynthetic efficiencies, and growth of rice seedlings under salt stress. Turk. J. Agric. Forestry 34(6), 455–479 (2010)
80. R.D. Mountain, D. Thirumalai, Molecular dynamics simulations of end-to-end contact formation in hydrocarbon chains in water and aqueous urea solution. J. Am. Chem. Soc. 125(7), 1950–7 (2003)
81. S.N. Timasheff, The control of protein stability and association by weak interactions with water: how do solvents affect these processes? Annu. Rev. Biophys. Biomol. Struct. 22(22), 67–97 (1993)
82. A. Gómezzavaglia, R. Fausto, Low-temperature solid-state FTIR study of glycine, sarcosine and N, N-dimethylglycine: observation of neutral forms of simple α-amino acids in the solid state. Phys. Chem. Chem. Phys. 5(15), 268–270 (2003)
83. S. Kumar, A.K. Rai, V.B. Singh, S.B. Rai, Vibrational spectrum of glycine molecule. Spectrochim. Acta Part A Mol. Biomol. Spectrosc. 61(11–12), 2741–2746 (2005)
84. N. Derbel, B. Hernández, F. Pflüger, J. Liquier, F. Geinguenaud, N. Jaïdane, Z.B. Lakhdar, M. Ghomi, Vibrational Analysis of amino acids and short peptides in hydrated media. I.L-glycine and L-leucine. J. Phys. Chem. B 111(6), 1470–1477 (2007)
85. G. Zhu, X. Zhu, Q. Fan, X. Wan, Raman spectra of amino acids and their aqueous solutions. Spectrochim. Acta Part A Mol. Biomol. Spectrosc. 78(3), 1187 (2011)
86. A. Oren, B.R. Elevi, N. Kandel, Z. Aizenshtat, J. Jehlička, Glycine betaine is the main organic osmotic solute in a stratified microbial community in a hypersaline evaporitic gypsum crust. Extremophiles 17(3), 445–451 (2013)
87. Y. Hayashi, Y. Katsumoto, I. Oshige, S. Omori, A. Yasuda, Comparative study of urea and betaine solutions by dielectric spectroscopy: liquid structures of a protein denaturant and stabilizer. J. Phys. Chem. B 111(40), 11858–63 (2007)
88. J.Y. Kim, S. Im, B. Kim, C. Desfrançois, S. Lee, Structures and energetics of Gly–(H$_2$O) 5: thermodynamic and kinetic stabilities. Chem. Phys. Lett. 451(4–6), 198–203 (2008)
89. A. Chaudhari, P.K. Sahu, S.L. Lee, Hydrogen bonding interaction in sarcosine–water complex using ab initio and DFT method. Int. J. Quantum Chem. 101(1), 97–103 (2005)
90. M. Civera, A. Fornili, M. Sironi, S.L. Fornili, Molecular dynamics simulation of aqueous solutions of glycine betaine. Chem. Phys. Lett. 367(1–2), 238–244 (2003)
91. T. Takayanagi, T. Yoshikawa, A. Kakizaki, M. Shiga, M. Tachikawa, Molecular dynamics simulations of small glycine–(H$_2$O)$_n$ (n = 2–7) clusters on semiempirical PM6 potential energy surfaces. J. Mol. Struct. (Thoechem) 869(1), 29–36 (2008)

92. A. Mukaiyama, Y. Koga, K. Takano, S. Kanaya, Osmolyte effect on the stability and folding of a hyperthermophilic protein. Proteins Structure Funct. Bioinform. **71**(1), 110–118 (2008)
93. A. Panuszko, P. Bruździak, E. Kaczkowska, J. Stangret, General mechanism of osmolytes' influence on protein stability irrespective of the type of osmolyte cosolvent. J. Phys. Chem. B **120**(43), 11159–11169 (2016)
94. Y.L. Rezus, H.J. Bakker, Destabilization of the hydrogen-bond structure of water by the osmolyte trimethylamine N-oxide. J. Phys. Chem. B **113**(13), 4038–44 (2009)
95. P. Bruździak, A. Panuszko, J. Stangret, Influence of osmolytes on protein and water structure: a step to understanding the mechanism of protein stabilization. J. Phys. Chem. B **117**(39), 11502–11508 (2013)
96. A. Panuszko, M. Śmiechowski, J. Stangret, *Fourier transform infrared spectroscopic and theoretical study of water interactions with glycine and its N-methylated derivatives.* J. Chem. Phys. **134**(11), 115104 (2011)
97. A. Kuffel, J. Zielkiewicz, The hydrogen bond network structure within the hydration shell around simple osmolytes: urea, tetramethylurea, and trimethylamine-N-oxide, investigated using both a fixed charge and a polarizable water model. J. Chem. Phys. **133**(3), 07B605 (2010)
98. A. Di Michele, M. Freda, G. Onori, M. Paolantoni, A. Santucci, P. Sassi, Modulation of hydrophobic effect by cosolutes. J. Phys. Chem. B **110**(42), 21077–21085 (2006)
99. K.J. Tielrooij, J. Hunger, R. Buchner, M. Bonn, H.J. Bakker, Influence of concentration and temperature on the dynamics of water in the hydrophobic hydration shell of tetramethylurea. J. Am. Chem. Soc. **132**(44), 15671–8 (2010)
100. P. Chettiyankandy, Effects of co-solutes on the hydrogen bonding structure and dynamics in aqueous N-methylacetamide solution: a molecular dynamics simulations study. Mol. Phys. **112**(22), 2906–2919 (2014)
101. H. Lee, J.H. Choi, P.K. Verma, M. Cho, Spectral graph analyses of water hydrogen-bonding network and osmolyte aggregate structures in osmolyte-water solutions. J. Phys. Chem. B **119**(45), 14402–14412 (2015)
102. K. Tielrooij, N. Garcia-Araez, M. Bonn, H. Bakker, Cooperativity in ion hydration. Science **328**(5981), 1006–1009 (2010)
103. J. Hunger, K.J. Tielrooij, R. Buchner, M. Bonn, H.J. Bakker, Complex formation in aqueous trimethylamine-N-oxide (TMAO) solutions. J. Phys. Chem. B **116**(16), 4783–95 (2012)
104. A.A. Bakulin, M.S. Pshenichnikov, H.J. Bakker, C. Petersen, Hydrophobic molecules slow down the hydrogen-bond dynamics of water. J. Phys. Chem. A **115**(10), 1821–9 (2011)
105. H. Fang, X. Liu, C.Q. Sun, Y. Huang, Phonon spectrometric evaluation of the solute-solvent interface in solutions of glycine and its N-methylated derivatives. J. Phys. Chem. B **122**(29), 7403–7408 (2018)
106. Y. Zhou, Y. Zhong, Y. Gong, X. Zhang, Z. Ma, Y. Huang, C.Q. Sun, Unprecedented thermal stability of water supersolid skin. J. Mol. Liq. **220**, 865–869 (2016)
107. X.J. Liu, M.L. Bo, X. Zhang, L. Li, Y.G. Nie, H. TIan, Y. Sun, S. Xu, Y. Wang, W. Zheng, C.Q. Sun, Coordination-resolved electron spectrometrics. Chem. Rev. **115**(14), 6746–6810 (2015)
108. C.Q. Sun, J. Chen, Y. Gong, X. Zhang, Y. Huang, (H, Li)Br and LiOH solvation bonding dynamics: molecular nonbond interactions and solute extraordinary capabilities. J. Phys. Chem. B **122**(3), 1228–1238 (2018)
109. E. Agabiti-Rosei, From macro- to microcirculation: Benefits in hypertension and diabetes. J. Hypertens. Suppl. Off. J. Int. Soc. Hypertens. **26**(3), 15–9 (2008)
110. B.P. Murphy, T. Stanton, F.G. Dunn, Hypertension and myocardial ischemia. Med. Clin. North Am. **93**(3), 681–695 (2009)
111. K.K. Gaddam, A. Verma, M. Thompson, R. Amin, H. Ventura, Hypertension and cardiac failure in its various forms. Med. Clin. North Am. **93**(3), 665–680 (2009)
112. E. Reisin, A.V. Jack, Obesity and hypertension: mechanisms, cardio-renal consequences, and therapeutic approaches. Med. Clin. North Am. **93**(3), 733–51 (2009)

113. W.B. White, Defining the problem of treating the patient with hypertension and arthritis pain. Am. J. Med. **122**(5 Suppl), 3–9 (2009)
114. L.D. Truong, S.S. Shen, M.H. Park, B. Krishnan, Diagnosing nonneoplastic lesions in nephrectomy specimens. Arch. Pathol. Lab. Med. **133**(2), 189–200 (2009)
115. R.E. Tracy, S. White, A method for quantifying adrenocortical nodular hyperplasia at autopsy: some use of the method in illuminating hypertension and atherosclerosis. Ann. Diagnostic Pathol. **6**(1), 20–9 (2002)
116. S. Mendis, P. Puska, B. Norrving, S. Mendis, P. Puska, B. Norrving, *Global atlas on cardiovascular disease prevention and control* (Geneva World Health Organization, Geneva, 2011)
117. A. Chockalingam, Impact of World Hypertension Day. Can. J. Cardiol. **23**(7), 517 (2007)
118. W.H. Organization, *A global brief on hypertension: silent killer, global public health crisis: World Health Day 2013* (2013)
119. W.G. Members, E.J. Benjamin, M.J. Blaha, S.E. Chiuve, M. Cushman, S.R. Das, R. Deo, S.D.D. Ferranti, J. Floyd, M. Fornage, Heart disease and stroke statistics—2017 update: a report from the American Heart Association. Circulation **121**(7), e46 (2010)
120. B. Williams, N.R. Poulter, M.J. Brown, M. Davis, G.T. Mcinnes, J.F. Potter, P.S. Sever, T.S. Mcg, Guidelines for management of hypertension: report of the fourth working party of the British Hypertension Society, 2004-BHS IV. J. Hum. Hypertens. **18**(3), 139 (2004)
121. P.K. Whelton, J. He, L.J. Appel, J.A. Cutler, S. Havas, T.A. Kotchen, E.J. Roccella, R. Stout, C. Vallbona, M.C. Winston, Primary prevention of hypertension: clinical and public health advisory from the national high blood pressure education program. JAMA **288**(15), 1882–1888 (2002)
122. K. Kurihara, Glutamate: from discovery as a food flavor to role as a basic taste (umami). Am. J. Clin. Nutr. **90**(3), 719S–722S (2009)
123. L. Baad-Hansen, B.E. Cairns, M. Ernberg, P. Svensson, Effect of systemic monosodium glutamate (MSG) on headache and pericranial muscle sensitivity. Cephalalgia **30**(1), 68–76 (2010)
124. Z. Shi, B. Yuan, A.W. Taylor, Y. Dai, X. Pan, T.K. Gill, G.A. Wittert, Monosodium glutamate is related to a higher increase in blood pressure over 5 years: findings from the Jiangsu Nutrition Study of Chinese adults. J. Hypertens. **29**(5), 846–853 (2011)
125. S. Mascoli, R. Grimm, C. Launer, Sodium chloride raises blood pressure in normotensive subjects. Hypertension, **17**(Suppl I) (1991)
126. S.N. Orlov, A.A. Mongin, Salt-sensing mechanisms in blood pressure regulation and hypertension. Am. J. Physiol. Heart Circ. Physiol. **293**(4), H2039–H2053 (2007)
127. G.R. Meneely, L.K. Dahl, Electrolytes in hypertension: the effects of sodium chloride. The evidence from animal and human studies. Med. Clinics North Am. **45**(2), 271 (1961)
128. R. Kwok, Chinese-restaurant syndrome. New England J. Med. **278**(14), 796 (1968)
129. T.R. Du, Y. Volsteedt, Z. Apostolides, Comparison of the antioxidant content of fruits, vegetables and teas measured as vitamin C equivalents. Toxicology **166**(1–2), 63–69 (2001)
130. S. Duffy, N. Gokce, M. Holbrook, A. Huang, B. Frei, J.F. Keaney, J.A. Vita, Treatment of hypertension with ascorbic acid. The Lancet **354**(9195), 2048–2049 (1999)
131. B.A. Mullan, I.S. Young, H. Fee, D.R. McCance, Ascorbic acid reduces blood pressure and arterial stiffness in type 2 diabetes. Hypertension **40**(6), 804–809 (2002)
132. J.P. Moran, L. Cohen, J.M. Greene, G. Xu, E.B. Feldman, C.G. Hames, D.S. Feldman, Plasma ascorbic acid concentrations relate inversely to blood pressure in human subjects. Am. J. Clin. Nutr. **57**(2), 213–217 (1993)
133. S.P. Juraschek, E. Guallar, L.J. Appel, M.E. Rd, Effects of vitamin C supplementation on blood pressure: a meta-analysis of randomized controlled trials. Am. J. Clin. Nutr. **95**(5), 1079–1088 (2012)
134. S. Kondo, K. Tayama, Y. Tsukamoto, K. Ikeda, Y. Yamori, Antihypertensive effects of acetic acid and vinegar on spontaneously hypertensive rats. Biosci. Biotechnol. Biochem. **65**(12), 2690–2694 (2001)

135. Y. Zhou, Y. Huang, Z. Ma, Y. Gong, X. Zhang, Y. Sun, C.Q. Sun, Water molecular structure-order in the NaX hydration shells (X = F, Cl, Br, I). J. Mol. Liq. **221**, 788–797 (2016)

136. C.Q. Sun, Y. Sun, *The Attribute of Water: Single Notion, Multiple Myths*. Springer Series in Chemical Physics, vol. 113 (SpringerVerlag, Heidelberg, 2016), 494pp

137. N. Peica, C. Lehene, N. Leopold, S. Schlücker, W. Kiefer, Monosodium glutamate in its anhydrous and monohydrate form: differentiation by Raman spectroscopies and density functional calculations. Spectrochim. Acta Part A Mol. Biomol. Spectrosc. **66**(3), 604–15 (2007)

138. T. Nakabayashi, K. Kosugi, N. Nishi, Liquid structure of acetic acid studied by Raman spectroscopy and Ab initio molecular orbital calculations. J. Phys. Chem. A **103**(43), 8595–8603 (1999)

139. P.C. Yohannan, V.H. Tresa, D. Philip, FT-IR, FT-Raman and SERS spectra of vitamin C. Spectrochim Acta A Mol. Biomol. Spectrosc. **65**(3–4), 802–804 (2006)

140. C. Ni, C. Sun, Z. Zhou, Y. Huang, X. Liu, Surface tension mediation by Na-based ionic polarization and acidic fragmentation: inference of hypertension. J. Mol. Liq. **259**, 1–6 (2018)

141. L.M. Burke, R.S. Read, Dietary supplements in sport. Sports Med. **15**(1), 43–65 (1993)

142. M. Baghbanbashi, G. Pazuki, A new hydrogen bonding local composition based model in obtaining phase behavior of aqueous solutions of sugars. J. Mol. Liq. **195**(4), 47–53 (2014)

143. B.L. Cantarel, P.M. Coutinho, C. Rancurel, T. Bernard, V. Lombard, B. Henrissat, The carbohydrate-active enzymes database (CAZy): an expert resource for glycogenomics. Nucleic Acids Res. **37**(1), D233–D238 (2009)

144. F. Franks, M. Jones, Biophysics and biochemistry at low temperatures. FEBS Lett. **220**(2), 391–391 (1986)

145. C.A. Oksanen, G. Zografi, The relationship between the glass transition temperature and water vapor absorption by poly (vinylpyrrolidone). Pharm. Res. Dordr **7**(6), 654–657 (1990)

146. A. Magno, P. Gallo, *Understanding the mechanisms of bioprotection: a comparative study of aqueous solutions of trehalose and maltose upon supercooling*. J. Phys. Chem. Lett. **2**(9), 977–982 (2011)

147. B.J. Sinclair, J.R. Stinziano, C.M. Williams, H.A. Macmillan, K.E. Marshall, K.B. Storey, Real-time measurement of metabolic rate during freezing and thawing of the wood frog, *Rana sylvatica*: implications for overwinter energy use. J. Exp. Biol. **216**(Pt 2), 292–302 (2013)

148. R.E.L. Jr, Insect cold-hardiness: to freeze or not to freeze. Bioscience **39**(5), 308–313 (1989)

149. S.N. Thompson, Trehalose-the insect 'blood' sugar. Adv. Insect Physiol. **31**(3), 205–285 (2003)

150. J.P. Costanzo, R.E. Lee, P.H. Lortz, Glucose concentration regulates freeze tolerance in the wood frog *Rana sylvatica*. J. Exp. Biol. **181**(1), 245–255 (1993)

151. J. Costanzo, R. Lee, M.F. Wright, Glucose loading prevents freezing injury in rapidly cooled wood frogs. Am. J. Physiol. **261**(6), R1549–R1553 (1991)

152. J.P. Costanzo, R.E. Lee Jr., M.F. Wright, Effect of cooling rate on the survival of frozen wood frogs, *Rana sylvatica*. J. Comp. Physiol. B **161**(3), 225–229 (1991)

153. J.P. Costanzo, R.E. Lee, M.F. Wright, Cooling rate influences cryoprotectant distribution and organ dehydration in freezing wood frogs. J. Exp. Zool. **261**(4), 373–378 (1992)

154. H. Kanno, M. Soga, K. Kajiwara, Linear relation between TH (homogeneous ice nucleation temperature) and Tm (melting temperature) for aqueous solutions of sucrose, trehalose, and maltose. Chem. Phys. Lett. **443**(4–6), 280–283 (2007)

155. M.E. Gallina, P. Sassi, M. Paolantoni, A. Morresi, and R.S. Cataliotti, Vibrational analysis of molecular interactions in aqueous glucose solutions. Temperature and concentration effects. J. Phys. Chem. B, **110**(17), 8856–8864 (2006)

156. C. Branca, S. Magazù, G. Maisano, S. Bennington, B. Fåk, Vibrational studies on disaccharide/H_2O systems by inelastic neutron scattering, Raman, and IR spectroscopy. J. Phys. Chem. B **107**(6), 1444–1451 (2003)

157. C. Branca, S. Magazu, G. Maisanoa, A. Mangionea, S.M. Benningtonb, J. Taylorb, INS investigation of disaccharide/H_2O mixtures. J. Mol. Struct. **700**(1), 229–231 (2004)

158. M. Paolantoni, P. Sassi, A. Morresi, S. Santini, *Hydrogen bond dynamics and water structure in glucose-water solutions by depolarized Rayleigh scattering and low-frequency Raman spectroscopy*. J. Chem. Phys. **127**(2), 024504 (2007)

159. S. Di Fonzo, C. Masciovecchio, A. Gessini, F. Bencivenga, A. Cesàro, Water dynamics and structural relaxation in concentrated sugar solutions. Food Biophys. **8**(3), 183–191 (2013)

160. M. Heyden, E. Bründermann, U. Heugen, G. Niehues, D.M. Leitner, M. Havenith, Long-range influence of carbohydrates on the solvation dynamics of water-answers from terahertz absorption measurements and molecular modeling simulations. J. Am. Chem. Soc. **130**(17), 5773–5779 (2008)

161. C. Branca, S. Magazu, G. Maisano, P. Migliardo, E. Tettamanti, Anomalous translational diffusive processes in hydrogen-bonded systems investigated by ultrasonic technique. Raman scattering and NMR. Physica B **291**(1), 180–189 (2000)

162. M.E. Elias, A.M. Elias, Trehalose + water fragile system: properties and glass transition. J. Mol. Liq. **83**(1), 303–310 (1999)

163. W. Yamamoto, K. Sasaki, R. Kita, S. Yagihara, N. Shinyashiki, Dielectric study on temperature-concentration superposition of liquid to glass in fructose-water mixtures. J. Mol. Liq. **206**(1), 39–46 (2015)

164. A. Lerbret, F. Affouard, P. Bordat, A. Hédoux, Y. Guinet, M. Descamps, Slowing down of water dynamics in disaccharide aqueous solutions. J. Non-Cryst. Solids **357**(2), 695–699 (2010)

165. K.N. Kirschner, R.J. Woods, Solvent interactions determine carbohydrate conformation. Proc. Natl. Acad. Sci. USA **98**(19), 10541 (2001)

166. T. Steiner, W. Saenger, Geometry of carbon-hydrogen.cntdot..cntdot..cntdot.oxygen hydrogen bonds in carbohydrate crystal structures. Analysis of neutron diffraction data. J. Am. Chem. Soc. **114**(26): 10146–10154, (1992)

167. Y.L. Huang, X. Zhang, Z.S. Ma, G.H. Zhou, Y.Y. Gong, C.Q. Sun, Potential paths for the hydrogen-bond relaxing with (H$_2$O)(N) Cluster Size. J. Phys. Chem. C **119**(29), 16962–16971 (2015)

168. C. Ni, Y. Gong, X. Liu, C.Q. Sun, Z. Zhou, The anti-frozen attribute of sugar solutions. J. Mol. Liq. **247**, 337–344 (2017)

169. X. Zhang, P. Sun, Y. Huang, Z. Ma, X. Liu, J. Zhou, W. Zheng, C.Q. Sun, Water nan-odroplet thermodynamics: quasi-solid phase-boundary dispersivity. J. Phys. Chem. B **119**(16), 5265–5269 (2015)

170. M. Mathlouthi, C. Luu, A.M. Meffroy-Biget, V.L. Dang, Laser-Raman study of solute-solvent interactions in aqueous solutions of d-fructose, d-glucose, and sucrose. Carbohyd. Res. **81**(2), 213–223 (1980)

171. A.M. Gil, P.S. Belton, V. Felix, Spectroscopic studies of solid α-α trehalose. Spectrochim Acta A **52**(12), 1649–1659 (1996)

172. S. Söderholm, Y.H. Roos, N. Meinander, M. Hotokka, Raman spectra of fructose and glucose in the amorphous and crystalline states. J. Raman Spectrosc. **30**(11), 1009–1018 (1999)

173. S.N. Wren, D.J. Donaldson, Glancing-angle Raman study of nitrate and nitric acid at the air–aqueous interface. Chem. Phys. Lett. **522**, 1–10 (2012)

174. C.Q. Sun, X. Zhang, X. Fu, W. Zheng, J.-L. Kuo, Y. Zhou, Z. Shen, J. Zhou, Density and phonon-stiffness anomalies of water and ice in the full temperature range. J. Phys. Chem Lett. **4**, 3238–3244 (2013)

Chapter 8
Multifield Coupling

Contents

Abstract Transiting the NaX/H_2O solutions from liquid into ice VI (at P_{C1}) and then into ice VII (P_{C2}) phase at 298 K needs excessive pressures with respect to the same sequence of phase transition for pure water. P_{C1} and P_{C2} vary simultaneously with the solute type in the Hofmeister series order: I > Br > Cl > F ~ 0. However, the P_{C1} grows faster than the P_{C2} with the increase of NaI/H_2O concentration, following the (P, T) path upwardly along the Liquid-VI phase boundary. The P_{C1} and P_{C2} meet then at the Liquid-VI-VII triple-phase junction at 3.3 GPa and 350 K. Observations confirmed that compression recovers the electrification-elongated O:H–O bond first

© Springer Nature Singapore Pte Ltd. 2019
C. Q. Sun, *Solvation Dynamics*, Springer Series in Chemical
Physics 121, https://doi.org/10.1007/978-981-13-8441-7_8

and then proceeds the phase transitions, which requires excessive energy for the same sequence of phase transitions. Heating enhances the effect of salting on bond relaxation but opposite on polarization that dictates the surface stress of the solution. It is also confirmed that molecular undercoordination disperses the quasisolid phase boundaries and the room-temperature ice-quasisolid phase transition needs excessive pressure. Polarization by salt solvation and skin undercoordination and boundary reflection transit the phonon abundance-lifetime-stiffness cooperatively. An extension of the HB and anti-HB or super-HB clarifies the energetic storage and structural stability for the spontenous and constrained explosion of energetic carriers.

Highlight

- *Compression does contrastingly to polarization or undercoordination on HB relaxation.*
- *Heating enhances the effect of salting on HB relaxation but opposite on polarization.*
- *Polarization and grain boundary reflection dictate the phonon energy dissipation lifetime.*
- *HB tension constrains but super-HB or anti-HB repulsion fosters explosion.*

8.1 Wonders of Multifield Effect

8.1.1 *P_C and T_C for Phase Transition*

Salt solvation modulates not only the solubility of dissolving biological molecules such as DNA and proteins but also the surface stress and solution viscosity and solubility [1–7]. Salt solvation varies the critical temperatures T_C and the critical pressures P_C, for phase transition. The gelation time, t_C, for energy accumulation also varies with the type and concentration of the solutes for transiting the colloid from sol to gel [8–10].

It is usual for other materials that mechanical compression shortens all the interatomic bonds and stiffens their Raman phonons such as carbon allotropes [11]. However, for water and ice, compression shortens the O–O distance and the O:H length but lengthens the H–O bond, towards O:H–O segmental length symmetrization transiting into the X-th phase at some 59 GPa pressure and in almost full temperature regime [12–15]. The O:H and the H–O are identical in length at 0.10 nm but have different bond energies at the VII-X and VIII-X phase boundaries [16]. Compression of the phase X shortens both the H–O and the O:H slightly and simultaneously [17]. Accordingly, compression stiffens the O:H phonon from below 400 cm^{-1} to its above but softens the H–O phonon from above 3200 cm^{-1} to its below [18–22], because of the O:H–O bond cooperativity and O–O repulsivity [23, 24].

The combination of salt solvation and mechanical compression makes the situation much more complicated [25]. Low-temperature and high-pressure Raman spectroscopy of ice transition from the phase VII to phase VIII and subsequently to the symmetrical phase X revealed that NaCl or LiCl solvation raises the P_C for transiting the solution ice from phase VII to phase X compared to pure ice [26, 27]. Solvation of LiCl in the LiCl/H_2O number ratio of 1/50 and 1/6 requires 30 and 85 GPa more pressures, respectively, with respect to that required for transiting pure water into ice X at 60 GPa. Meanwhile, salt hydration not only stiffens the ω_H but also lengthens the O–O distance and shortens the ω_H phonon lifetime [28, 29].

Compression elevated P_C for the VII-VIII and the VIII-X phase transition is attributed to the O:H compression with an association of H–O elongation through the intrinsic Coulomb repulsion between electron pairs on adjacent oxygen anions [14, 15]. Another opinion for this kind of phase transition is that the pressure degenerates the symmetrical double-well potentials into one located midway between oxygen ions, terminating the proton quantum transitional tunneling, which locates the proton eventually in the fixed position [12]. The third view is that salt hydration elevated P_C for the VII-X and VIII-X phase transition is explained on the base of two-component phase diagram for salted-water. The presence of salt hinders proton ordering and O:H–O segmental length symmetrization [26–28].

8.1.2 Salting and Heating

Likewise, anion polarization also affects the T_C for phase transition of the methyl-cellulose aqueous solutions [8, 30]. The transition was characterized by enthalpies and the specific heat (C_p) under constant pressure. Differential scanning calorimetric (DGA) measurements revealed that NaCl solvation exhibits the salt-out and NaI solvation the salt-in effect. Increasing NaCl concentration lowers the temperature of the C_p peak but raises the peak intensity in comparison to NaI solvation that does contrastingly on the evolution of the C_p peak temperature and its intensity.

An IR absorption spectroscopic investigation [31] revealed that NaCl addition has the same effect to heating on shifting the ω_H from 3200 to 3450 cm^{-1} and shifting the ω_{B1} for the $\angle$O:H–O bending vibration from 600 to 470 cm^{-1} for salting and to 530 cm^{-1} for heating. The ω_{B1} absorbance at 470 cm^{-1} for salting is about 25% that for heating, evidencing the salting enhanced polarization and structure ordering to limit the IR absorbance. The contact angle measurements further confirmed the salt ionic polarization raises but thermal fluctuation depresses the liquid surface stress. As an indicator of surface stress, the contact angle increases proportionally to the extent of polarization and decreases with thermal fluctuation or network disruption by H↔H formation.

However, little attention had been paid to the joint effect of salt solvation, confinement, and mechanical compression at the ambient temperature [25] or the joint effect of salt solvation and heating under the ambient pressure [31] or heating on the O:H–O bond cooperative relaxation. In 2014, Zhang et al. [31] resolved the joint effect of NaCl solvating and heating using IR absorption. Subsequently, Zeng et al. [25, 32] systematically examined the P_C for mechanical icing of the Na(F, Cl, Br, I) solutions and the concentration dependence of NaI/H$_2$O solution over the full frequency range for the Liquid-VI-VII phase transition with multiple findings. These exercises clarify how the O:H–O bond responds to the joint effect of charge injection by solvation, thermal excitation, and mechanical compression, and how the solute-solute interaction mediate the O–O repulsion and the critical pressure and temperature for phase transition.

This section shows that heating has the same, but mechanical compression has an opposite effect of salt solvation, heating and confinement on the O:H–O bond relaxation. However, heating enhances the thermal fluctuation to depress the solution surface stress, but salt salvation and geometric confinement do it contrastingly by polarization. Phase transition requires excessive energy to recover the salvation- or confined-elongated O:H–O bond by raising the P_{C1} and P_{C2} simultaneously and the ΔP_C changes with the solute type in the order of I > Br > Cl > F ~ 0. However, NaI concentration performs differently with solute type because of the anion-anion repulsion comes into play at higher concentrations [33]. The P_{C1} grows faster than the P_{C2}. The P_{C1} proceeds along the P-T path of the Liquid-VI boundary and the P_{C2} along the VI-VII boundary. P_{C1} and P_{C2} then merge at the liquid-VI-VII triple-phase junction of 3.3 GPa and 350 K.

8.2 P-T-E Joint Effect on Bond Energy

8.2.1 Compression, Confinement, and Electrification

Electrification means that the electric fields of ions align, cluster, stretch, and polarize the surrounding H$_2$O molecular dipoles to form hydration shells without charge sharing between the solute and the solvent molecules, which is equivalent to applying an E field of regularly by dispersed point charges. The electrification stretches the H$_2$O dipoles by elongating the O:H nonbond and softens its phonon and relaxes the H–O bond contrastingly because of the O:H–O bond cooperativity. Therefore, solute electrification and solution confinement reduce the molecular size and enlarges their separations with an association of strong polarization of the nonbonding electron lone pairs, which forms the supersolid phase to be responsible for the skin hydrophobicity, viscoelasticity, stress, and solubility, etc. [24].

Mechanical icing of salt solutions proceeds in two steps [25]. Firstly, salt solvation stores energy into the H–O bond by contraction and the amount of energy storage varies with the solute type and concentration. Secondly, mechanical compression then has to reverse the relaxed O:H–O bond to the right O:H and H–O energies of pure water for the phase transition [14]. Compression stores energy to the O:H nonbond and release energy from the initially contracted H–O by H–O elongation. Moving along the pressure path at 298 K in the phase diagram, one will go through phases of Liquid, ice VI, ice VII, and cross their boundaries, towards phase X at even higher pressures. For the deionized water, the Liquid-VI and the VI-VII phase transitions occur at pressure dropping from 1.33 to 1.14 (1.33→1.14) GPa and at 2.23→2.17 GPa, respectively. The pressure sharp falls as indicated by "→" at the phase boundary results from the structural relaxation and O–O repulsion attenuation [25, 27, 34].

8.2.2 T_{xC} and P_{Cx} Versus Bond Energy

O:H–O bond relaxation and electrostatic polarization dictate the detectable quantities such as phonon frequencies ω_x, O 1s binding energy shift, and the T_{xC} and P_{Cx} for the phase transition denoted order x, as well as macroscopic properties such as hydrophobicity, toughness, slipperiness, viscoelasticity [24]. Quantities of immediate concern are the ω_x and the T_{xC} [24]:

$$\begin{cases} \omega_x \propto \sqrt{E_x/\mu_x}/d_x \\ T_{xC} \propto \sum_{x=H,L} E_{xC} \end{cases} \tag{8.1}$$

The O:H–O bond cooperativity means that if one segment becomes shorter, it will be stiffer, and its characteristic phonon undergoes a blue shift; the other segment of the O:H–O bond relaxes contrastingly. The ω_x depends intrinsically on the reduced mass μ_x, segmental length d_x and energy E_x. The d_x and E_x depend functionally on perturbation such as molecular undercoordination, etc. These sequential correlations ensure the advantage of the DPS in direct estimating the segmental length and energy from measurements.

The P_{Cx} and the T_C for a phase transition is correlated to the O:H–O bond energy E_{xC} (X = L, H), for instance:

$$\begin{aligned} T_C &\propto \sum_{L,H} E_{xC} \\ &= \begin{cases} \sum_{L,H} \left(E_{x0} - s_x \int_{P_0}^{P_{C0}} p\frac{dd_x}{dp} dp \right) & \text{(a, neat } H_2O) \\ \sum_{L,H} \left(E_x - s_x \int_{P_0}^{P_C} p\frac{dd_x}{dp} dp \right) & \text{(b, solution)} \end{cases} \end{aligned} \tag{8.2}$$

$E_{H0} < E_H$ means that the H–O bond for pure water is weaker than it is in the salt solution under the same pressure (ambient $P_0 = 100$ kPa ~ 0) [25, 32]. The integrals are energies stored into the bonds by compression. The summation is over both segments of the O:H–O bond. When the pressure increases from P_0 to the P_{C0} for neat water and to the P_C for the salted, phase transition occurs. At transition, the bond energy equals to the difference between the two terms in the bracket. The change of the cross-section s_x of the specific bond of d_x length is assumed insignificant at relaxation with ignorance of the Poisson ratio. To raise the critical pressure from P_{C0} to P_C at the same critical temperature ($\Delta T_C = 0$ in the present situation), one needs excessive energy which is the difference between the bond energy of the salted and the neat water, therefore, (8.2) yields,

$$\Delta E_x - s_x \left(\int_{P_0}^{P_C} p \frac{dd_x}{dp} dp - \int_{P_0}^{P_{C0}} p \frac{dd_x}{dp} dp \right) = 0$$

where $\Delta E_x = E_x - E_{x0}$ is the energy stored into the O:H–O bond by salt solvation; compression from P_{C0} to P_C recovers the salt-induced distortion, and then phase transition occurs. This expression indicates that, if one wants to overcome the effect of solute electrification, or molecular undercoordination, on the bond distortion for phase transition, one should increase the pressure from P_{C0} to P_C under the same critical temperature, say at 298 K, presently. The ΔE_x varies with the solute type and concentration in different manners, as the solute-solute interaction changes with solute concentration. The ΔE_x also changes with the feature size of confinement by hydrophobic interface.

The segmental length d_x varies with pressure in trends of [14],

$$\frac{dd_L}{dp} < 0 \quad \frac{dd_H}{dp} > 0 \quad \text{and} \quad P_C > P_{C0},$$

which defines uniquely the segmental deformation energies derived by salt solvation,

$$\Delta E_x = s_x \left(\int_{P_0}^{P_C} p \frac{dd_x}{dp} dp - \int_{P_0}^{P_{C0}} p \frac{dd_x}{dp} dp \right) \begin{cases} > 0 \ (H{-}O) \\ < 0 \ (O{:}H) \end{cases} \qquad (8.3)$$

Indeed, ionic polarization strengthens the H–O bond but weakens the O:H non-bond. The resultant of the ΔE_L loss and the ΔE_H gain governs the ΔP_C for the phase transition of the aqueous solutions. Figure 8.1 shows the energy stored in the two segments of the O:H–O bond by salt solvation that raises the critical pressure from P_{C0} to P_C. The blue-shaded zone stands for the O:H energy gain but the green-shaded area corresponds to the H–O energy loss under compression of a salt solution.

8.3   Compression Icing of Salt Solutions

8.3.1   Hofmeister P_{Cx} for NaX/H$_2$O Icing

High-pressure Raman examination of the NaX and concentrated Na/H$_2$O Liquid-VI and the VI-VII phase transition confirmed consistently the above predictions [25]. Figures 8.2 and 8.3 show the H$_2$O and typical NaI/H$_2$O phonon cooperative relaxation at 298 K. Table 8.1 records the variation of the P_{Cx} for each type of the solutions. The abrupt changes of the spectral shapes correspond to the P_{Cx} for phase transition with structural relaxation. The P_C sharp fall at phase boundary indicates that the structural

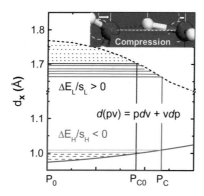

Fig. 8.1 Compression stores energy into water by shortening the O:H and lengthening the H–O bond (dark and red shaded areas) for phase transition at P_{C0}. Compression a salt solution into ice requires excessive energy ΔE_H (<0, green shaded) and ΔE_L (>0, blue shaded) to recover the initially deformed H–O and O:H requires additional pressure P_C—from P_{C0} (reprinted with permission from [25])

relaxation weakens the O–O repulsion. Raman spectra for the room-temperature salt solution Liquid-VI and then VI-VII icing revealed the following [25]:

(1) Three pressure zones from low to high correspond to phases of Liquid, ice VI, and ice VII, toward phase X [24].
(2) Compression shortens the O:H nonbond and stiffens its ω_L phonon but does the opposite to the H–O bond over the full pressure range except for the P_{Cx} at phase boundaries.
(3) At transition, the gauged pressure drops, resulting from geometric restructuring that weakens the O–O repulsion; both the O:H and the H–O contract abruptly when cross the phase boundaries.
(4) Most strikingly, the P_{Cx} increases with the anion radius or the electronegativity difference between Na and X, following the Hofmeister series order: $I > Br > Cl > F \approx 0$.

8.3.2 NaI Concentration Resolved P_{Cx}

Figures 8.4 and 8.5 show the concentration dependence of the P_{Cx} for the Liquid-VI and VI-VII transition for the concentrated NaI/H$_2$O solutions at 298 K. Results show consistently that compression shortens the O:H nonbond and stiffens its phonons but the H–O bond responds to pressure contrastingly throughout the course unless at the phased boundaries. At higher concentrations, say 0.05 and 0.10, the skin 3450 cm^{-1} mode are more active in responding to pressure, which evidence the preferential skin occupancy of the I$^-$ anions that enhances the local electric filed. The P_{Cx} abrupcy

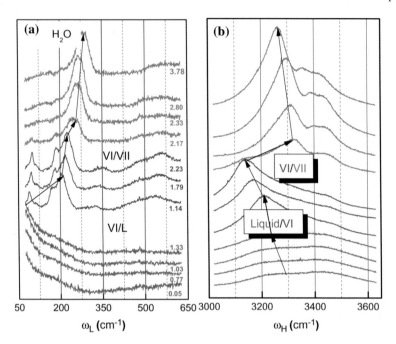

Fig. 8.2 a O:H and **b** H–O phonon cooperative relaxation of pure H_2O under 298 K mechanical compression. Liquid-VI and VI-VII phase transition takes place at P_C associated with fine shifts and pressure sharp falls at the phase boundaries (reprinted with copyright permission from [25])

Table 8.1 O:H–O bond relaxation dynamics of the NaX solutions transiting from Liquid to VI and then to VII phases under compression and ambient temperature

		H_2O	NaF	NaCl	NaBr	NaI
$\Delta\eta$ ($\eta_{Na} = 0.9$)		–	3.1	2.1	1.9	1.6
R ($R_{Na+} = 0.98$ Å)		–	1.33	1.81	1.96	2.20
Liquid	$\Delta\omega_L$	>0				
VI	$\Delta\omega_H$	<0				
VII						
L→VI	P_{C1}	1.33→1.14	1.45→1.13	1.59→1.36	1.56→1.51	1.94→1.74
boundary	$\Delta\omega_L$	>0				
	$\Delta\omega_H$	<0				
VI→VII	P_{C2}	2.23→2.17	2.22→2.06	2.35→2.07	2.79→2.71	3.27→2.98
boundary	$\Delta\omega_L$	>0				
	$\Delta\omega_H$	>0				

η is the elemental electronegativity

The critical pressures P_{C1} and P_{C2} vary with the solute type in the Hofmeister series order. An anion of larger radius and lower electronegativity raises more the P_{Cx}

The pressure abruption at transition indicates the weakening of the inter-oxygen Coulomb repulsion, shortening the O:H and the H–O spontaneously at phase boundaries

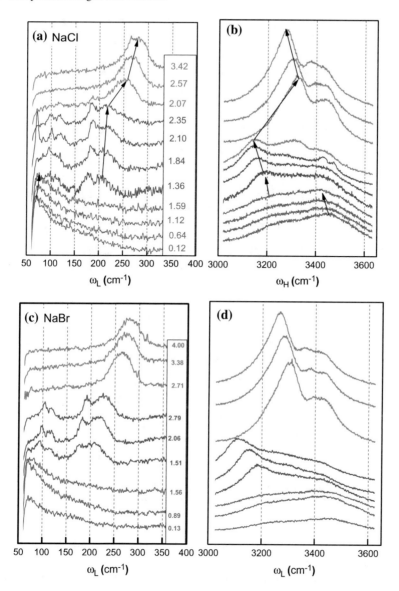

Fig. 8.3 O:H–O phonon cooperative relaxation of the (**a, b**) NaCl/H$_2$O and (**c, d**) NaBr/H$_2$O solutions of 0.016 molar concentration under room-temperature mechanical compression. Liquid-VI and VI-VII phase transitions occur at P$_C$ associated with abruptions at the phase boundaries (reprinted with permission from [25])

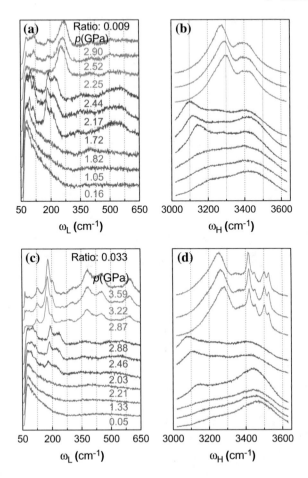

Fig. 8.4 (**a, c**) O:H and (**b, d**) H–O phonon frequency shift of the concentrated NaI solutions and the critical pressures at 298 K (reprinted with permission from [32])

at phase boundaries shows the weakening of the O–O repulsion due to structure transition from one phase to the other.

The high-pressure Raman spectra from the concentrated NaI/H$_2$O solutions revealed the following:

(1) The P$_{C1}$ for the Liquid-VI transition increases faster than the P$_{C2}$ with NaI concentration till its maximum at 3.0 GPa and 0.10 concentration. The P$_{C1}$ and P$_{C2}$ approach and eventually meet at the triple phase junction at 3.3 GPa and 350 K.

(2) The P$_{C2}$ for VI-VII transition changes insignificantly with concentration, keeping almost at the VI-VII boundary in the phase diagram, which contrasts with the trend of solution type.

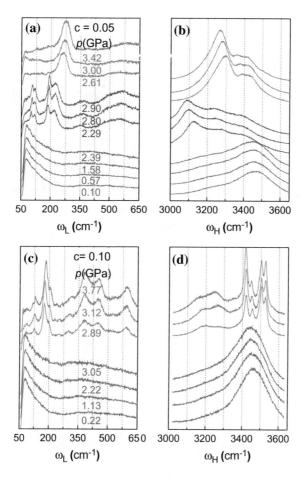

Fig. 8.5 Spectral transition from the (**a, b**) Liquid-VI and VI-VII series to the (**c, d**) Liquid-VI-VII triple phase junction when the NaI/H$_2$O concentration reaches 0.1 molar ratio (reprinted with permission from [32])

(3) The concentration trend of P_{C1} along the L-VI boundary is equivalent to the simultaneous compressing and heating in the phase diagram.

The discrepancy between the solute type and the concentration on the critical pressures pathway for phase transition arises from the involvement of anion-anion repulsion that weakens the electric field at higher concentrations. Solute type determines the nature and the extent of the initial electrification; concentration increase reduces the local electric field and the extent of the initial H–O bond energy storage.

Furthermore, the P_{C2} is less sensitive than the P_{C1} to the change of solute concentration. One can imagine that the highly-compressed O:H–O bond is less sensitive to the local electric field of the hydration shells. They could be harder to deform further than those less-deformed under the same pressure.

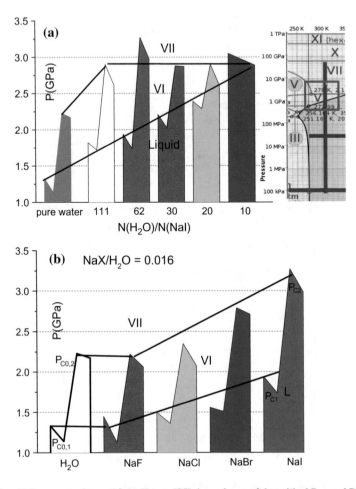

Fig. 8.6 **a** NaI concentration and **b** NaX type [25] dependence of the critical P_{C1} and P_{C2} for the Liquid-VI and the VI-VII phase transition. Inset a shows the water phase diagram with the pressure path at 298 K. Indicated also the pressure path along the Liquid-VI boundary for the concentrated NaI solutions to the triple phase joint at 3.3 GPa and 350 K. Solute capabilities for raising the critical pressures at 0.016 molar ratio follow the Hofmeister order of I > Br > Cl > F = 0. The involvement of the anion-anion repulsion discriminates the solute concentration from the type effect. The L-VI boundary is more sensitive than the VI-VII boundary to the solute concentration change

One may note that, the O:H–O bond is very sensitive to the environment such as pressure holding time, temperature, and phase precipitation, the measurement may not be readily reproducible, but the trends of measurements and the physical origin retain. The maximal P_{C2}, 3.27 > 3.05 > 2.23 GPa, for the 0.016 NaI/H_2O solution [25], the 0.10 NaI/H_2O, and the deionize water, demonstrates clearly the Hofmeister effect, the presence of the anion-anion repulsion, on the critical pressures for the room-temperature phase transition, as compared in Fig. 8.6 and Table 8.2.

Table 8.2 NaI/H$_2$O concentration dependence of the critical pressures for transiting Liquid into phase VI and then to phase VII under compression (unit in GPa)

	0	0.009	0.016	0.033	0.050	0.100
H$_2$O/NaI	–	111	62	30	20	10
P$_{C1}$	1.33→1.14	1.82→1.72	1.94→1.74	2.21→2.03	2.39→2.29	3.05→2.89
P$_{C2}$	2.23→2.17	2.90→2.61	3.27→2.98	2.88→2.87	2.90→2.61	

Arrows denote the critical pressure drops when transition occurs due to structure relaxation. See Fig. 8.4 for example

8.3.3 O:H–O Bond Length and Phonon Relaxation

Figure 8.7a shows that the ω_H phonon in ice and LiCl solution undergoes redshift under compression [28]. The extent of the H–O bond softening varies with its sites in the solution. The H–O stiffness of the solution relaxes less than it is in pure ice [35, 36], because of its initial contraction by ionic polarization [25, 32]. The initially shorter and stiffer H–O bonds under polarization are hardly compressible than the H–O bonds in pure water and ice.

Figure 8.7b compares the mean O–O distance (O:H–O length) in pure ices, in LiCl- and NaCl-contained ices [29]. The O:H–O in the hydration shells of Li$^+$ and Na$^+$ is always longer than it is in the pure ice. The close situations for the pure ice, LiCl- and NaCl-contained ice suggest these signals arise from those O:H–O bonds outside the hydration shells. The O:H–O segmental length symmetrization could hardly happen even at 180 GPa for the O:H–O bonds with the hydration shells, in contrasting to the O:H–O symmetrization at 60 GPa pressure [14].

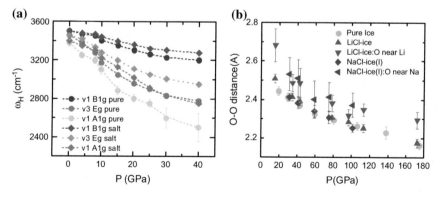

Fig. 8.7 **a** Raman H–O stretching mode in pure and LiCl-contained ices (0.08 mol concentration [28]) and **b** the mean O–O distance in ices (squares), LiCl-contained ices (solid circles), and in NaCl ices (solid triangles). ν1 B1 g > ν3 Eg > ν1 A1 g features the components of the H–O Raman spectral peak. Open circles and triangles correspond to the mean O–O distances near the cation (reprinted with permission from [29])

These observations confirm the opposite effect of mechanical compression and ionic polarization on the O:H–O bond cooperative relaxation. Compression shortens and stiffens the O:H nonbond while lengthens and softens the H–O bond, but the ionic polarization does it oppositely in salt solutions.

8.3.4 Salt Impact Freezing: Impact and Immersion

It is amazing that freezing more likely when the ionic catalyst impinges into or collides with the air–water interface. Much research in contact freezing has been motivated by its relevance for ice initiation in Earth's atmosphere and the snow ice-quasisolid transition [37]. Silver iodide, sand, and clay can trigger freezing at a higher temperature in the collision mode than in the immersion mode. Collision of salt and sugar can initiate freezing at -11.0 and -13.5 °C, respectively [38], which is slightly higher than pure water freezing at -15.0 °C. Niehaus and Cantrell [37] demonstrated that collisions of six soluble substances (KCl, KI, NaCl, NaI, NaOH, and KOH) with moderately supercooled water raise the freezing temperatures within 10 ns time scale, as shown in Table 8.3, rather than at -34 °C for homogeneous ice nucleation of supercooled water. They examined the effect of particle size and density, impact velocity (~3 m/s), and collision kinetic energy on the temperature of freezing and suggested that the freezing behavior depends on the collision itself, as the freezing catalysts would modulate the freezing point upon dissolving into the bulk. In other words, the collision freezing nuclei needs not be effective as immersing the salt into the solution.

The kinetic energy from a mechanical impulse at contact is identified as the key factor in reducing the energy barrier for nucleation, enhancing the probability of a phase transition [37].

Salt impinging into water induces cooling in the surrounding liquid as heat is absorbed as the bonds within the salt are broken and the resulting ions hydrated. If the water is cooled below the eutectic point for the water-salt system, freezing is possible with the salt as a substrate. The eutectic point in a phase diagram indicates the chemical composition and temperature corresponding to the lowest T_m of a mixture of components. Alternatively, water may be cooled below its homogeneous freezing

Table 8.3 Threshold temperatures for the initial freezing (T_0), 80% freezing (T_{80}), and the eutectic temperature ($T_{eutectic}$) of the alkali salts tested for collision freezing activity [37]

Salt	T_0 ($-$°C)	T_{80} ($-$°C)	$T_{eutecic}$ ($-$°C)
NaI	7	13	31.5
KI	8	12	23.2
NaOH	11	15	28.0
KOH	11	15	62.8
NaCl	12	15	21.2
KCl	12	13	10.8

limit, inducing freezing before the ions from the dissolving crystal have diffused into the region that has been supercooled to that point.

Multiple mechanisms are available to describe the collision freezing. The typical ones include subcritical ice embryos adsorbed to the surface of incoming particles [39]; momentary reduction in the energy barrier between water and ice as a result of the heat of wetting [40]. The collision freezing is also explained as intrinsic reduction in the free energy barrier at a three phase contact line [41, 42], and the slat endothermic dissolution in the atmosphere [43].

Solute electrification plays its role in both collision and immersion modes to modulate the T_N of the solution. The O:H bond energy determines the critical T_N for transforming the quasisolid supercooling phase to homogenous ice, or the inverse. Mechanical collision and ionic electrification relax the O:H–O bond in opposite manners. Collision provides an impulse having the same effect of compression to shortens and stiffen the O:H nonbond momently, causing instant icing at elevated T_N; ionic polarization elongates and softens the O:H nonbond. These events modulate the T_N oppositely—collision raises but electrification depresses the T_N. In fact, the salt particle collision freezing combines discrete steps of mechanical icing [34] and the Hofmeister solute electrification [31]:

(1) First, collision of the speedy salt particle with the supercooling quasisolid droplet provides impulse that promotes local transformation of quasisolid water into ice. This process is momently or temporally pressure P dependent. The impulse $f\Delta t = \Delta P$ and $P = f/S$ with f being the force acting on the S area of the droplet surface for Δt period.

(2) Secondly, when the salt particle is dissolved, ion electrification effect comes into play, which lowers the T_N and raises the T_m. This process is solute type and concentration dependent.

(3) The competition of mechanical collision and ion electrification matters the QS boundaries and the sequence order of these two stimuli matters the freezing differently.

Figure 8.8 illustrates two examples of collision regelation. Supersolid nanocrystals turn into ice by collision and supercooled water turns into ice at the ambient. Ice forms preferentially on the upper front of the positively curved surface of wings for the fast-moving airplane. At a height of tens of thousands of meters, the temperature is about $-40\ ^\circ$C or even lower. Water molecules form the supercooled cloud or fog clusters in the supersolid states at a freezing temperature lower than $-100\ ^\circ$C. Molecular undercoordination and low pressure share the same effect on O:H–O bond relaxation, dispersing outwardly the QS boundary. However, the supersolid nanocrystals colliding inelastically with the upper front of the fast-moving wings provide impact that lowers the dew point and the T_N of the accumulated crystals, the supersolid crystals then turn to be bulk ice. The competition of the molecular undercoordination, lowered pressure, and the heavy impact determines the T_N for the phase transition. No such phenomenon could be seen from the bottom flat surface of the wings, as no collision takes place.

It is often to observe that clapping the bottle filled with supercooled water can turn the water into ice. The gentle contact between the supercooled water and ice provides an impact that shifts the QS boundary inwardly to raise the T_N.

8.4 Ionic Polarization and Site-Resolved Thermal Excitation

Figure 8.9 compares the FTIR ω_x DPS of concentrated NaCl/H$_2$O solutions with the DPS for water under thermal excitation [31]. The FTIR absorption covers the wavenumbers from 400 to 3800 cm^{-1} show more bulk information than Raman scattering does. Results show that the ω_{B1} (~600 cm^{-1} corresponding to the $\angle$O:H–O bending mode) undergoes redshift and the ω_H (3150–3450 cm^{-1}) blue shift upon being heated or NaCl solvated. The ω_H shifts from 3170 gradually to the skin value of 3450 cm^{-1} at 25 mass% salting [44–47] and heating to 328 K. The latter is companied with a shoulder at 3650 cm^{-1} that corresponds to the H–O dangling bond under heating [48]. The broader H–O phonon peak FWHM results from thermal fluctuation.

Meanwhile, salting softens the ω_{B1} from 600 to 520 cm^{-1} but heating softens the ω_{B1} from to 470 cm^{-1}. A 50 cm^{-1} difference in the ω_{B1} shift indicates the O:H is softer at heating than it is under polarization. The ω_{B1} absorption peak intensity for salted water is much weaker than that of the heated water, which indicates that the salted water molecules are in the higher structure order. The polarized and ordered structure reflects more of the incident light so the absorbance is lower than it is by the heated water in less molecular structure order. The higher structure order of the salted water also gives narrower linewidth of the ω_H peak.

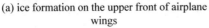

(a) ice formation on the upper front of airplane wings   (b) Ice formation of supercooled water

Fig. 8.8 QS boundary dispersion by mechanical perturbation **a** Supersolid nanocrystals transit into ice on the upper front of the fast-moving airplane's wings at lower temperatures and **b** supercooled water transits into ice by the ignorable, mechanical packing impulse under the ambient temperature (public domain)

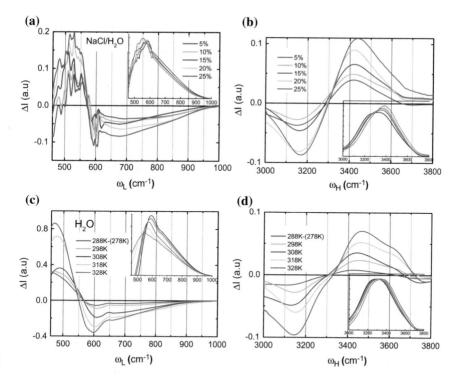

Fig. 8.9 FTIR DPS for the (**a, c**) ω_{B1} and (**b, d**) the ω_H modes of water as a function of (**a, b**) salinity (mass% NaCl) and (**c, d**) heating temperature with respect to the referential spectrum collected from deionized water at 278 K. Insets show the raw spectra. Heating shifts the ω_{B1} from 600 to 470 cm^{-1} while salting shifts the ω_{B1} to 520 cm^{-1}. Heating and salting shift the ω_H from 3150 to the same 3450 cm^{-1} but heating creases a shoulder at 3650 cm^{-1} corresponding to the skin H–O bond thermal fluctuation

Phonon softening/stiffening indicates bond stiffness loss/gain. NaCl salting and water heating effect the same on the ω_x relaxation despite the 50 cm^{-1} $\Delta\omega_{B1}$ difference, which resolve the polarization effect on the O:H length and $\angle$O:H–O bending. This trend agreement indicates that salt solvation indeed modulates the Coulomb repulsion between ions [25]; however, the O:H–O bond in liquid water is subject to the segmental specific heat disparity which specifies that the O:H nonbond undergoes thermal expansion and the H–O bond thermal contraction in the liquid phase [49]. Heating and salting have the similar effect on the O:H–O bond cooperative relaxation but completely different mechanisms behind the similar observations.

Figure 8.10 compares the full-frequency Raman spectra for deionized water and 0.1 NaI/H$_2$O solution heated from 278 to 358 K. Indeed, salting enhances the heating effect on the ω_x shift. The ω_H is decomposed into three components of bulk (3200 cm^{-1}), hydration shell (~3500 cm^{-1}), and the H–O dangling bond (3610 cm^{-1}). The supersolid hydration shell component overlaps the supersolid skin of the ordinary water at 3450 cm^{-1}.

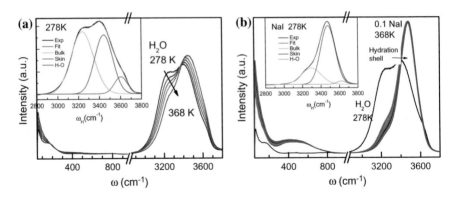

Fig. 8.10 Full-frequency Raman spectra for **a** the deionized water and **d** 0.10 NaI/H₂O solution heated from 278 to 368 K with respect to the spectrum collected from 278 K water. Decomposition of the ωH spectra into the bulk, skin/hydration-shell, and H–O dangling bond components shown in the insets (**c, d**) manifests that effects of heating and salting enhance each other on the blue shift of the H–O phonon frequency and abundance

One can examine the thermal stability of the three components of the ω_H peak upon heating and salting. Examination of the thermal stability of the H–O bond [48] by heating from 5 to 95 °C revealed that heating stiffens the skin H–O phonon component from 3443 cm^{-1} by 0.38%, and the bulk phonon from 3239 cm^{-1} by 1.48%, but softens the H–O dangling bonds from 3604 cm^{-1} by −0.40%. Heating also varies the respective molecular structural order and phonon abundance by −8.4/−19.7%, 4.9/−2.6%, and 13.0/137.0% for the bulk, the skin, and the dangling bond components, respectively. Observations also evidence that heating has the same effect of molecular undercoordination on O:H–O bond relaxation but opposite on the structure order and surface stress.

The convoluted ω_x DPS in Fig. 8.11 shows that the ω_x undergoes opposite shift because of H–O contraction and O:H elongation by both heating and salting. NaI addition turns a fraction of the ordinary bulk water into the ionic hydration shells. The FWHM for the hydration shell is narrower than that of the bulk component because of the polarization-slowed molecular dynamics [31] in the hydration shells. Heating enhances the solute electrification effect by a different mechanism—O:H elongation and H–O contraction because of the O:H–O bond segmental specific heat disparity [49].

Figure 8.12 refines the molecular-site and temperature resolved O:H–O bond relaxation in terms of H–O bond stiffness, structural order, and phonon abundance in comparison to those for deionized water. Table 8.4 summarizes the observations. While the abundance of the hydration shells undergoes net gain and the H–O dangling bond net loss, the ordinary bulk component evolves to symmetrical gain and loss at 318 K and above. This observation results from the effect of heating that enhances the blue shift of the H–O phonons for the bulk mode. However, the H–O free radicals and the bulk feature change a little. The skin of water and the ionic hydration shell

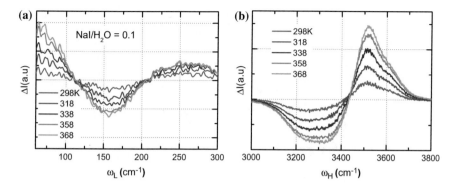

Fig. 8.11 Temperature dependence of the convoluted (**a, b**) ω_x DPS for 0.1 NaI/H$_2$O solutions with respect to the spectrum for the solution measured at 278 K. Heating asymmetrizes the peak shape towards higher frequencies (reprinted with permission from [48])

are thermally extraordinarily stable due to their supersolidity nature [31, 48]. The change of the H–O radical is almost negligible. The abundance gain may not equal its loss for a particular component, but the overall transition of the H–O phonons should conserve, as the bonds transits from one component to another upon excitation.

Figure 8.13 shows the NaI concentration and temperature dependent contact angle for the NaI/H$_2$O solutions on glass substrate. Heating lowers but salting raises, the contact-angle though both relax the ω_x phonon in the same manner. Heating softens the O:H nonbond through thermal fluctuation, which shifts the $\Delta\omega_{B1}$ to 470 cm^{-1} but slating does the same through polarization shifting the $\Delta\omega_{B1}$ to 520 cm^{-1}. Therefore, the DPS is very powerful in resolving the effect of ionic polarization and thermal fluctuation on the phonon frequency relaxation and the fluctuation and polarization on the surface stress of solutions and liquid water [44].

8.5 Multifield P$_C$ and T$_C$ Mediation Ice-QS Transition

8.5.1 *QS Phase Boundary Mediation*

Table 8.5 features the multifield effect on the O:H–O segmental length and stiffness, Debye temperatures, quasisolid (QS) phase boundary (close to T$_m$ and T$_N$) dispersion. The H–O energy E$_H$ governs the T$_m$ and the O:H energy E$_L$ dictates the T$_N$. Compression and QS heating share the same effect, and molecular undercoordination and liquid heating share the contrasting effect on O:H–O bond relaxation and QS phase boundary dispersion. These regulations aid one to understand the confinement-compression modulated pressures and temperatures for the Solid/QS and QS/Liquid transition of water and ice.

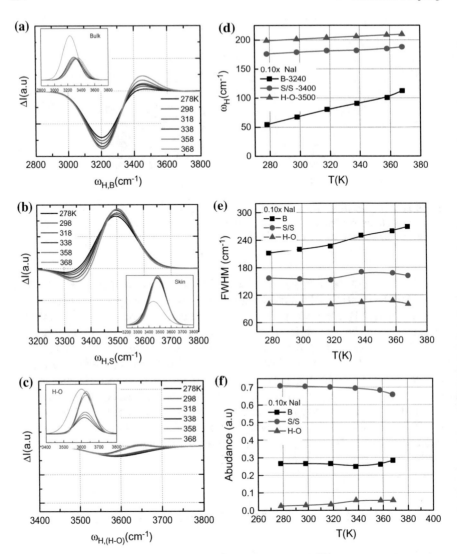

Fig. 8.12 NaI/H$_2$O solution molecular-site and temperature dependent phonon relaxation dynamics (**a–c**) in terms of (**d**) H–O bond stiffness, (**e**) molecular fluctuation order, and (**f**) phonon abundance. B represents for the bulk, S/S for the skin and hydration shell, and H–O the dangling bond radicals (reprinted with permission form [48])

Table 8.4 Molecular-site and temperature resolved O:H–O bond relaxation dynamics in terms of H–O bond stiffness, molecular fluctuation, and phonon abundance (reprinted with permission form [48])

		Bulk	Skin	H–O radical	Notes
ω$_H$ (cm^{-1})	278 K water	3239.0	3442.6	3603.0	• Joint heating and salting neutralize the heating softening and salting stiffening of the H–O radicals and makes them stable
	278 K NaI	3294.6	3476.0	3618.7	• The H–O stiffness in the hydration shell is thermally more stable than that in the water skin
	298	3307.4	3478.9	3621.2	• The bulk H–O stiffness increases abruptly at 320 K due to the joint effect
	318	3320.4	3481.6	3623.8	
	328	3333.8	3483.3	3624.7	
	338	3330.6	3481.9	3626.0	
	355	3340.6	3485.0	3628.6	
	368	3352.5	3488.0	3629.7	
FWHM (cm^{-1})	278 K water	217.1	167.5	121.8	Molecules in the hydration shells are less mobile than water skin and insensitive to temperature
	278 K NaI	212.0	157.0	100.2	• At 320 K and above, the bulk molecules become more active but the H–O less mobile
	298	220.0	155.0	98.7	
	318	226.7	152.7	99.6	
	328	236.4	150.6	102.3	
	338	250.4	170.4	104.4	
	358	259.9	167.9	107.9	
	368	269.2	162.2	100.2	
Abundance	278 K water	0.532	0.385	0.083	• At ≥320 K, the shell reduces due to thermal fluctuation • The abundance of H–O radicals drops slightly when heated

(continued)

Table 8.4 (continued)

	Bulk	Skin	H–O radical	Notes
278 K NaI	0.267	0.708	0.025	
298	0.266	0.705	0.029	
318	0.266	0.701	0.033	
328	0.283	0.681	0.036	
338	0.249	0.695	0.056	
358	0.262	0.683	0.055	
368	0.284	0.660	0.056	

Table 8.5 O:H–O segmental cooperative relaxation in length, vibration frequency, Debye temperature and the critical temperatures for phase transition with respect to $d_{L0} = 1.6946$ Å, $d_{H0} = 1.0004$ Å, $\omega_{H0} = 3200$ cm^{-1}, $\omega_{L0} = 200$ cm^{-1} at 277 K [24, 50]

	Δd_H	$\Delta\omega_H \propto \Delta\Theta_{DH}$	$\Delta E_H \propto \Delta T_m$	Δd_L	$\Delta\omega_L \propto \Delta\Theta_{DL}$	$\Delta E_L \propto \Delta T_N$	References
Compression	>0	<0		<0	>0 QS boundary disperses inwardly (Fig. 8.15d)		[25]
QS heating							[49]
Liquid heating	<0	>0		>0	<0 QS boundary disperses outwardly		
Undercoordination							[51]

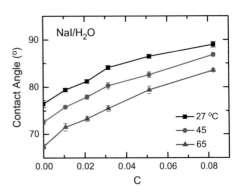

8.5.2 Regelation: Ice-QS and Ice-Vapor Transition

Prior to discussing the room-temperature solution compression icing, let us look at the QS phase boundary inward depression by compression—Regelation—lowered T_m but raised T_N. Using laser confocal microscopy and Michelson interferometry, Chen and co-workers [52] examined the QS (or quasi-liquid, QL) skins of single crystal and polycrystalline ice films. The latter has large amount of grain boundaries and defects. They found that the QS exist stably on the polycrystalline ice even at -16.2 °C and the QS on ice single crystals disappear at temperature lower than -2.4 ± 0.5 °C. In addition, they also found that critical vapor pressure above which the QS can grow is always higher than the solid–vapor equilibrium curve, indicating that the QS can be formed in a metastable state by the deposition of supersaturated water vapor.

The O:H–O specific heat disparity defines, and stimulus perturbation disperses the QS boundary, resulting in the supercooling and superheating phenomenon, as illustrated in Fig. 8.14a. Compression disperses the QS boundary inwardly, resulting in the T_N elevation and the T_M depression. Electric polarization and molecular undercoordination disperses the QS boundary outwardly. The QS/Liquid phase boundary meets the $dT/dP < 0$ ($dd_H/dP > 0$) criterion of H–O relaxation dominance for the T_m (T_{C1}) depression and ice regelation. The QS/Solid and Solid/Vapor Phase boundaries meet the $dT/dP > 0$ ($dd_L/dP < 0$) criterion of O:H relaxation dominance for the critical temperature T_N (T_{C2}) for ice homogenous nucleation and the critical temperature T_V for evaporation [52]. The measured QS/Solid and Vapor/Solid phase boundaries match to both the Solid–Vapor and the Liquid–Vapor equilibrium curves of water ice [53, 54].

Compared with the P-T curve for ice regelation [55]—compression depresses the melting point, T_m, the QS/Solid and Vapor/Solid phase boundaries couple the T-P curve of Liquid-QS transition, as compared in Fig. 8.14b. Compression up to a pressure of 220 MPa lowers the T_m for the Liquid-QS transition of bulk water from 273 to 250 K. This process is fully reversible, known as Regelation—ice melts under compression and freezes again when the pressure is relieved [34].

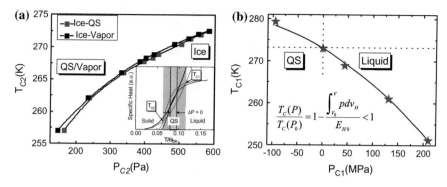

Fig. 8.14 Comparison of the **a** QS/Solid and Vapor/Solid phase boundaries for polycrystalline skin [52] and the **b** Liquid/QS phase boundary for water ice [34]. Inset a illustrates the O:H–O segmental specific heat derived phases and the T_{C1} from melting and T_{C2} for homogeneous ice nucleation. Inset **b** formulates the T_{C1} dependence on the pressure with E_{H0} and V_H being the H–O bond energy and volume. O:H contraction by compression raises the $T_{C2}(T_N)$ and H–O elongation lowers the $T_{C1}(T_m)$ [34]. Electric polarization and molecular undercoordination have the opposite effect of mechanical compression on the T_C shift [56, 57]

8.5.3 Low-Dimensional Ice-QS Transition: $\Delta T_C = 0$

Figure 8.15a–c show that the atomic force microscopy investigation [58] revealed the reversible transition from two-dimensional (2D) ice into a QS phase confined between the hydrophobic graphene and muscovite mica by compression. This observation exemplified the joint effect of compression and molecular undercoordination (called confinement) on the QS boundary dispersion, as illustrated in Fig. 8.15d. Figure 8.16 shows two facts as consequence of compression-undercoordination-QS heating on the P_C and T_N for the Solid-QS transition or homogeneous ice formation:

1. At room temperature, the critical pressure P_C for the confined Solid-QS transition at T_N amounts at 6 GPa that is much higher than the P_C at 1.33 GPa for bulk Liquid-QS transition at T_m and the P_C at 3.5 GPa for NaI solution of 0.1 molar concentration [25, 32].
2. The P_C compensates substrate temperature. The P_C drops as the T increases. The ice and QS phase coexistence line appears at temperature between 293 and 333 K, which is higher than the QS phase between 258 and 277 K for bulk water at the ambient pressure [49]. Heating lengthens the H–O bond in the QS phase and compression always does so without discrimination of phase.

Figure 3.1c, d (see Chap. 3) showed the QS phase boundary defined by the superposition of the O:H and the H–O specific heat curves and the specific-heat defined structural phases of mass density oscillation [49]. Corresponding to the density extremes, or close to the T_m for liquid-QS and T_N for solid-QS phase transition, the QS boundaries are retractable by the Debye temperature and the vibration frequency shift, $\Theta_{Dx} \propto \omega_x$, co correlation.

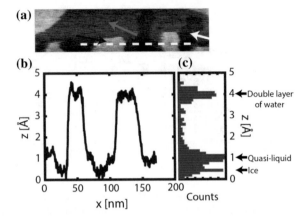

Fig. 8.15 Height distribution of confined water under an external applied pressure higher than 6 GPa. **a** AFM topographic image (230×60 nm^2) of a melted fractal (~ 10 GPa). Three levels are present in the height profiles, namely, (i) the ice layer (denoted with a white arrow), (ii) the quasi-liquid layer (blue arrow), and (iii) the double water layer (black arrow). **b** Cross section across the white dashed line marked in panel (**a**). The height difference between the fractal and the double layer of water is about 3.6 Å. The QS layer is $\sim 70 \pm 5$ pm higher than the ice layer. **c** Histogram of the cross section in (**b**), showing three distinct peaks corresponding to the three different layers (reprinted with copyright permission from [58] and [59])

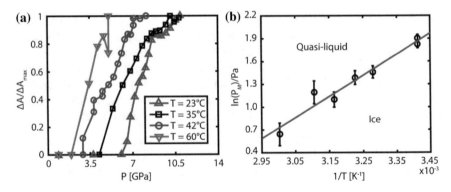

Fig. 8.16 P-T compensation in QS/Solid phase transition of confined water. **a** Ice patch areal change as a function of pressure at different substrate temperatures. At elevated temperatures, the pressure needed to initiate melting decreases. **b** P-T correlation for the QS/Solid ΔT_{C2} transition (reprinted with copyright permission from [58])

Raman spectroscopy measurements of the phase boundaries of water confined within isolated carbon nanotubes of different diameters [60] revealed an exquisite sensitivity to diameter and larger temperature elevations of the freezing transition (by as much as 100 °C) than ever known. The T_m and T_N transitions of the water filled single-walled carbon nanotubes (SWCNTs) were marked by 2–5 cm^{-1} shifts in the radial breathing mode frequency, showing reversible melting bracketed to 105–151 °C and 87–117 °C for 1.05 and 1.06 nm SWCNT, respectively. Phase changes for 1.44 and 1.52 nm CNTs, occur between 15–49 °C and 3–30 °C, respectively, whereas the T_N was observed for the 1.15 nm nanotube between −35 and 10 °C. These observations offered further evidence for the outward dispersion and upward shift of the QS phase by nanoconfinement.

Molecular undercoordination disperses the QS boundary outwardly, which has raised the T_m for a monolayer and the skin of bulk to 325 K [61] and 310 K [44] and lowers the T_N for a 4.4, 3.4, 1.4 and 1.2 nm sized droplet to 242 K [62], 220 K [62], 205 K [63] and 172 K [64], respectively. For a cluster of 18 molecules or less, the T_N is about 120 K [65]. Ionic polarization has an identical effect of molecular undercoordination on the O:H–O bond relaxation and the QS phase dispersion [49]. This fact explains why the Ice-QS phase coexists between 299 and 333 K under compression, instead of 258 and 277 K for bulk water.

According to Table 8.5 regulations, one needs to shorten and stiffen the O:H non-bond by compression to raise the T_N for translating the room-temperature confined ice into the QS phase at 299 K and below, as observed [58]. Compression up to 6 GPa raises the T_N from far below to the room-temperature of confined ice. On the other hand, within the QS phase, the H–O bond follows the regular rule of thermal expansion - heating lengthens the H–O bond and shortens the O:H nonbond, which eases the O:H–O compression. Therefore, QS heating lowers the P_C for the Solid-QS transition.

8.5.4 QS Droplet Compressibility

Figure 8.17 shows the P and T dependence of the compressibility, κ_T, of micrometer sized water droplets in the temperature range of 18–310 K [66]. The pressure is elevated from 1 to 1700 bar (10^5 Pa) for the power-law fitting to the measurements, showed as solid lines. The scattered datum are measurements. One can see from the measurements, the temperature and pressure compensate each other to keep the κ_T as a constant, which means that one wants to have the same κ_T at a higher pressure he must lower the temperature. The situation is the same as presented in Fig. 8.16b where the phase transition from ice to the QS at the boundary is equivalent to the present κ_T. Consistency between Figs. 8.16b and 8.17 indicates the supercooling state is in the QS phase. In the QS phase, cooling shortens the H–O bond and lengthens the O:H nonbond but compression always does contrastingly, so the compression and heating compensate each other to keep the constant κ_T value.

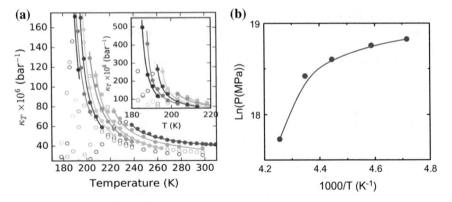

Fig. 8.17 Isothermal compressibility (scattered solid points), κ_T, power-law simulating to pressures of 1, 500, 1000, 1200, 1400, 1500, 1600 and 1700 bar, shown in **a** from right to left solid curves, respectively. The inset shows κ_T at a de-magnified scale for the lower part of the temperature range. **b** P-T compensation at a constant value of $\kappa_T = 60 \times 10^6$ Pa^{-1} in (**a**) (reprinted with permission from [66])

8.5.5   *Water Droplet Freezing*

Researchers have performed experimental studies of pure water droplets in a vacuum generated as a train of droplets in a liquid jet. The droplet does not freeze immediately after the temperature falls below the freezing point, i.e., they are supercooled substantially. Sellberg et al. observed water droplets with diameters of 9–12 μm up to 5 ms after generation by ultrashort X-ray scattering, pointing out that they freeze in the temperature range of 227–232 K as estimated by numerical simulation [67]. Goy et al. reported that a fraction of water droplets with an initial diameter of 6.38 μm stay in the liquid phase down to 230 K [68]. A smaller droplet of Goy et al. should freeze at a lower temperature. The discrepancy exists because the cooling rate should be higher for a smaller droplet due to its large surface/volume ratio and because the homogeneous ice nucleation rate is lower for a smaller one as it is proportional to the volume of the liquid. Ando et al. [69] examined the freezing of droplets of pure water with diameters of 49–71 μm in a vacuum. They measured the fraction of frozen for 200 droplets as a function of time (a freezing curve) using laser scattering at each time. In addition to the experiment, numerical simulation was performed to obtain temporal evolution of the droplet temperature (a cooling curve) to understand the size-dependent freezing time, as shown in Fig. 8.18.

Results show that a larger droplet takes a longer time to be frozen at relatively high temperatures. For the 49.2 μm droplets, the frozen fraction F_{ice} at 7.0 ms was ~5% (~10 out of 200 droplets), whereas it increased rapidly and exceeded 95% at 7.9 ms. For 65.6 μm droplets, F_{ice} exceeded 5% at 9.7 ms, and reached 95% at 10.7 ms. Complete measurement for 71.0 μm droplets was limited by the experimental setup. The largest droplets required the longest time to be frozen. In contrasting, the freezing

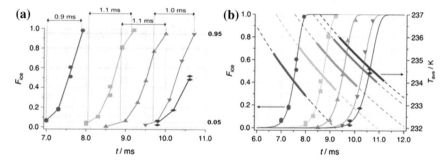

Fig. 8.18 Droplet size dependence of the freezing time and temperature as **a** measured and **b** simulated for, l to r, 49.2(1.0 ms), 58.0(1.2), 60.2(1.1), 65.6(1.0) and 71.0(1.6) μm droplets. The thick lines are the temperature ranges derived from simulation (reprinted with permission from [69])

temperature increases slightly with the diameter of the droplets. The thick solid parts of the cooling curves represent the ranges of the mass-averaged temperature T_{ave}, where F_{ice} increases from 1 to 99%. The range of T_{ave} is raised by about 0.7 K as the diameter of the droplet is increased from 49.2 to 69.4 μm, which indicates the trend that smaller droplets freeze at lower temperatures. These observations further confirm the effect of molecular undercoordination on the QS dispersion and the water droplet freezing - droplets having high fraction of undercoordinated molecules freeze at lower temperature and faster.

8.6 Phonon Abundance-Lifetime-Stiffness Cooperativity

8.6.1 NaBr Solvation and Water Molecular Undercoordination

Using the ultrafast IR spectroscopy, Park and co-workers [70–72] studied the hydrogen bond dynamics of water in NaBr salt solutions and in reverse micelles by measuring the spectral diffusion and relaxation lifetime. The measurements performed on NaBr solutions as a function of concentration revealed that the hydrogen bond dynamics becomes slower as the NaBr concentration increases. The most pronounced change is in the longest time scale dynamics which are related to the global rearrangement of the hydrogen bond structure. Complete hydrogen bond network randomization slows by a factor of 3 in the 6 M NaBr solution compared to that in bulk water. They examined the bond dynamics of water in the nanoscopically confined environments by encapsulating water molecules in ionic head group (AOT) and nonionic head group (Igepal CO 520) reverse micelles as a function of size. The relaxation dynamics deviate significantly from bulk water when the size of the reverse micelles is smaller than several nm and become nonexponential and slower as the size of the reverse micelles decreases. In the smallest reverse micelles, the relaxation is almost

20 times slower than that in bulk water. They confirmed that confinement by an interface to form a nanoscopic water pool is a primary factor governing the dynamics of nanoscopic water rather than the presence of charged groups at the interface.

Ultrafast spectroscopy and MD computations [73, 74] showed consistently that both salt solvation and water confinement not only slow down MD dynamics characterized by the phonon lifetime but also shift the phonon frequency to higher frequencies. However, one may wonder why they happen in the same manner and what intrinsically dictates the chemical and physical properties of the confined water and the salt solution. The answer is the chemical bond [75] and the valence electrons [50, 76].

According to Pauling [75], the nature of the chemical bond bridges the structure and property of a crystal and molecule. Therefore, bond formation and relaxation and the associated energetics, localization, entrapment, and polarization of electrons mediate the macroscopic performance of substance accordingly [76]. O:H–O bond segmental disparity and O–O repulsivity form the soul dictating the extraordinary adaptivity, cooperativity, recoverability, and sensitivity of water and ice [50]. Therefore, one must focus on the hydrogen bond relaxation and electron polarization in the skin region or in the hydration volume and an interplay of the multiscale approaches is necessary.

Salt solvation [33] and molecular undercoordination [51, 77, 78] share the same effect on intramolecular covalent bond contraction, intermolecular nonbond elongation, and charge polarization, associated with high-frequency phonon blueshift and low-frequency phonon redshift. The polarization results in the supersolidity in the ionic solute hydration volume [33, 79] and in the covering sheet of water ice [51, 80]. Higher viscosity, lower diffusivity, slower molecular dynamics, and longer H–O or D–O phonon lifetime do feature the supersolidity.

During the ultrafast IR spectroscopy detection, one switches off/on the pump/probe simultaneously and monitors the population time decay that corresponds to the rate of vibration energy dissipation hindered by the surrounding coordination environment such as defects, impurities, and the energy states of electrons as well as the molecular mobility that is subject to the viscosity of the solution. Ultrafast IR spectroscopy detection [70–72] revealed indeed that 10% NaBr solvation slows the hydrogen bond network relaxation in $5\%D_2O + 95\%H_2O$ solvent by a factor of 3 associated with D–O bond stiffness transition from 2480 to 2560–2580 cm^{-1}. Br$^-$ $\leftrightarrow$ Br$^-$ repulsion weakens the local electric field of the Br$^-$ anion, which softens the DPS from 2580 to 2560 cm^{-1} at higher solute concentrations [81]. The structure order characterized by the DPS linewidth and the transition of phonon abundance increase with the NaBr concentration. However, water molecules confined in the reverse micelles relax even slower than those in the ionic hydration volume associated with blue shift of the D–O phonons from 2480 to 2600 cm^{-1}.

An integration of the DPS shown in Fig. 8.19a and b give rise to the fraction of bonds transiting from the mode of water to polarization in the Na$^+$ and Br$^-$ hydration volume and the skin of the droplets. Figure 8.19c and d compare the solute concentration and the droplet size resolved fraction coefficients [56]:

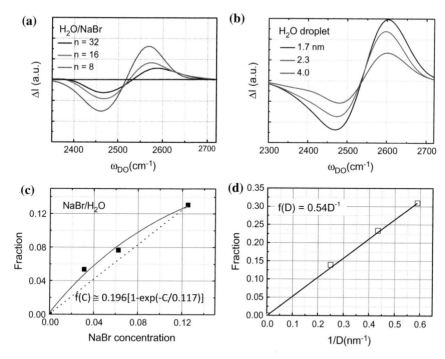

Fig. 8.19 D–O phonon DPS (**a**, **b**) and the fraction coefficients (**c**, **d**) for the concentrated NaBr solutions (**a**, **c**) and (**b**, **d**) the sized water nanodroplets. The fraction coefficient $f_{NaBr}(C) \propto 1-\exp(-C/C_0)$ follows the general form for salt solutions [33] and **d** the $f_{droplet}(D) \propto D^{-1}$ follows universally the skin/body volume ratio for nanostructures [83] (reprinted with permission from [88])

$$f_{NaBr}(C) \propto 0.196\left[1 - \exp(-C/0.117)\right]$$
$$f_{droplet}(D) = 0.54D^{-1}$$

The $f_{NaBr}(C)$ follows the general form for salt solutions [33, 79, 82] and the $f_{droplet}(D)$ follows the general form of the skin/body volume ratio for nanostructures [83].

The $f_{NaBr}(C)$ results from the invariant Na^+ hydration volume and the variant Br^- hydration volume [33]. The Na^+ solute forms a constantly sized hydration droplet without responding to interference of other ions because its hydrating H_2O dipoles fully screen its electric field, which contributes to the linear term of the Jones–Dole viscosity [33, 84]. However, the number inadequacy of the highly ordered hydration H_2O dipoles partially screens the large Br^-. The Br^- then interacts repulsively with other Br^- anions, which weakens its electric field. The $f_{Br}(C)$ approaches saturation at higher solute concentration consequently.

Contrastingly, molecular undercoordination shares the same effect of electronic polarization. The $f_{droplet}(D) = 0.54D^{-1}$ indicates that the H–O bond contraction dictates the performance of nanodroplets. The $f_{droplet}(D)$ clarifies the presence of the supersolid covering sheet of a constant thickness ΔR. From Fig. 8.19d, one can estimate the covering shell thickness ΔR for a spherical droplet of $V \propto R^3$. The fraction $f(R) \propto \Delta V/V = \Delta N/N = 3\Delta R/R$, and the skin shell thickness $\Delta R = f(R)R/3 = 0.90$ Å is exactly the dangling H–O bond length [24] at a flat surface featured at 3610 cm^{-1} wavenumber of vibration [51]. The next shortest H–O bond is about 0.95 Å featured at 3550 cm^{-1} [85]. The DPS distills only the first hydration shell [86] and the outermost layer of a surface [87] without discriminating the transition region between the ordinary water and the supersolid hydration shell or the covering sheet.

8.6.2 Lifetime Resolved by Confinement and Polarization

Figure 8.20a and b compare the phonon (intensity) population decay for the NaBr solutions and water nanodroplets. The orientation relaxation (or vibration energy dissipation) lifetime increases from 2.6 to 3.9 and 6.7 ps as the water transit into solution with concentration increasing from 32 to 16 and 8 H$_2$O per NaBr solute. The relaxation time increases from 2.7 to 50 ps when the size is reduced from 28 to 1.7 nm [70–72]. Water molecular dynamics in the 4 nm diameter nanopool of the neutral and the ionic reverse micelles is almost identical, which confirms that confinement by the cell-water interface is a primary factor governing the dynamics of nanoscopic water rather than the presence of charged groups at the interface. The interface plays the dominant yet unclear role in determining the hydrogen bonding dynamics, whereas the chemical nature of the interface plays a secondary role [89]. Compared to the "frozen" shell, the core water dynamics is much faster but still not as fast as in the bulk case [73], which is suggested as arises from curvature-induced frustration as the dynamics becomes slower with the reduction of droplet size [74].

Likewise, a proton mobility investigation [90] revealed that proton-charge diffusion slows down significantly and linearly with decreasing size, see Fig. 8.20c, being in the form of $Y = 1.585D$. For water droplet smaller than 1 nm, the diffusivity is about two orders lower than it is in bulk water. The lower drift motion mobility is attributed to the more rigid hydrogen-bond network of nanoconfined water or the high viscosity of the droplet. On the other hand, it is harder to break the even shorter and stronger H–O bond for the undercoordinated water molecules for detecting H$^+$ drift motion [24]. Table 8.6 correlates the relaxation time constant τ and the fraction coefficients $f(x)$ for bond transition from the mode of ordinary water to the supersolid states in the hydration shells and in the covering sheet of the droplets.

Figure 8.20d shows the linear $\tau \propto f$ relation with a slope as a function of $d\tau/df \cong 30$ for the NaBr solution, which clarifies that the phonon lifetime is proportional to the number fraction of O:H–O bonds being polarized. The supersolidity of the hydration volume hinders phonon propagation and elongates its lifetime.

An extrapolation of the $\tau(f)$ for the polarization to the droplet regime leads to the excessive $\tau(f)_{droplet} - f_{droplet}d\tau/df_{solution} = \tau(f)_{confinement}$, which shall discriminate the geometric confinement of the droplet boundary, as illustrated in Fig. 8.20 from the effect of polarization. Therefore, the population decay of D–O phonons in

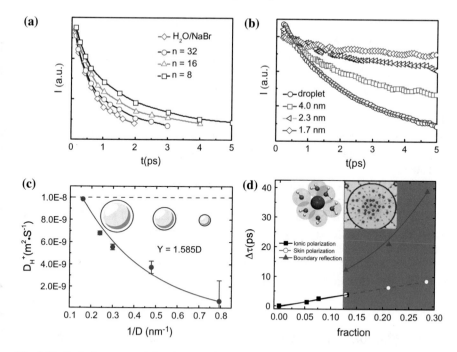

Fig. 8.20 D–O phonon population decay profiles (**a, b**) [70], (**c**) droplet-size resolved D$^+$ diffusivity [90] and **d** the D–O phonon lifetime as a function of the fraction coefficient with discrimination of the polarization (red circles) and confinement effect (triangles) for solution and nanodroplets, $\Delta\tau_{solution} = \tau_{solution} - \tau_{bulk}$ and $\Delta\tau_{confinement} = \tau_{droplet} - \tau_{bulk} - f_{droplet}(d\tau/df)_{solution}$. Insets in d illustrate the droplet confined in a reverse micelle [70] and the supersolid solute hydration shell (reprinted with permission from [88])

Table 8.6 Comparison of the D–O bond orientational relaxation time and the fraction of bonds transferred from the mode of pure water to the polarized states in NaBr solutions and water nanodroplets

Sample		τ(ps) [70]	ω_M(DPS)	f(x)	$\tau_{polarization}$	$\tau_{confinement}$	$\Delta R = f(R)R/3$
Pure water		2.6	0	–	0	–	–
H$_2$O/NaBr	32	3.9	2586	0.0535	1.3	–	–
	16	5.1	2574	0.0766	2.5	–	–
	8	6.7	2568	0.1306	3.9		
Nanograin (nm)	4.0	18	2600	0.1285	3.86	12.32	0.090
	2.3	30	2597	0.2108	6.32	21.08	0.090
	1.7	50	2602	0.2839	8.52	38.84	0.090

For solution, $d\tau/df \approx 30$ arises from the polarization. The excessive $\tau_{confinement} = \tau_{droplet} - \tau_{bulk} - f_{droplet}(d\tau/df)_{solution}$ arises from the geometric phonon confinement by the droplet grain boundaries that prohibiting phonon propagation (reprinted with permission from [88])

the nanodroplet consists two parts. One is the skin supersolidity and the other the geometric confinement.

The striking difference between salt solution and the droplet is that the phonon lifetime τ and the fraction coefficients. The slope $d\tau/df \approx 30$ remains constant for the NaBr solutions, but the $d\tau/df$ remains no constant for the water droplet. The τ values for the droplets are much greater than that for the NaBr solutions though both salt solvation and molecular undercoordination derive the same supersolidity [33, 51]. The DPS peak frequency difference of 2600-2580 cm^{-1} may indicate that the supersolidity of the droplet skin is higher than that of the salt hydration shell.

Besides the effect of polarization, the phonon spatial confinement the grain boundary reflection shall inhibit the phonon propagation or vibration energy dissipation. The phonon propagates radially away and partially transmitted and reflected at the boundary. The superposition of the propagating and reflecting waves form then the weakly time dependent standing wave because of the weaker intensity of the reflecting waves. Ideally, the lifetime of a standing wave is infinity. Such a standing wave is weakly time dependent and has thus a longer lifetime.

8.6.3 Phonon and Electronic Evidence of Skin Supersolidity

An electron is the smallest anion. A free electron introduced into a polar solvent, such as water [91] or ammonia [92], may be trapped by locally oriented solvent molecules. In water, an "equilibrated" hydrated electron can be transiently confined within a roughly spherical cavity defined by six OH bonds oriented toward the negative charge distribution in the so-called Kevan geometry [93, 94], which has the same effect of a Y^+ cation albeit its polarity on forming the screened hydration shell. The hydrated electron is an important reagent in condensed-phase chemistry and molecular biology, as it participates in radiation chemistry, electron transfer, DNA damage, and charge-induced reactivity [95]. Thus, research investigating the dynamics of this species, whether in the presence or absence of other reagents, and how water networks accommodate an excess electron has attracted considerable attention in the theoretical and experimental physical chemistry communities [96–98].

Hydrated electrons provide a probe for the cluster size and molecular site resolved information on their vertical bound energy and emission lifetime. Using the ultrafast pump-probe photoelectron spectroscopy, Verlet et al. [97] discovered that an excess electron can bound to the surface of a water cluster and to the ambient water/air interface. The key feature to the experimental approach is generating solvated electrons by a short pump pulse of 267 nm light and recording photoelectron spectra using a time delayed 38.7 eV (32 nm) high harmonic probe pulse. The internally solvated electron bound energy for a $(D_2O)_{50^-}$ cluster is centered at -1.75 eV and the surface localized states are centered at -0.90 eV. These two states vary with the cluster size and from $(D_2O)_{50^-}$ to $(H_2O)_{50^-}$ slightly. For the bulk water, the electron bound energy is -1.6 eV at the surface and it is -3.2 eV at the bulk interior [99–102].

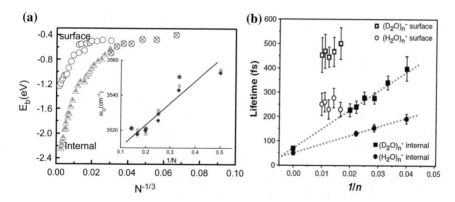

Fig. 8.21 **a** $(H_2O)_{n^-}$ cluster size and molecular site resolved vertical bound energy of the hydrated electron [100–102] with inset showing the size resolved H–O phonon frequency [103–106]; **b** The lifetime of hydrated electrons at the surface and interior of $(H_2O)_{n^-}$ and $(D_2O)_{n^-}$ clusters [97]. The relaxation time scales are given after electronic excitation of the surface isomers at 0.75 eV (open symbols) and internal isomers at 1.0 eV (solid symbols) of $(H_2O)_{n^-}$ (circles) and $(D_2O)_{n^-}$ (squares)

Figure 8.21a shows that the hydrated electron can reside at the water/air interface, but remain below the dividing surface, within the first nanometer [99–102]. Comparatively, the neutral $(H_2O)_N$ size-resolved H–O vibration frequency (inset a) and photoelectron lifetime (Fig. 8.21b) shifts linearly with the inverse of cluster size [103–106]. The electrons on a crystalline water ice surface can exist up to minutes. Long living states even exist when the surface electron is only separated by 1.2 bilayers of water molecules from the metal substrate and thus from an apparent recombination channel [107]. The lifetime of the surface states is much longer than that of the internal states. The unexpectedly long lifetime of solvated electrons bound at the water surface is attributed to a free-energy barrier that separates surface from the interior states [108].

These cluster size and molecular site resolved electron bound energy, phonon stiffness, and electron phonon lifetimes confirm consistently the molecular undercoordination induced supersolidity. The H–O phonon lifetime is proportional to its vibration frequency [109]. The polarization not only shortens the H–O bond and stiffens its phonon but also lengthens the O:H nonbond and softens the O:H phonon. The polarization lowers the bound energy of the hydrated electrons. The extent of polarization increases with decreases of cluster size and the surface polarization is more significant than it is at the interior of the cluster. The invariant photoelectron lifetime at the water cluster surface coincides with the extraordinary thermal stability of the supersolid skin [48].

8.7   Explosion: Intermolecular Tension and Repulsion

8.7.1   Aquatic Unconstrained Explosion

Alkali metals (Y) and molten alkali halides (YX) can rigorously react with water at room temperature, causing massive explosion [110–112]. It was thought that the reaction produces sodium hydroxide, hydrogen, and heat. The heat ignites the hydrogen and the cationic Coulomb fission initiates the explosion at earlier stage [111]. When contacting water, alkali atoms erupt each its valence electron into the water to form hydroxide or turn to be hydrated. Video clips in Fig. 8.22 show that Na solvation undergoes four stages in a few seconds [110]: first, the bulk Na is dissolved gradually and dyes the water into purple, secondly, the Na specimen releases gas, thirdly, the released gas catches fire, and finally explodes abruptly.

The mixture of Na and K forms a liquid alloy at room temperature. The K/Na alloy reacts with water explosively. A group of researchers [111, 113] observed using high-speed camera that the K/Na droplet shoots spikes out just 0.4 ms after it enters the water. Figure 8.23 shows typical photos captured 0.2–0.4 ms after the contact between the allot droplet and water. This explosion process is too fast to attribute to

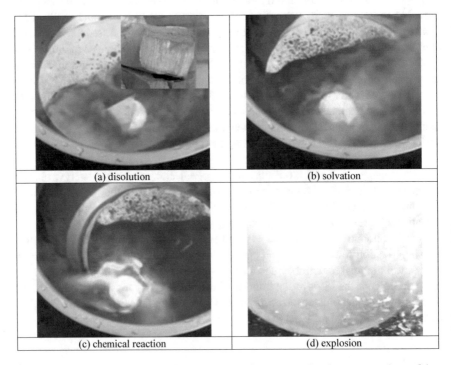

Fig. 8.22 Sodium solvation caused four-stage explosion: water color change, gas release, firing, and explosion (public domain [110])

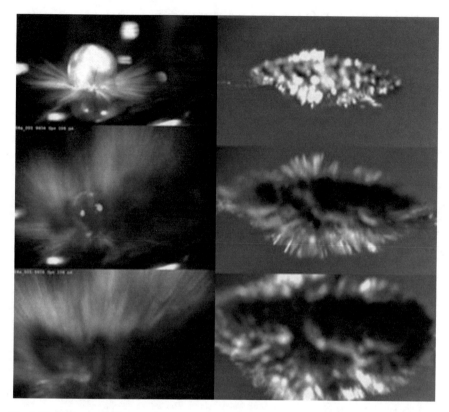

Fig. 8.23 High-speed camera (left) sideview and (right) bottom-view of the Na/K liquid alloy exploding in liquid water (reprinted with permission from [111])

the heat repellant. Replacing H_2O with liquid ammonia (at $-77.8\ °C$) can form less-apparent transient spikes but does not foster an explosion. Replacing the K/Na alloy with a drop of molten aluminum (at ~1000 °C), produces even no sparks. Instead, a Leidenfrost effect [114] occurred in which steam kept the liquid aluminium and water separate on a 100 ms timescale.

Computer simulations [111] suggested that alkali metal atoms at the surface of a small cluster each lose its valence electron within picoseconds. These electrons dissolve in water and react in pairs to form molecular hydrogen and hydroxide. The positively charged ions rapidly repel each other. Such a Coulomb fission [115] initiates the explosion. The out shooting metal spikes continue the Coulomb fission Rayleigh instability [116], generate new surface area that drives the subsequent reactions.

Likewise, molten NaCl ($T_m = 801\ °C$) causes massive explosion when getting into liquid water at the ambient temperature, as shown in Fig. 8.24 captured from a video clip [113]. However, molten Na_2CO_3 ($T_m = 851\ °C$) and molten H_3BO_3 ($T_m = 171\ °C$) do not cause any explosion when reacting with water though the melting tem-

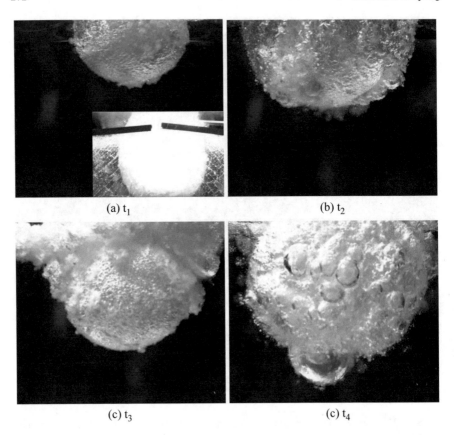

(a) t_1 (b) t_2

(c) t_3 (c) t_4

Fig. 8.24 High-speed camera captured sodium chloride explosion in liquid water at a sequence of time: $t_4 > t_3 > t_2 > t_1$. The explosion breaks the fish tank shown in inset **a** (reprinted with permission of [113])

perature of the Na_2CO_3 is higher than that of NaCl. Pouring molten NaCl on ice causes explosion leaving an opening behind the ice. Pouring molten Cu ($T_m = 1083$ °C) generates spikes due to the Leidenfrost effect and the molten aluminium (660 °C) just melts the ice see Fig. 8.25 for comparison.

Unfortunately, mechanisms for the aquatic explosion of the alkali metals and molten alkali halides remain open for investigation. Questions arise as following:

(a) Is there any correlation between alkali metals and alkali halides on aquatic explosion?
(b) How does aqueous solvation get rid the valence electrons of the alkali metals?
(c) Why is ammonia different from water in fostering the explosion?
(d) Why do molten Na_2CO_3 ($T_m = 851$ °C) and molten H_3BO_3 ($T_m = 171$ °C) not explode when dropping into water?

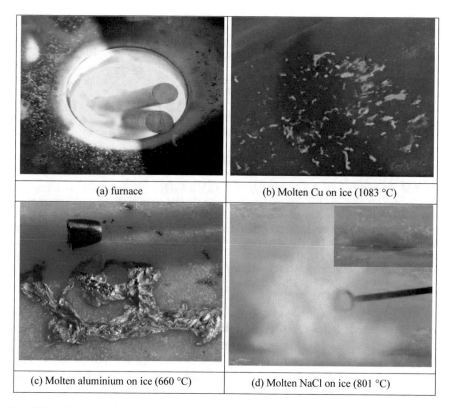

(a) furnace	(b) Molten Cu on ice (1083 °C)
(c) Molten aluminium on ice (660 °C)	(d) Molten NaCl on ice (801 °C)

Fig. 8.25 Interaction between the molten substance and ice. **a** The furnace used to melt the samples. **b** Liquid Cu shoots spikes and shows the Leidenfrost effect. **c** Liquid Al lands on ice softly, and **d** Molten NaCl explodes leaving a hole behind the ice as the inset shows (Tencent video, public domain)

8.7.2 Constrained Explosion: TNT and Cyclo-N_5- Complex

The energetic CHNO molecular crystals and full-N $cyclo$-N_5- complexes undergo constrained explosion. The explosion creates shock waves used in weapon of killing and in launching satellite. The sensitivity, energy density and detonation velocity are key factors for consideration. The sensitivity determines the ignition easiness and safety in storage and transportation. The energy density and detonation velocity determine the impact power of the explosive. Controlling these contradictory factors is the challenge for practical application [117].

The coupling of the inter- and intramolecular interactions is crucial to the performance of the CNHO molecular crystals and $clyco$-N_5- complexes. The orbital hybridization happens to the O and N upon the crystal and complex formation. Each O creates two lone pairs and N creates one upon their electronic orbital hybridization. The lone pairs ':' and protons 'H⁺' interact with their alike or unlike to form

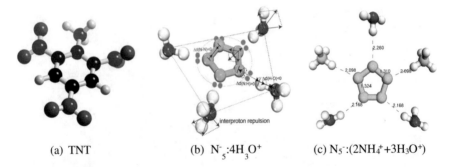

(a) TNT   (b) N$_5^-$:4H$_3$O$^+$ (c) N$_5^-$:(2NH$_4^+$+3H$_3$O$^+$)

Fig. 8.26 Molecular motif for **a** TNT (2C$_7$H$_5$N$_3$O$_6$), **b** full-N *cyclo*-N$_5^-$:cH$_3$O$^+$ with the force diagram, and (c) the *cyclo*-N$_5^-$: (3H$_3$O$^+$ + 2NH$_4^+$) in the (N$_5$)$_6$(H$_3$O)$_3$(NH$_4$)$_4$Cl complex. At the critical concentration c = 3–5, the circumferential H↔H repulsion stretches the radial N:H–O, which not only stabilize the *cyclo*-N$_5^-$:cH$_3$O$^+$ complex but also store energy by shortening all H-O and N-N bonds. The N–N bond shortens from 1.38 to 1.31–1.32 Å. The N:H–O tension also lengthens the N:H nonbond to 2.10 Å for the N:H–N and to a value varying from 2.17–2.26 Å for the N:H–O bonds compared with the O:H length of 1.70 Å for water (reprinted with permission from [118, 119] and courtesy of Dr. Zhang Lei)

the hydrogen bond X:H–Y, H↔H anti-HB, or X:⇔:Y super-HB. These unexpected nonbond interactions govern the performance of explosive crystals in terms of the sensitivity, energy density and detonation velocity. One can tune the number ratio of the HBs to the super-HBs or the anti-HBs to adjust the sensitivity and energy density of an explosive energy carrier.

Figure 8.26 shows example motifs for (a) the N-based TNT and (b) the full-nitrogen *cyclo*-N$_5^-$. For the TNT (2C$_7$H$_5$N$_3$O$_6$) motif, there are 10 H$^+$ and 2 × (3 + 2×6) = 30 lone pairs ":". According to the nonbonding counting rules, in Sect. 3.6, the motif will form 20 HBs and 20 super–HBs with neighboring motifs. No H↔H anti-HB is generated. Two typical *cyclo*-N$_5^-$ complexed are the *cyclo*-N$_5^-$:4H$_3$O$^+$ [118] and the *cyclo*-N$_5^-$: (3H$_3$O$^+$ + 2NH$_4^+$) in the (N$_5$)$_6$(H$_3$O)$_3$(NH$_4$)$_4$Cl complex [119] optimized using quantum calculations. The *cyclo*-N$_5^-$ can only be stabilized in acidic conditions [120–126].

Figure 8.26b illustrates the forces balancing intermolecular interaction and energy storage in the cyclic N–N covalent bonds and in the H–O in the H$_3$O$^+$. The N atoms in the *cyclo*-N$_5^-$ ring undergo the sp^2 orbital hybridization with each creating one lone pair and one unpaired π electron. The lone pairs and the unpaired π electrons form the dual aromaticity (π and σ) of attraction and repulsion along the ring. At a critical concentration of 3 ~ 5 H$_3$O$^+$ surrounding the *cyclo*-N$_5^-$, the radial N:H–O is stretched by the resultant forces of the circumferential H$^+$↔H$^+$ repulsion. The N:H–O tension shortens the H–O bond and lengthens the N:H, which reduces the inter-lone-pair repulsion along the central σ-aromatic *cyclo*-N$_5^-$ ring, resulting in the N–N bond contraction and energy gain.

As shown in Fig. 8.26c, the N–N bond shortens from 1.38 to 1.31–1.32 Å, the N:H nonbond lengthens to 2.10 Å for the N:H–N and varies from 2.17–2.26 Å for the N:H–O bonds [32, 59] with respect to the O:H nonbond of 1.70 Å at 4 °C. Therefore, the circumferential H$^+$↔H$^+$ repulsion, radial N:H–O or N:H–N bond tension, and

H–O/N and N–N covalnet bond contraction raise the crystal to a metastable state to store excessive energy. This configuration also clarifies the reason for the *cyclo*-N$_5$-stabilization under acidic conditions [44, 60–63].

8.7.3 Formulation: Super-HB and HB Competition

From the perspectives of O:H–O bond relaxation, H↔H anti-HB and O:⇔:O super-HB repulsion, one may consider the non-constrained and constrained explosion. Table 8.7 formulates the reaction dynamics and intermolecular interactions involved in the aquatic explosive alkali metals and alkali halides and in the CHNO type molecular crystals and cyclo-N$_5$- complex. The presence of anti-HB or super-HB repulsion and HB tension are the keys to the spontenous and constrained explosion.

As it can be seen, the HB tension constrains and the anti-HB or super-HB repulsion fosters explosion, which reconciles the sponteneous aquatic explosion of alkali metals and molten alkali halides and the contrained explosion of CNHO molecular crystals. The absence of the repulsive interaction does not fosters explosion. In the aquatic explosion, H$_2$O dipoles tend to dissolve a polar crystal or a polar molecule into charged constituents. The YX salt and YOH base are easy to be dissolved because they are polar molecules/crystals. Therefore, $2YX + 2H_2O \Rightarrow 2HX + 2YOH + H_2$ and $2YOH + 4H_2O \Rightarrow Y^+ \leftrightarrow Y^+ + 2(HO:\Leftrightarrow:H_2O)$ transitions occur upon solvation, producing the interaction repulsion and the O:⇔:O super-HB. Dissolution of the neutral alkali metal requires a process of oxidation. The metal atom donates its valence electron to O by $Y + H_2O \Rightarrow H + YOH \Rightarrow H + Y^+ + OH^-$ replacement reaction, rather than erupts the electron into the solvent to become free. Neutral atom oxidation turns the Y into the YOH and then proceeds the same YOH solvation, yielding the identical $[Y^+ \leftrightarrow Y^+ + 2(H_2O:\Leftrightarrow:OH^-)]$ + heat product with either H$_2$↑ or HX↑ vapor and H$_2$O steam.

Alkali metal or the alkali halide undergo unconstrained explosion at solvation because of the absence of X:H–Y stretching. The intermolecular X:⇔:Y or H↔H repulsion stretches the X:H–Y. Both the repulsion and X:H–Y stretching shorten the intramolecular covalent bonds. The entire bonding network is in the energetically metastable state. The compressively deformed covalent bond not only stores excessive energy but also emits energy abruptly when breaks. Once the X:H or the X:⇔:Y is broken by mechanical or thermal shock wave excitation, the entire bonding network collapses and explosion occurs.

However, the absence of super-HB does not foster alkali metal-liquid ammonia explosion. Alkali metal solvation produces insufficient number of lone pairs. The reaction $Y + NH_3 \Rightarrow H + NH_2Y \Rightarrow Y^+ + (NH_2)^- + H$ turns out the (NH$_2$)$^-$ with two pairs of lone pairs. The (NH$_2$)$^-$:(NH$_2$)$^-$ stabilizes the (NH$_2$)$^-$·NH$_3$ or the NH$_3$·NH$_3$ interactions, instead. Each NH$_3$ molecule has three protons and one lone pair, which endows the NH$_3$ to interact with its tetrahedrally structured neighbors through two H↔H anti-HBs and two N:H–N bonds. The H↔H anti-HBs destabilize the ammonia with lower melting temperature. Furthermore, water dissolves the molten Y$_2$CO$_3$ and

Table 8.7 Interactions involved in the aquatic explosive alkali metals and alkali halides and in the CHNO and cyclo-N_5^-: cH_3O^+ complex

Non-constrained explosion	2Y + nH$_2$O (ambient T)	⇑	2YOH + H$_2$↑ + heat + (n-2)H$_2$O	(1) [Y$^+$↔Y$^+$ + 2(O:⇔:O)] Coulomb repulsivity drives explosion
		⇑	[Y$^+$↔Y$^+$ + 2(H$_2$O:⇔:OH$^-$)] + heat + (n-4)H$_2$O + H$_2$↑	(2) Thermal enhancement of repulsivity
				(3) Preferential solvation of polar molecules/crystals
	2YX(molten) + nH$_2$O	⇑	2YOH + (n-2)H$_2$O + heat + 2HCl↑	(4) O:⇔:O compression heats up solution
		⇑	[Y$^+$↔Y$^+$ + 2(H$_2$O:⇔:OH$^-$)] + heat + (n-4)H$_2$O + 2HX↑	(5) Y$^+$↔Y$^+$ repulsion results in Rayleigh disability
Constrained explosion	CHNO	⇑	[X:⇔:Z + X:H–Z] + ⋯	Competition between HB stretching, and anti-HB or super-HB repulsion dictates the sensitivity, energy density, and explosion velocity
	cyclo-N$_5$− complex	⇑	[H↔H + X:H–Z] + ⋯	
Inexplosive	2Y + nNH$_3$ (−78 °C)	⇑	[Y$^+$↔Y$^+$ + **H$_2$N$^-$:H$_2$N$^-$**] + (n-2)NH$_3$ + H$_2$↑	Absence of super-HB repulsion fosters no explosion

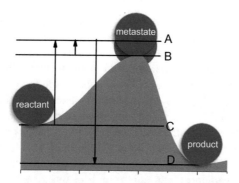

Fig. 8.27 Energy diagram for the "sensitivity-energy density-detonation velocity" of energetic substance subjecting to the constrained (BA: X:H–Y tension) and unconstrained (molten alkali halides) explosion. AD: energy ejection by bond dissociation; CA: energy storage by intramolecular bond compression though X:⇔:Y or H↔H repulsion and X:H–Y tension that is also related to the sensitivity

molten H_3BO_3 into the respective $2Y^+ + (HCO_3)^-$ and $H^+ + (H_3BO_3)^-$. The dipolar anions foster only short-range interactions.

Thus, the $[Y^+↔Y^+ + 2(H_2O:⇔:OH^-) + heat]$ reconciles the aquatic explosion of alkali metals and alkali halides. The $Y^+↔Y^+$ Coulomb fission initiates and the O:⇔:O super-HB fosters the explosion. The exothermic oxidation-hydration lengthens the O:⇔:O and O:H nonbonds but shortens the H–O bonds to store energy. When the O:⇔:O or the O:H is broken, the H–O bonds erupt energy of explosion, which is the same process of energetic crystal explosion.

Figure 8.27 illustrates the energy transition of energetic substance. The X:H–O tension serves as the constraint to stabilize the system and the X:⇔:Y or the H↔H repulsions store excessive energy by intramolecular bond compression. Once the constraint X:H breaks, the system will collapse erupting energy. The product is in gaseous state of lower binding energy.

8.8 Summary

Observations thus verified expectations of the multiple field effect on the O:H–O relaxation, critical pressures and temperatures for phase transition, and the surface stress from the perspectives of O:H–O bond cooperative relaxation, structure order fluctuation and electron polarization. The same trend of phonon relaxation induced by salting and heating shows that they share the same mechanism of Coulomb modulation of O–O interaction by varying ion size, ion charge, and O–O distance. Heating lengthens the O–O distance by O:H expansion and H–O contraction but salting elongates the O–O separation by polarization. Larger ions with lower ionicity weaken the Coulomb repulsion and stiffen the ω_H more than the otherwise, which is exactly the situation of expectation. However, heating and salting effects contrastingly on surface stress because of their functionality of O:H–O polarization and thermal fluctuation. The superposition of the specific heat curves is the key to dictating the O:H–O bonding thermodynamics and phase transition under various stimuli.

NaX/H$_2$O solutions phase transition from Liquid to ice VI (at P$_{C1}$) and then to ice VII (P$_{C2}$) phase at 298 K requires excessive pressures to recover the electrification-deformed O:H–O bond. P$_{C1}$ and P$_{C2}$ vary simultaneously with the solute type in the Hofmeister series order: I > Br > Cl > F ~ 0. However, the P$_{C1}$ increases faster than the P$_{C2}$ with the NaI/H$_2$O concentration. The (P, T) follows the path upwardly along the Liquid–VI phase boundary and then meet at the Liquid-VI-VII triple-phase junction at 3.3 GPa and 350 K. Discrepancy between the solute type and solute concentration arises from the involvement of inter-anion repulsion that governs the critical pressures and temperatures for phase transition at higher solute concentrations. Molecular undercoordination has the same effect of ionization, which disperses the QS boundary contrastingly to the effect of compression.

Oxidation-solvation-hydration of alkali metals and alkali halides fosters their aquatic explosion that can be formulated by [Y$^+$$\leftrightarrow$Y$^+$ + 2(H$_2$O:$\Leftrightarrow$:OH$^-$) + heat] + (H$_2$/HX)↑. The Y$^+$$\leftrightarrow$Y$^+$ Coulomb fission initiates and the O:$\Leftrightarrow$:O super-HB fosters the explosion. Compared with the alkali metals and alkali halides, the CHNO explosives show constrained explosion with tunable sensitivity and energy density. The competition between the super-HB repulsion and HB attraction determines sensitivity and the intramolecular bond contraction determines energy density. Lack of HB attraction makes the alkali metal and alkali halide unconstrained explosion.

A combination of the DPS, ultrafast phonon and photoelectron spectroscopies has reconciled the phonon abundance-lifetime-stiffness cooperative transition dynamics upon NaBr solvation and nanodroplet confinement, and hence bridges the perspectives of molecular dynamics and O:H–O bond cooperative relaxation. Water droplet size reduction shares the same effect of salt polarization on the covalent bond shortening and stiffening and O:H nonbond elongation and polarization at its covering sheet. The polarization raises the viscosity, slows down the diffusivity and molecular dynamics, and prolongs the phonon/photoelectron lifetime by slowing the vibrational energy dissipation. Besides the supersolidity contribution, the geometric confinement lengthens the phonon lifetime in nanodroplet because of the formation of the weakly time dependent standing waves.

References

1. L.M. Levering, M.R. Sierra-Hernández, H.C. Allen, Observation of hydronium ions at the air–aqueous acid interface: vibrational spectroscopic studies of aqueous HCl, HBr, and HI. J. Phys. Chem. C **111**(25), 8814–8826 (2007)
2. L.M. Pegram, M.T. Record, Hofmeister salt effects on surface tension arise from partitioning of anions and cations between bulk water and the air-water interface. J. Phys. Chem. B **111**(19), 5411–5417 (2007)
3. R.K. Ameta, M. Singh, Surface tension, viscosity, apparent molal volume, activation viscous flow energy and entropic changes of water + alkali metal phosphates at T = (298.15, 303.15, 308.15) K. J. Mol. Liq. **203**: 29–38 (2015)

4. P. Lo Nostro, B.W. Ninham, Hofmeister phenomena: an update on ion specificity in biology. Chem. Rev, **112**(4): 2286–322 (2012)
5. C.M. Johnson, S. Baldelli, Vibrational sum frequency spectroscopy studies of the influence of solutes and phospholipids at vapor/water interfaces relevant to biological and environmental systems. Chem. Rev. **114**(17), 8416–8446 (2014)
6. X.P. Li, K. Huang, J.Y. Lin, Y.Z. Xu, H.Z. Liu, Hofmeister ion series and its mechanism of action on affecting the behavior of macromolecular solutes in aqueous solution. Prog. Chem. **26**(8), 1285–1291 (2014)
7. E.K. Wilson, Hofmeister Still Mystifies. C&EN Arch. **90**(29), 42–43 (2012)
8. Y.R. Xu, L. Li, P.J. Zheng, Y.C. Lam, X. Hu, Controllable gelation of methylcellulose by a salt mixture. Langmuir **20**(15), 6134–6138 (2004)
9. M. van der Linden, B.O. Conchúir, E. Spigone, A. Niranjan, A. Zaccone, P. Cicuta, Microscopic origin of the Hofmeister effect in gelation kinetics of colloidal silica. J. Phys. Chem. Lett. 2881–2887 (2015)
10. F. Aliotta, M. Pochylski, R. Ponterio, F. Saija, G. Salvato, C. Vasi, Structure of bulk water from Raman measurements of supercooled pure liquid and LiCl solutions. Phys. Rev. B **86**(13), 134301 (2012)
11. W.-T. Zheng, C.Q. Sun, Underneath the fascinations of carbon nanotubes and graphene nanoribbons. Energy Environ. Sci. **4**(3), 627–655 (2011)
12. M. Benoit, D. Marx, M. Parrinello, Tunnelling and zero-point motion in high-pressure ice. Nature **392**(6673), 258–261 (1998)
13. D. Kang, J. Dai, Y. Hou, J. Yuan, Structure and vibrational spectra of small water clusters from first principles simulations. J. Chem. Phys. **133**(1), 014302 (2010)
14. C.Q. Sun, X. Zhang, W.T. Zheng, Hidden force opposing ice compression. Chem Sci **3**, 1455–1460 (2012)
15. S. Chen, Z. Xu, J. Li, The observation of oxygen-oxygen interactions in ice. New J. Phys. **18**(2), 023052 (2016)
16. Y. Huang, X. Zhang, Z. Ma, Y. Zhou, G. Zhou, C.Q. Sun, Hydrogen-bond asymmetric local potentials in compressed ice. J. Phys. Chem. B **117**(43), 13639–13645 (2013)
17. X. Zhang, S. Chen, J. Li, Hydrogen-bond potential for ice VIII-X phase transition. Sci Rep **6**, 37161 (2016)
18. Y. Yoshimura, S.T. Stewart, M. Somayazulu, H. Mao, R.J. Hemley, High-pressure x-ray diffraction and Raman spectroscopy of ice VIII. J. Chem. Phys. **124**(2), 024502 (2006)
19. P. Pruzan, J.C. Chervin, E. Wolanin, B. Canny, M. Gauthier, M. Hanfland, *Phase diagram of ice in the VII-VIII-X domain. Vibrational and structural data for strongly compressed ice VIII.* J. Raman Spectrosc. **34**(7–8): 591-610 (2003)
20. M. Song, H. Yamawaki, H. Fujihisa, M. Sakashita, K. Aoki, Infrared absorption study of Fermi resonance and hydrogen-bond symmetrization of ice up to 141 GPa. Phys. Rev. B **60**(18), 12644 (1999)
21. Y. Yoshimura, S.T. Stewart, M. Somayazulu, H.K. Mao, R.J. Hemley, Convergent Raman features in high density amorphous ice, ice VII, and Ice VIII under Pressure. J. Phys. Chem. B **115**(14), 3756–3760 (2011)
22. Y. Yoshimura, S.T. Stewart, H.K. Mao, R.J. Hemley, In situ Raman spectroscopy of low-temperature/high-pressure transformations of H2O. J. Chem. Phys. **126**(17), 174505 (2007)
23. N. Mishchuk, V. Goncharuk, On the nature of physical properties of water. J. Water Chem. Technol. **39**(3), 125–131 (2017)
24. Y.L. Huang, X. Zhang, Z.S. Ma, Y.C. Zhou, W.T. Zheng, J. Zhou, C.Q. Sun, Hydrogen-bond relaxation dynamics: resolving mysteries of water ice. Coord. Chem. Rev. **285**, 109–165 (2015)
25. Q. Zeng, T. Yan, K. Wang, Y. Gong, Y. Zhou, Y. Huang, C.Q. Sun, B. Zou, Compression icing of room-temperature NaX solutions (X = F, Cl, Br, I). Phys. Chem. Chem. Phys. **18**(20), 14046–14054 (2016)
26. G.N. Ruiz, L.E. Bove, H.R. Corti, T. Loerting, Pressure-induced transformations in LiCl–H_2O at 77 K. Phys. Chem. Chem. Phys. **16**(34), 18553–18562 (2014)

27. S. Klotz, L.E. Bove, T. Strässle, T.C. Hansen, A.M. Saitta, The preparation and structure of salty ice VII under pressure. Nat. Mater. **8**(5), 405–409 (2009)
28. L.E. Bove, R. Gaal, Z. Raza, A.A. Ludl, S. Klotz, A.M. Saitta, A.F. Goncharov, P. Gillet, Effect of salt on the H-bond symmetrization in ice. Proc Natl Acad Sci U S A **112**(27), 8216–8220 (2015)
29. Y. Bronstein, P. Depondt, L.E. Bove, R. Gaal, A.M. Saitta, F. Finocchi, Quantum versus classical protons in pure and salty ice under pressure. Phy. Rev. B **93**(2), 024104 (2016)
30. Y. Xu, C. Wang, K.C. Tam, L. Li, Salt-assisted and salt-suppressed sol-gel transitions of methylcellulose in water. Langmuir **20**(3), 646–652 (2004)
31. X. Zhang, T. Yan, Y. Huang, Z. Ma, X. Liu, B. Zou, C.Q. Sun, Mediating relaxation and polarization of hydrogen-bonds in water by NaCl salting and heating. Phys. Chem. Chem. Phys. **16**(45), 24666–24671 (2014)
32. Q. Zeng, C. Yao, K. Wang, C.Q. Sun, B. Zou, Room-Temperature NaI/H_2O Compression Icing: solute–solute interactions. PCCP **19**, 26645–26650 (2017)
33. C.Q. Sun, J. Chen, Y. Gong, X. Zhang, Y. Huang, (H, Li)Br and LiOH solvation bonding dynamics: molecular nonbond interactions and solute extraordinary capabilities. J. Phys. Chem. B **122**(3), 1228–1238 (2018)
34. X. Zhang, P. Sun, Y. Huang, T. Yan, Z. Ma, X. Liu, B. Zou, J. Zhou, W. Zheng, C.Q. Sun, Water's phase diagram: from the notion of thermodynamics to hydrogen-bond cooperativity. Prog. Solid State Chem. **43**, 71–81 (2015)
35. A.F. Goncharov, V.V. Struzhkin, M.S. Somayazulu, R.J. Hemley, H.K. Mao, Compression of Ice to 210 Gigapascals: Infrared Evidence for a Symmetric Hydrogen-Bonded Phase. Science **273**(5272), 218–220 (1996)
36. A.F. Goncharov, V.V. Struzhkin, H.-K. Mao, R.J. Hemley, Raman spectroscopy of dense H_2O and the transition to symmetric hydrogen bonds. Phys. Rev. Lett. **83**(10), 1998 (1999)
37. J. Niehaus, W. Cantrell, *Contact Freezing of Water by Salts*. J. Phys. Chem. Lett. **6**(17), 3490–3495 (2015)
38. N.R. Gokhale, J.D. Spengler, Freezing of freely suspended, supercooled water drops by contact nucleation. J. Appl. Meteorol. **11**(1), 157–160 (1972)
39. W.A. Cooper, A possible mechanism for contact nucleation. J. Atmos. Sci. **31**(7), 1832–1837 (1974)
40. N. Fukuta, A study of the mechanism of contact ice nucleation. J. Atmos. Sci. **32**(8), 1597–1603 (1975)
41. C. Gurganus, A.B. Kostinski, R.A. Shaw, Fast imaging of freezing drops: no preference for nucleation at the contact line. The J. Phys. Chem. Letters **2**(12), 1449–1454 (2011)
42. R.A. Shaw, A.J. Durant, Y. Mi, Heterogeneous surface crystallization observed in undercooled water. J. Phys. Chem. B **109**(20), 9865–9868 (2005)
43. R.G. Knollenberg, A laboratory study of the local cooling resulting from the dissolution of soluble ice nuclei having endothermic heats of solution. J. Atmos. Sci. **26**(1), 115–124 (1969)
44. X. Zhang, Y. Huang, Z. Ma, Y. Zhou, W. Zheng, J. Zhou, C.Q. Sun, A common supersolid skin covering both water and ice. Phys. Chem. Chem. Phys. **16**(42), 22987–22994 (2014)
45. T.F. Kahan, J.P. Reid, D.J. Donaldson, Spectroscopic probes of the quasi-liquid layer on ice. J. Phys. Chem. A **111**(43), 11006–11012 (2007)
46. Q. Sun, Raman spectroscopic study of the effects of dissolved NaCl on water structure. Vib. Spectrosc. **62**, 110–114 (2012)
47. M. Baumgartner, R.J. Bakker, Raman spectroscopy of pure H_2O and NaCl-H_2O containing synthetic fluid inclusions in quartz—a study of polarization effects. Mineral. Petrol. **95**(1–2), 1–15 (2008)
48. Y. Zhou, Y. Zhong, Y. Gong, X. Zhang, Z. Ma, Y. Huang, C.Q. Sun, Unprecedented thermal stability of water supersolid skin. J. Mol. Liq. **220**, 865–869 (2016)
49. C.Q. Sun, X. Zhang, X. Fu, W. Zheng, J.-L. Kuo, Y. Zhou, Z. Shen, J. Zhou, Density and phonon-stiffness anomalies of water and ice in the full temperature range. J. Phys. Chem. Letters **4**, 3238–3244 (2013)

50. C.Q. Sun, Y. Sun, The attribute of water: single notion, multiple myths. Springer Ser. Chem. Phys. **113**, 494 pp (2016)

51. C.Q. Sun, X. Zhang, J. Zhou, Y. Huang, Y. Zhou, W. Zheng, Density, elasticity, and stability anomalies of water molecules with fewer than four neighbors. J. Phys. Chem. Letters **4**, 2565–2570 (2013)

52. J. Chen, K. Nagashima, K.-I. Murata, G. Sazaki, Quasi-liquid layers can exist on polycrystalline ice thin films at a temperature significantly lower than on ice single crystals. Cryst. Growth Des. **19**, 116–124 (2018)

53. D.M. Murphy, T. Koop, Review of the vapour pressures of ice and supercooled water for atmospheric applications. Q. J. Royal Meteorol. Soc.: J. Atmos. Sci. Appl. Meteorol. Phys. Oceanogr. **131**(608), 1539–1565 (2005)

54. D. Sonntag, Import new values of the physical constants of 1986, vapour pressure formulations based on the ITS-90, and psychrometer formulae. Z. Meterol. **70**, 340 (1990)

55. X. Zhang, Y. Huang, P. Sun, X. Liu, Z. Ma, Y. Zhou, J. Zhou, W. Zheng, C.Q. Sun, Ice regelation: hydrogen-bond extraordinary recoverability and water quasisolid-phase-boundary dispersivity. Sci Rep **5**, 13655 (2015)

56. C.Q. Sun, Perspective: Supersolidity of Undercoordinated and Hydrating Water. Phys. Chem. Chem. Phys. **20**, 30104–30119 (2018)

57. C.Q. Sun, Aqueous charge injection: solvation bonding dynamics, molecular nonbond interactions, and extraordinary solute capabilities. Int. Rev. Phys. Chem. **37**(3–4), 363–558 (2018)

58. K. Sotthewes, P. Bampoulis, H.J. Zandvliet, D. Lohse, B. Poelsema, Pressure induced melting of confined ice. ACS Nano **11**(12), 12723–12731 (2017)

59. X. Zhang, P. Sun, Y. Huang, Z. Ma, X. Liu, J. Zhou, W. Zheng, C.Q. Sun, Water nanodroplet thermodynamics: quasi-solid phase-boundary dispersivity. J. Phys. Chem. B **119**(16), 5265–5269 (2015)

60. K.V. Agrawal, S. Shimizu, L.W. Drahushuk, D. Kilcoyne, M.S. Strano, Observation of extreme phase transition temperatures of water confined inside isolated carbon nanotubes. Nat. Nanotechnol. **12**(3), 267 (2017)

61. H. Qiu, W. Guo, Electromelting of confined monolayer ice. Phys. Rev. Lett. **110**(19), 195701 (2013)

62. M. Erko, D. Wallacher, A. Hoell, T. Hauss, I. Zizak, O. Paris, Density minimum of confined water at low temperatures: a combined study by small-angle scattering of X-rays and neutrons. PCCP **14**(11), 3852–3858 (2012)

63. F. Mallamace, C. Branca, M. Broccio, C. Corsaro, C.Y. Mou, S.H. Chen, The anomalous behavior of the density of water in the range 30 K < T < 373 K. Proc. Natl. Acad. Sci. U.S.A. **104**(47), 18387–18391 (2007)

64. F.G. Alabarse, J. Haines, O. Cambon, C. Levelut, D. Bourgogne, A. Haidoux, D. Granier, B. Coasne, Freezing of water confined at the nanoscale. Phys. Rev. Lett. **109**(3), 035701 (2012)

65. R. Moro, R. Rabinovitch, C. Xia, V.V. Kresin, Electric dipole moments of water clusters from a beam deflection measurement. Phys. Rev. Lett. **97**(12), 123401 (2006)

66. A. Späh, H. Pathak, K.H. Kim, F. Perakis, D. Mariedahl, K. Amann-Winkel, J.A. Sellberg, J.H. Lee, S. Kim, J. Park, K.H. Nam, T. Katayama, A. Nilsson, Apparent power-law behavior of water's isothermal compressibility and correlation length upon supercooling. Phys. Chem. Chem. Phys. **21**, 26–31 (2019)

67. J.A. Sellberg, C. Huang, T.A. McQueen, N.D. Loh, H. Laksmono, D. Schlesinger, R.G. Sierra, D. Nordlund, C.Y. Hampton, D. Starodub, D.P. DePonte, M. Beye, C. Chen, A.V. Martin, A. Barty, K.T. Wikfeldt, T.M. Weiss, C. Caronna, J. Feldkamp, L.B. Skinner, M.M. Seibert, M. Messerschmidt, G.J. Williams, S. Boutet, L.G. Pettersson, M.J. Bogan, A. Nilsson, Ultrafast X-ray probing of water structure below the homogeneous ice nucleation temperature. Nature **510**(7505), 381–384 (2014)

68. C. Goy, M.A. Potenza, S. Dedera, M. Tomut, E. Guillerm, A. Kalinin, K.-O. Voss, A. Schottelius, N. Petridis, A. Prosvetov, Shrinking of rapidly evaporating water microdroplets reveals their extreme supercooling. Phys. Rev. Lett. **120**(1), 015501 (2018)

69. K. Ando, M. Arakawa, A. Terasaki, Freezing of micrometer-sized liquid droplets of pure water evaporatively cooled in a vacuum. Phys. Chem. Chem. Phys. **20**(45), 28435–28444 (2018)
70. S. Park, D.E. Moilanen, M.D. Fayer, Water dynamics: the effects of ions and nanoconfinement. J. Phys. Chem. B **112**(17), 5279–5290 (2008)
71. D.E. Moilanen, N.E. Levinger, D.B. Spry, M.D. Fayer, Confinement or the nature of the interface? Dynamics of nanoscopic water. J. Am. Chem. Soc. **129**(46), 14311–14318 (2007)
72. M.D. Fayer, Dynamics of water interacting with interfaces, molecules, and ions. Acc. Chem. Res. **45**(1), 3–14 (2011)
73. A.A. Bakulin, D. Cringus, P.A. Pieniazek, J.L. Skinner, T.L. Jansen, M.S. Pshenichnikov, Dynamics of water confined in reversed micelles: multidimensional vibrational spectroscopy study. J. Phys. Chem. B **117**(49), 15545–15558 (2013)
74. P.A. Pieniazek, Y.-S. Lin, J. Chowdhary, B.M. Ladanyi, J. Skinner, Vibrational spectroscopy and dynamics of water confined inside reverse micelles. J. Phys. Chem. B **113**(45), 15017–15028 (2009)
75. L. Pauling, *The Nature of the Chemical Bond*. 3rd ed. (Cornell University press, Ithaca, NY, 1960)
76. C.Q. Sun, Relaxation of the chemical bond. Springer Ser. Chem. Phys. **108**, 807 pp (2014)
77. Q. Zeng, J. Li, H. Huang, X. Wang, M. Yang, Polarization response of clathrate hydrates capsulated with guest molecules. J. Chem. Phys. **144**(20), 204308 (2016)
78. F. Yang, X. Wang, M. Yang, A. Krishtal, C. Van Alsenoy, P. Delarue, P. Senet, Effect of hydrogen bonds on polarizability of a water molecule in (H_2O) N (N = 6, 10, 20) isomers. Phys. Chem. Chem. Phys. **12**(32), 9239–9248 (2010)
79. X. Zhang, Y. Xu, Y. Zhou, Y. Gong, Y. Huang, C.Q. Sun, HCl, KCl and KOH solvation resolved solute-solvent interactions and solution surface stress. Appl. Surf. Sci. **422**, 475–481 (2017)
80. X. Zhang, X. Liu, Y. Zhong, Z. Zhou, Y. Huang, C.Q. Sun, Nanobubble skin supersolidity. Langmuir **32**(43), 11321–11327 (2016)
81. X. Zhang, Y. Zhou, Y. Gong, Y. Huang, C. Sun, Resolving H(Cl, Br, I) capabilities of transforming solution hydrogen-bond and surface-stress. Chem. Phys. Lett. **678**, 233–240 (2017)
82. Y. Zhou, Y. Gong, Y. Huang, Z. Ma, X. Zhang, C.Q. Sun, Fraction and stiffness transition from the H–O vibrational mode of ordinary water to the HI NaI, and NaOH hydration states. J. Mol. Liq. **244**, 415–421 (2017)
83. C.Q. Sun, Size dependence of nanostructures: Impact of bond order deficiency. Prog. Solid State Chem. **35**(1), 1–159 (2007)
84. Y. Chen, H.I.I. Okur, C. Liang, S. Roke, Orientational ordering of water in extended hydration shells of cations is ion-specific and correlates directly with viscosity and hydration free energy. Phys. Chem. Chem. Phys. **19**(36), 24678–24688 (2017)
85. B. Wang, W. Jiang, Y. Gao, Z. Zhang, C. Sun, F. Liu, Z. Wang, Energetics competition in centrally four-coordinated water clusters and Raman spectroscopic signature for hydrogen bonding. RSC Adv. **7**(19), 11680–11683 (2017)
86. Y. Gong, Y. Zhou, C. Sun, Phonon spectrometrics of the hydrogen bond (O:H–O) segmental length and energy relaxation under excitation, B.o. intelligence, Editor. China (2018)
87. X.J. Liu, M.L. Bo, X. Zhang, L. Li, Y.G. Nie, H. TIan, Y. Sun, S. Xu, Y. Wang, W. Zheng, C.Q. Sun, Coordination-resolved electron spectrometrics. Chem. Rev. **115**(14), 6746–6810 (2015)
88. Y. Peng, Y. Yang, Y. Sun, Y. Huang, C.Q. Sun, Phonon abundance-stiffness-lifetime transition from the mode of heavy water to its confinement and hydration. J. Mol. Liq. **276**, 688–693 (2019)
89. E.E. Fenn, D.B. Wong, M. Fayer, Water dynamics at neutral and ionic interfaces. Proc. Natl. Acad. Sci. **106**(36), 15243–15248 (2009)
90. T.H. van der Loop, N. Ottosson, T. Vad, W.F. Sager, H.J. Bakker, S. Woutersen, Communication: slow proton-charge diffusion in nanoconfined water. J. Chem. Phys. **146**(13), 131101 (2017)

91. E.J. Hart, J. Boag, Absorption spectrum of the hydrated electron in water and in aqueous solutions. J. Am. Chem. Soc. **84**(21), 4090–4095 (1962)

92. W. Weyl, Ann. Phys. **197**, 601 (1863)

93. L. Kevan, Solvated electron structure in glassy matrixes. Acc. Chem. Res. **14**(5), 138–145 (1981)

94. M. Boero, M. Parrinello, K. Terakura, T. Ikeshoji, C.C. Liew, First-principles molecular-dynamics simulations of a hydrated electron in normal and supercritical water. Phys. Rev. Lett. **90**(22), 226403 (2003)

95. K.R. Siefermann, B. Abel, The hydrated electron: a seemingly familiar chemical and biological transient. Angew. Chem. Int. Ed. **50**(23), 5264–5272 (2011)

96. A. Bragg, J. Verlet, A. Kammrath, O. Cheshnovsky, D. Neumark, Hydrated electron dynamics: from clusters to bulk. Science **306**(5696), 669–671 (2004)

97. J. Verlet, A. Bragg, A. Kammrath, O. Cheshnovsky, D. Neumark, Observation of large water-cluster anions with surface-bound excess electrons. Science **307**(5706), 93–96 (2005)

98. J.M. Herbert, M.P. Coons, The hydrated electron. Annu. Rev. Phys. Chem. **68**, 447–472 (2017)

99. D. Sagar, C.D. Bain, J.R. Verlet, Hydrated electrons at the water/air interface. J. Am. Chem. Soc. **132**(20), 6917–6919 (2010)

100. J. Kim, I. Becker, O. Cheshnovsky, M.A. Johnson, Photoelectron spectroscopy of the 'missing' hydrated electron clusters $(H_2O)^-$ n, n = 3, 5, 8 and 9: isomers and continuity with the dominant clusters n = 6, 7 and $\geq$ 11. Chem. Phys. Lett. **297**(1–2), 90–96 (1998)

101. J.V. Coe, S.M. Williams, K.H. Bowen, Photoelectron spectra of hydrated electron clusters vs. cluster size: connecting to bulk. Int. Rev. Phys. Chem. **27**(1), 27–51 (2008)

102. A. Kammrath, G. Griffin, D. Neumark, J.R.R. Verlet, Photoelectron spectroscopy of large (water)[sub n][sup −] (n = 50–200) clusters at 4.7 eV. J. Chem. Phys. **125**(7), 076101 (2006)

103. J. Ceponkus, P. Uvdal, B. Nelander, Intermolecular vibrations of different isotopologs of the water dimer: experiments and density functional theory calculations. J. Chem. Phys. **129**(19), 194306 (2008)

104. J. Ceponkus, P. Uvdal, B. Nelander, On the structure of the matrix isolated water trimer. J. Chem. Phys. **134**(6), 064309 (2011)

105. J. Ceponkus, P. Uvdal, B. Nelander, Water tetramer, pentamer, and hexamer in inert matrices. J. Phys. Chem. A **116**(20), 4842–4850 (2012)

106. V. Buch, S. Bauerecker, J.P. Devlin, U. Buck, J.K. Kazimirski, Solid water clusters in the size range of tens-thousands of H_2O: a combined computational/spectroscopic outlook. Int. Rev. Phys. Chem. **23**(3), 375–433 (2004)

107. U. Bovensiepen, C. Gahl, J. Stahler, M. Bockstedte, M. Meyer, F. Baletto, S. Scandolo, X.-Y. Zhu, A. Rubio, M. Wolf, A dynamic landscape from femtoseconds to minutes for excess electrons at ice–metal interfaces. J. Phys. Chem. C **113**(3), 979–988 (2008)

108. K.R. Siefermann, Y. Liu, E. Lugovoy, O. Link, M. Faubel, U. Buck, B. Winter, B. Abel, Binding energies, lifetimes and implications of bulk and interface solvated electrons in water. Nat. Chem. **2**(4), 274 (2010)

109. S.T. van der Post, C.S. Hsieh, M. Okuno, Y. Nagata, H.J. Bakker, M. Bonn, J. Hunger, Strong frequency dependence of vibrational relaxation in bulk and surface water reveals sub-picosecond structural heterogeneity. Nat. Commun. **6**, 8384 (2015)

110. M.M.A. Science, *Sodium in Water Explosion—Chemical Reaction* (2016)

111. P.E. Mason, F. Uhlig, V. Vaněk, T. Buttersack, S. Bauerecker, P. Jungwirth, Coulomb explosion during the early stages of the reaction of alkali metals with water. Nat. Chem. **7**, 250 (2015)

112. M. Schiemann, J. Bergthorson, P. Fischer, V. Scherer, D. Taroata, G. Schmid, A review on lithium combustion. Appl. Energy **162**, 948–965 (2016)

113. P.E. Mason, *Pouring Molten Salt into Water—Explosion!* (2017)

114. J.D. Bernardin, I. Mudawar, A cavity activation and bubble growth model of the Leidenfrost point. J. Heat Transfer **124**(5), 864–874 (2002)

115. D. Duft, T. Achtzehn, R. Müller, B.A. Huber, T. Leisner, Coulomb fission: Rayleigh jets from levitated microdroplets. Nature **421**(6919), 128 (2003)

116. L. Rayleigh, XX. On the equilibrium of liquid conducting masses charged with electricity. Lond. Edinb. Dublin Philos. Mag. J. Sci. **14**(87), 184–186 (1882)

117. C. Zhang, F. Jiao, L. Hongzhen, Crystal engineering for creating low sensitivity and highly energetic materials. Cryst. Growth Des. **18**, 5713–5726 (2018)

118. L. Zhang, C. Yao, S.-L. Jiang, Y. Yu, C.Q. Sun, J. Chen, Stabilization of the dual-aromatic pentazole cyclo-N_5^- anion by acidic entrapment. J Phys Chem Lett **10**, 2378–2385 (2019)

119. C. Jiang, L. Zhang, C. Sun, C. Zhang, C. Yang, J. Chen, B. Hu, Response to comment on "Synthesis and characterization of the pentazolate anion cyclo-N_5—in $(N_5)_6(H_3O)_3(NH_4)_4Cl$". Science **359**, 8953–8955 (2018)

120. C. Zhang, C. Sun, B. Hu, C. Yu, M. Lu, Synthesis and characterization of the pentazolate anion cyclo-N_5^- in $(N_5)_6(H_3O)_3(NH_4)_4Cl$. Science **355**(6323), 374 (2017)

121. C. Zhang, C. Yang, B. Hu, C. Yu, Z. Zheng, and C. Sun, A Symmetric $Co(N_5)_2(H_2O)_4.4 H_2O$ High-Nitrogen Compound Formed by Cobalt(II) Cation Trapping of a Cyclo-N_5^- Anion. Angew Chem. Int. Ed. **56**(16), 4512 (2017)

122. Y. Xu, Q. Wang, C. Shen, Q. Lin, P. Wang, M. Lu, A series of energetic metal pentazolate hydrates. Nature **549**(7670), 78 (2017)

123. W. Zhang, K. Wang, J. Li, Z. Lin, S. Song, S. Huang, Y. Liu, F. Nie, Q. Zhang, Stabilization of the pentazolate anion in a zeolitic architecture with $Na_{20}N_{60}$ and $Na_{24}N_{60}$ nanocages. Angew. Chem. Int. Ed. **57**(10), 2592 (2018)

124. C. Sun, C. Zhang, C. Jiang, C. Yang, Y. Du, Y. Zhao, B. Hu, Z. Zheng, K.O. Christe, Synthesis of AgN_5 and its extended 3D energetic framework. Nat. Commun. **9**(1), 1269 (2018)

125. Y. Xu, Q. Lin, P. Wang, M. Lu, Stabilization of the pentazolate anion in three anhydrous and metal-free energetic salts. Chem. Asian J **13**(8), 924 (2018)

126. P. Wang, Y. Xu, Q. Lin, M. Lu, Recent advances in the syntheses and properties of polynitrogen pentazolate anion cyclo-N_5^- and its derivatives. Chem. Soc. Rev. (2018)

Chapter 9
Concluding Remarks

Contents

Abstract A combination of the O:H–O bond cooperativity, segmental DPS strategy, and contact-angle detection, etc., has enabled systematic quantification and clarification of the hydration bonding dynamics for HX acids, YOH bases and H_2O_2 hydrogen peroxide, YX, ZX_2 and complex NaT salts, alcohols, organic acids, aldehydes, and sugars. Advancement of the theoretical and experimental strategies has enabled resolution of the solute capabilities of transiting the O:H–O bonds from the mode of ordinary water into the hydrating states in terms of phonon abundance, bond stiffness, and fluctuation order, and electron polarization. O:H vdW formation, H↔H point fragilization, O:⇔:O point compression, and ionic or dipolar polarization form the basic elements for molecular nonbond interactions. Nonbond-bond cooperativity and solute-solvent interfacial structure distortion govern the performance of the solutions in terms of surface stress, solution viscosity, molecular diffusivity, phonon lifetime, solution temperature, phase boundary dispersion, critical pressures and temperatures for phase transition under mutifield excitation.

Highlight

- *O:H, H↔H, O:⇔:O, ionic or dipolar polarization govern molecular interactions.*
- *Intermolecular nonbond and intramolecular bond cooperativity activates a solution.*
- *DPS resolves O:H–O bonds transition from the mode of ordinary water to hydration.*

- *Molecular and O:H–O bonding dynamics substantiates profoundly solvation dynamics.*

9.1   Attainments and Limitations

9.1.1   Degrees of Freedom Due Charge Injection

Table 9.1 summarizes the possible ways of charge injection and their functionalities on the O:H–O bond network and solution properties. These ways of change injection add new degrees of freedoms to the considered variables of pressure, temperature, coordination, electric and magnetic fields. It is therefore to clarify the degrees of freedom as each of them follows its basic rule in reaction.

9.1.2   Major Findings

An extension of the conventional phonon spectroscopy to the DPS has enabled resolution of the fraction and stiffness of O:H–O bonds transiting from the mode of ordinary water to their hydration, which amplifies solvation study from the molecular

Table 9.1 Ways of charge injection and their functionalities on the O:H–O bond network and solution properties

Injected charge	Interaction		Remarks
Free electrons	Site resolved bound energy		Energy and emission lifetime
Proton H^+	H_3O^+; $H \leftrightarrow H$	Point beaker	Neither H^+ or ":" stay freely
Lone pairs ":"	OH^-; $O: \Leftrightarrow :O$	Point compressor	
Y^+ ions	Polarization; O:H–O stretching, clustering, aligning	Fully screened	Weak interionic interaction
Y^{2+} ions		Partly screened	Medium interionic interaction
X^- anion			High interionic interaction
Z^- complex anion		Dipolar interaction	Eccentric charge distribution
Molecular dipoles	Interface $O: \Leftrightarrow :O$; $H \leftrightarrow H$; O:H		":"and H^+ neighboring interaction interface structure distortion
Solvent	Oriented H_2O dipole shielding		Weakens solute interaction

dynamics in temporal and spatial domains to hydration bonding energetic dynamics. Intermolecular nonbond and intramolecular bond cooperative relaxation and charge polarization dictate the performance of aqueous solutions. Table 9.2 summarizes observations of the O:H–O segmental cooperative relaxation in length, vibration frequency, and surface stress upon external excitation with respect to standard situation at 277 K.

9.1.3 Limitations

As proven, cooperation of the solvent and solute intramolecular H–O bond relaxation and the intermolecular nonbond interaction govern the solution properties. It is unrealistic to discriminate one from the other of the inter- and the intra-molecular interactions. For instance, the intramolecular H–O bond determines the critical pressures and temperatures for QS-Liquid transition, melting temperature, exothermic and endothermic reaction of aqueous solutions. The intermolecular nonbond interactions such as the ionic polarization, O:H vdW bond, H↔H inter-proton and O:⇔:O inter-lone-pair repulsion stem the solvent H–O bond relaxation. Molecular motion, diffusion, structure fluctuation, or even evaporation dissipates energy caped at a few percentiles of the H–O bond energy. One can hardly detect the H↔H and the O:⇔:O repulsion but can probe their consequences on the intramolecular bond relaxation. Therefore, these weak and repulsive intermolecular interactions and their cooperativity with intramolecular covalent bond relaxation challenge capabilities of the classical thermodynamics, DFT or MD quantum computations, or the time-dependent phonon spectroscopy.

It would be much revealing to extend the solvation study from the perspective of molecular motion to bonding and nonbonding dynamics and expand the capabilities of the concurrently used approaches to cover comprehensively the intra- and intermolecular cooperative interactions. It is also essential to treat the solutions as strongly correlated, fluctuating crystal-like structures with solute impurities. One has to check if charge injection breaks the solvent conservation of the 2N number of protons and lone pairs and O:H–O configuration besides its relaxation.

The classical thermodynamics is an extension of statistics for gaseous molecules without attending to the intra- and intermolecular coupling. Thermodynamics in terms of the Gibbs free energy, entropy, enthalpy as a $\Delta G(P, T, \ldots)$ direct function of external stimulus does not include the individual bond or nonbond relaxation or the energy absorption, emission or dissipation. The weak and repulsive molecular interactions contribute insignificantly to the total energy, but they do dictate the performance of aqueous solutions.

DFT is temperature invariant and it hardly attends to the atomistic anisotropy, short-range, abrupt, and strong localization. The MD calculations and time-dependent spectroscopy treat a molecule as the basic and independent structure unit performing in the temporal and spatial domains with little involvement of the inter- and the intra-molecular coupling and electronic polarization. Therefore, the O:H–O cooperativity and charge polarization could be the right choice, which involves heavy multidisciplinary efforts.

Table 9.2 O:H–O segmental cooperative relaxation in length, vibration frequency, and surface stress with respect to $d_{L0} = 1.6946$ Å, $d_{H0} = 1.0004$ Å, $\omega_{H0} = 3200$ cm^{-1}, $\omega_{L0} = 200$ cm^{-1}, $\gamma_S = 72.5$ J m^{-2} at 277 K upon excitation by heating, compression, molecular undercoordination (skin, cluster, droplet, nanobubble) and acid, base, salt, and organic molecular solvation [1, 2]

		Δd_H	$\Delta \omega_H$	Δd_L	$\Delta \omega_L$	$\Delta \gamma_S$	Remark	References
Liquid water	Heating	<0	>0	>0	<0	<0	d_L elongation; d_H contraction; thermal fluctuation	[3]
	Under-coordination					>0	d_H contraction; d_L elongation; polarization; supersolidity	[4]
	Compression	>0	<0	<0	>0	–	d_L compression; d_H elongation	[5]
Aqueous solution	YX salt	<0	>0	>0	<0	>0	Y^+ and X^- polarization	[6–8]
	HX acid					<0	$H\leftrightarrow H$ fragilization; X^- polarization	[9]
	YOH base	>0	>0	>0	>0	>0	O:⇔:O compression; Y^+ polarization;	[10]
	H$_2$O$_2$	<0	<0	<0	<0		solute H–O bond contraction	[11]
	Alcohol	>0	<0	>0	<0	$T_N < 0$	Intermolecular interaction; solute dipolar induction	[12]
	Sugar						Solute dipolar interface distortion	[13]
	Aldehyde	>0	<0	<0	>0	<0	Strong network destruction	[14]
	Organic acid						Network destruction	

9.2 Regulations for Aqueous Charge Injection

Consistency in predictions and experimental observations uncovered the mechanisms behind the solvation dynamics, solute capabilities, and solute-solvent and solute-solute interactions, which follow the following regulations.

9.2.1 Conservation Rule Broken

(1) The sp^3-orbital hybridization occurs to O, N, Cl, S atoms upon reaction, which offers the anionic cluster of the Na-based complex salts and organic molecular solutes with multiple ":" lone pairs. The solute tetrahedron bond configuration allows only single bond formation between electronegative elements within the anionic clusters or organic crystals.

(2) A water specimen consisting of N oxygen atoms reserves its 2N protons H^+ and 2N lone pairs ":" and the O:H–O configuration unless excessive H^+ or ":" is involved upon solvation. Molecules follow the rule of O:H–O configuration conservation without significant rotation. The $\angle$O:H–O containing angle and its segmental length and energy are subject to relaxation and fluctuation, which determines the geometric structures and the properties of water and ice.

(3) Solute-solvent interacts through O:H vdW, H↔H and O:⇔:O repulsion, and dipolar induction without intermolecular covalent or ionic bond formation.

(4) HX acid, YOH base, and YX salt dissolve into H^+ proton, Y^+ cation, OH^- hydroxide, and X^- halogenic anions. A molecular crystal or liquid dissolves into solute dipoles surrounded with H^+ and lone pairs. Ions prefer the interstatial hollow sites interacting with four nearest and six next-nearst neigbours of oriented H_2O molecules through anitropic polarization.

(5) Introduction of the excessive H^+ or ':' breaks the 2N number and O:H–O configuration conservation rule pertained to liquid water with generation of the H↔H point breaker and the O:⇔:O point compressor by $(H_3O^+, OH^-)\cdot 4H_2O$ formation. The repulsive O:⇔:O super–HB and H↔H anti–HB do not stand alone. They must be accompanied with the O:H–O bond attraction for equilibrating molecular interactions.

(6) Neither regular bond nor electron exchange is involved in the solute-solvent and solute-solute molecular interactions. The solute-solvent interaction is realized through the O:H vdW bond, H↔H anti–HB, O:⇔:O super–HB, ionic or dipolar polarization associated with hydrating H_2O dipolar screening, causing O:H–O cooperative relaxation and local structure distortion.

(7) The O:⇔:O super–HB point compressor has the same or even stronger effect of mecanical compression. The O:⇔:O compression lengthens its neighboring H–O bond and softens its phonon from above 3100 cm^{-1} to its below. O:⇔:O polarization also raise the surface stress.

(8) The H↔H anti–HB point breaker has the same effect of thermal fluctuation that disrupts the solution network and its surface stress. The H^+ forms firmly the H_3O^+ without hopping, tunneling or polarizing the neighboring H_2O molecules. The H–O bond due H_3O^+ and OH^- contracts because of its local structural network termination; the H↔H fragilization elongates slightly the neighboring H–O bond through O:H compression.

(9) The H^+ neither polarizes its neighbors nor freely hops between water molecules but forms the H_3O^+ and H↔H anti–HB, as the H–O bond energy is at 4.0 eV level, needing 121.6 nm laser irradiation to break.

(10) Solvation dissolves an organic crystal into molecular dipoles covered inhomogeneously with ":" and H^+.

(11) Nonbond-bond cooperativity dictates the phonon frequency shift and fingerprints the performance of solutions such as viscosity and surface stress.

9.2.2 Polarization and Nonbonding Repulsion

(12) X^- and Y^+ stay freely forming the close/separated contacted ion pair (CIP or SIP) depending on their electronegativity difference. The Y^+ and X^- serve each as a charge center to align, cluster, stretch and polarize the O:H–O bonds to form the supersolid hydration shells without breaking the rule of conservation.

(13) Hydrated electrons, however, serve as probes to calibrate the molecular-site and cluster-size resolved HB polarization and the photoelectron lifetime of energy dissipation.

(14) Ionic polarization shortens the H–O bond and stiffens its phonon from 3200 to ~3500 cm^{-1} but the O:H nonbond responds to polarization contrastingly shifting its phonon from 200 to 75 cm^{-1}, which raises the solution surface stress, solution viscosity, depresses the T_N for homogeneous ice formation and elevates the T_m for quasisolid melting. The softer O:H oscillator ensures the adaptivity of water ice, dominating the slipperiness of ice.

(15) The X^- solute capability of O:H–O transition and polarization follows the Hofmeister I > Br > Cl > F = 0 series order; the Y^+ cations perform similarly in O:H–O transition but different in electronic polarization or stress elevation in the order of: Na > K > Li > Rb > Cs.

(16) The bond-order-deficiency shortens the solute H–O bonds and creates the vibration features at 3550 cm^{-1} due H_2O_2 and at 3610 cm^{-1} due OH^- and the dangling H–O free radicals in water.

(17) Polarization determines surface stress and viscosity of alkali halide solutions and basic solutions, which follows the Jones–Dole viscosity, but H↔H fragilation disrupts surface stress and viscosity of acid solutions.

9.2.3 Bond Transition Mediates Solution Properties

(18) The fraction coefficient $f_H(C) = 0, f_{OH}(C) \propto f_Y(C) \propto C$, and $f_X(C) \propto 1 - \exp(-C/C_0)$ feature the solute capabilities of transiting the fraction number of HBs in the concentrated solutions. The division of f(C)/C implies the hydration shell size that is sensitive to interference by other solutes because the hydration H_2O dipoles screen the solute electric field.

(19) The fraction coefficient $f_H(C) \equiv 0$ because a H^+ proton does not polarize its neighboring H_2O molecules, which enabled discrimination of the $f_Y(C)$ from $f_X(C)$ in the YX aqueous solutions: $f_{YX}(C) = f_Y(C) + f_X(C)$, and $f_X(C) = f_{HX}(C)$.

(20) The fraction coefficient $f_Y(C) \propto C$ for the tiny Y^+ alkali cation and large organic molecular solutes because of the invariance of the local electric field that determines the hydration shell size. H_2O dipoles in the hydration shells fully-screen the cation electric filed from being interfered by other solutes. Cation-solute interactions are absent in the relevant alkali halide solutions.

(21) The fraction coefficient $f_X(C) \propto 1 - \exp(-C/C_0)$ towards saturation fingerprints the anion-anion repulsion that weakens the local electric field and reduces shell size at higher concentrations. The number deficiency of the oppositely, orderly aligned hydrating H_2O dipoles screen-partly the X^- electric field. Therefore, anion-anion interaction exists in the aqueous solutions.

(22) The quasilinear $f(C) \propto C$ for OH^- and H_2O_2 is related to the solvent H–O elongation and solute H–O contraction. OH^- shows stronger ability than H_2O_2 of bond transition, surface stress construction, and exothermic solvation.

(23) O:H–O bond segmental phonon relaxation determines the critical pressures and temperatures for phase transition. The O:H energy determines the T_N and the H–O energy determines the T_m of the solution. The O:H and the H–O phonon frequency shifts modulate the Debye temperature of the respective specific heat curve and disperses the quasisolid phase boundary cooperativity, which offsets the T_N and T_m upon solvation.

(24) Both the solution viscosity and the surface stress of an alkali halide solution follow the fraction coefficient for O:H–O bonds transition from the water mode to the hydration shells: $\Delta\eta(C)/\eta(0) = AC^{1/2} + BC \propto 1 - \exp(-C/C_0)$, which clarifies that cation polarization contributes to the linear term and the anion polarization and anion-anion repulsion to the nonlinear term of Jones–Dole viscosity.

(25) The remnant energy of H–O bond exothermic elongation by O:⇔:O or H↔H repulsion and the solute H–O contraction by bond-order-deficiency heats up the basic and H_2O_2 or alcohol solutions. The solution temperature T(C) correlates directly to the f(C). The H–O bond emits 150% O:H cohesive energy of 0.1 eV that caps the energy dissipation by molecular motion or even evaporation.

(26) The concentration resolved viscosity of some molecular solutions and divalent halide solutions show positive curvature at higher concentrations, which do not

follow the Jones-Dole convention or trends of the fraction of hydrogen bond transition into the hydration mode.

9.2.4   Multifield Effect

(27) Salt hydration raises the critical pressures for the room-temperature Liquid-VI and VI-VII phase transitions in the Hofmeister series order according to the electronegativity difference $\Delta\eta$ and anionic radius R.

(28) NaX solvation raises the room-temperature P_{C1} for solution VI-VII transition and P_{C2} for L-VI transition simultaneously by amounts following the $I > Br > Cl > F = 0$ ($P_{C1} > P_{C2}$). Recovering the O:H–O deformation by solvation for phase transition needs more energy and so the critical P_C at constant temperature.

(29) Concentrated NaI solvation shifts contrastingly the P_{C2} up along the Liquid-VI phase boundary and the P_{C1} and P_{C2} merge at the triple-phase junction of 350 K and 3.3 GPa. The involvement of anion-anion repulsion weakens the anionic local electric field and the lessens the O:H–O bond deformation. The highly-deformed O:H–O bond at P_{C1} responds insensitively to compression.

(30) Saltation and heating have the same effect on O:H–O relaxation but contrastingly on surface stress. Thermal fluctuation disrupts but ionic polarization constructs the solution network and the surface stress though both stimuli shift similarly the O:H–O segmental phonons.

(31) Molecular undercoordination and compression disperse the quasisolid phase boundary oppositely. A six-GPa compression raises the T_N (low-temperature quasisolid phase boundary dominated by O:H energy) to 298 K from its far below for the Solid-QS transition of water confined between graphene and silica.

(32) Heating eases the compression and reduces the P_C for the undercoordinated Solid-QS phase transition. Within the QS phase, the H–O bond follows the regular rule of thermal expansion—heating lengthens the H–O bond and shortens the O:H nonbond spontaneously, compensating the effect of compression that always lengthens the H–O bond.

(33) H–O bonds in the hydrogen shells and at water skin are shorter and stiffener and they are less sensitive to temperatures; at temperature outside the QS phase, the O:H nonbond follows the regular rule of thermal expansion but H–O bond thermal contraction because the O:H–O cooperativity and specific heat disparity.

(34) Phonon abundance-lifetime-stiffness transits cooperatively upon salt solvation and confinement. Phonon lifetime depends linearly on the fraction of bond transition due to ionic and undercoordination polarization and phonon skin reflection of nanodroplet, which reconciles the approaches of molecular dynamics and O:H–O bond cooperative transition upon solvation and confinement.

(35) The combination of the X:H–Y tension and X:⇔:Y or H↔H repulsion foresters the sensitivity, energy density, and detonation velocity of energetic

substance explosion. The X:H–Y tension determines the sensitivity of the constrained explosion and the X:⇔:Y or H↔H repulsion shortens the intramolecular bonds to store energy.

(36) The O:H segment phonons (200 cm^{-1}) are sensitive to perturbation in its respective order, which could provide reference to tune the THz wave frequency towards understanding the life processes such as signaling, messaging, coupling, and regulating.

9.2.5 Advantage of the DPS Strategy

(37) DPS minimizes artefacts of detection by spectral peak area normalization. Artefacts such as cross-section of mode reflectivity and scattering efficiency may contribute to the spectral intensity and peak shape but never the nature for the fraction and stiffness transition of bonds from the reference water to the hydrating states.

(38) DPS resolves the transition of the solvent O:H–O bonds from the mode of ordinary water to the closest hydrating shells, bond-order-deficiency resolved solute H–O bond contraction, and the solute-solute interactions.

(39) DFT computations further verify the expected configuration and the X·H–O and O:H–O segmental relaxation in the hydration shells.

(40) The contact angle variation with solute concentration and experimental conditions offers direct information of HB network polarization or disruption. Polarization enhances the viscosity, stiffness, and supersolidity of the solution skin and hydration shells.

(41) The $f_{droplet}(D) \propto D^{-1}$ for droplet clarifies the core-shell configuration of water droplet of D size. For salt solutions, the phonon relaxation time $\tau_{solution}(C) \propto f_{solution}(C)$ but for nanodroplet, the $\tau_{droplet}(D)$ results from skin supersolidity and geometric confinement inhibiting vibration energy dissipation. Evidencing the essentiality of intermolecular nonbond and intramolecular bond cooperativity, exercises not only lead to consistent insight into the hydration interface bonding thermodynamics but also exemplify the efficient yet straightforward DPS towards solvation phonon spectrometrics.

(42) The intrinsic multifield is much more significant than the nonlinear effect on the spectral features and the solution properties so one must focus on correlation between the perturbation-bond-property transition in solvation upon charge injection.

Evidencing the essentiality of O:H–O bond polarization, H↔H anti–HB fragilization, O:⇔:O super–HB compression, and bond-order-deficiency induced solute H–O contraction, exercises lead to the hitherto consistent insight into the solvation bonding dynamics, solute capabilities, and solute-solvent, solute-solute interactions complementing conventional pursuit of solute motion mode and dynamics, hydration shell size, phonon relaxation time, etc.

9.3 Future Directions

From what we have experienced in the presently described exercises, one can be recommended the following ways of thinking and strategies of approaching for efficient investigation, and complementing to conventional approaches:

(1) The nonbonding electron lone pairs pertained to N, O, F and their neighboring elements in the Periodic Table form the primary element being key to our life, which should receive deserved attention. It is related to DNA folding and unfolding, regulating, and messaging. NO medication and CF_4 anticoagulation in synthetic blood are realized through lone pairs interaction with living cells. The lone pair forms the O:H and the O: $\Leftrightarrow$:O interactions together with the H$\leftrightarrow$H determines the molecular interactions. However, the presence and functionality of the localized and weak lone pairs interactions have been oversighted. Without lone pairs, neither O:H–O bond nor oxidation could be possible; molecular interaction equilibrium could not be realized. Extending the knowledge about lone pairs and their functionality of polarization to catalysis, solution-protein, drug-cell, liquid-solid, colloid-matrix, interactions and even energetic explosives and other molecular crystals would be even more fascinating.

(2) The key to the O:H–O bond is the O–O Coulomb coupling. Without such a coupling none of the cooperative relaxation or the mysterious of water ice and aqueous solutions such as ice floating, ice slipperiness, regelation, supercooling/heating or the negative thermal expansion, warm water cooling faster. Unfortunately, the O–O coupling has been long overlooked in practice. Extending to the general situation containing lone pairs and X:A–Y bond, the impact would be tremendously propounding to the universe where we are living. For instance, a combination of O: $\Leftrightarrow$:O or other forms of repulsion and O:H–O(N) elongation could shorten the intramolecular covalent bonds for energy storage in the energetic materials such as full-N and CNHO-based explosives because O:H–O bond elongation stiffens the H–O bond. Negative thermal expansion arises from segmental specific heat disparity of the X:H–Y and the superposition of the specific heat curves, such as graphite with discrepant inter- and intra-layer interactions.

(3) It is necessary to think about water and solvent matrix as the highly ordered, strongly correlated, and fluctuating crystals, particularly, the supersolid phase caused by hydration and molecular undercoordination rather than the amorphous or multiphase structures. Water holds the two-phase structure in the core shell configuration, rather than the randomly domain-resolved mixture of density patches. Liquid water and the matrix of aqueous solutions must follow the conservation rules for the 2N number of protons and lone pairs and for the O:H–O configuration despite its segmental length and energy relaxation unless excessive H^+ or lone pairs are introduced.

(4) One can consider the solvation as a process of charge injection with multiple interactions. Charge injection in the form of hydrated electrons, protons, lone pairs, cations, anions, molecular dipoles mediate the HB network and properties of a solution. Protons and lone pairs cannot stay alone but they are attached to

a H_2O molecule to form the H_3O^+ or OH^- tetrahedron, respectively, breaking the conservation rule of water ice. Drift or Brownian motion of H_3O^+ or OH^- may happen under electric or thermal fluctuation. It would be comprehensive to consider the electrostatic polarization and hydrating H_2O molecular screening, O:H, H↔H, O:⇔:O, solute-solute interactions and their variation with solute type and concentration.

(5) Multifield perturbation supplies the basic degrees of freedom that mediate the solution HB network and properties by relaxing and transiting the O:H–O bond and electrons. H–O bond relaxation exchanges energy while the O:H relaxation or molecular motion dissipates energy capped at its cohesive energy about 0.01 eV. As the ground of defect and surface sciences and nanoscience atomic/Molecular undercoordination is a prominent degree of freedom that should receive deserved attention.

(6) The concepts of quasisolid (quasiliquid) of negative thermal extensity due to O:H–O bond segmental specific disparity, supersolidity due to molecular under-coordination and electric polarization, the H↔H anti-HB due to excessive protons, and the O:⇔:O super-HB due to excessive lone pairs are essential to describe the multifield effect on the performance of water ice and aqueous solutions. The quasisolid phase boundary dispersion by perturbation determines the solution O:H–O bond network thermodynamic behavior.

(7) A combination of the multiscale theory such as classical thermodynamics, DFT ad MD quantum computations, nuclear quantum effect, and the O:H–O cooperativity and polarization notion to overcome limitation of them independently. Thermodynamics deals with the system from the perspective of statistics of a collection of neutral particles and related the system energy directly to external stimulus. DFT deals with electrons as in terms of wavefunction with spatial distribution probability under proper interaction potentials; MD takes a H_2O molecule as the basic structural unit of polarizable or non-polarizable dipoles with attention more to the intermolecular O:H interaction, drift motion, spatial and temporal performance, and refers the O:H as the hydrogen bond that is incomplete. As the key elements to the current described progress, the O:H–O polarization and segmental cooperative relaxation, the O:H–O segmental specific heats derived five phases over the full temperature range under ambient pressure, the charge injection derived nonbond interactions and the perturbation dispersed phase boundaries, continue challenging the limitations of available theories and computational methods.

(8) Integrating the inter- and intramolecular O:H–O cooperativity and nonbonding electron polarization would be much more appealing. Molecular motion dissipates energy caped at the O:H scale that is only less than 5% of the H–O energy. The H–O absorbs or emits energy through relaxation. Any detectible quantities are functional dependence on the physically elemental variables of length L, mass m, and time t (such as energy $[E] = [L^2/(mt^2)]$, frequency $[\omega^2] = [E/L^2]$, critical temperature for phase transition $[T_C] = [E]$, elasticity $[B] = [E/L^3]$, etc.) and their variations with external perturbation. Therefore, it would be more efficient to focus on the structural geometry and energy exchange of

the O:H–O bond responding to perturbation, as the key driver of solvation study and molecular engineering and science.

(9) Focusing on the bond-electron-phonon-property correlation and interplaying the spatially- and temporarily-resolved electron/phonon/photon spectrometrics would substantiate the advancement of related studies. Combining the spatially resolved electron/phonon DPS and the temporarily resolved ultrafast pump-probe spectroscopies not only distill the phonon abundance-stiffness-fluctuation due to the conditioning liquid but also fingerprint the electron/phonon energy dissipation and the ways of interactions. Molecular residing time or drift motion under a certain coordination environment fingerprints the way of energy dissipation but these processes could hardly give direct information of energy exchange under perturbation. Polarization, entrapment, or defect edge reflection and absorption determine the energy dissipation. Embracing the emerged O:H–O bond segmental disparity and cooperativity and the specific heat difference would be even more revealing.

Understanding may extend to water-protein interaction, biochemistry, environmental and pharmaceutical industries. Hydrophobic interface is the same to free surface. Charge injection by salt and other solute solvation provide the local electric fields. As the independent degrees of freedom, molecular undercoordination and electric polarization are ubiquitous to our daily life and living conditions. Knowledge developed could contribute to the science and society. It would be very promising for one to keep mind open and always on the way to developing experimental strategies and innovating theories toward resolution to the wonderful world.

References

1. C.Q. Sun, Y. Sun, *The Attribute of Water: Single Notion, Multiple Myths*. Springer Series in Chemical Physics, vol. 113 (Springer, Heidelberg, 2016), 494p
2. Y.L. Huang, X. Zhang, Z.S. Ma, Y.C. Zhou, W.T. Zheng, J. Zhou, C.Q. Sun, Hydrogen-bond relaxation dynamics: resolving mysteries of water ice. Coord. Chem. Rev. **285**, 109–165 (2015)
3. C.Q. Sun, X. Zhang, X. Fu, W. Zheng, J.-L. Kuo, Y. Zhou, Z. Shen, J. Zhou, Density and phonon-stiffness anomalies of water and ice in the full temperature range. J. Phys. Chem. Lett. **4**, 3238–3244 (2013)
4. C.Q. Sun, X. Zhang, J. Zhou, Y. Huang, Y. Zhou, W. Zheng, Density, elasticity, and stability anomalies of water molecules with fewer than four neighbors. J. Phys. Chem. Lett. **4**, 2565–2570 (2013)
5. Q. Zeng, T. Yan, K. Wang, Y. Gong, Y. Zhou, Y. Huang, C.Q. Sun, B. Zou, Compression icing of room-temperature NaX solutions (X = F, Cl, Br, I). Phys. Chem. Chem. Phys. **18**(20), 14046–14054 (2016)
6. X. Zhang, Y. Xu, Y. Zhou, Y. Gong, Y. Huang, C.Q. Sun, HCl, KCl and KOH Solvation resolved solute-solvent interactions and solution surface stress. Appl. Surf. Sci. **422**, 475–481 (2017)
7. Y. Zhou, Y. Huang, Z. Ma, Y. Gong, X. Zhang, Y. Sun, C.Q. Sun, Water molecular structure-order in the NaX hydration shells (X = F, Cl, Br, I). J. Mol. Liq. **221**, 788–797 (2016)
8. Y. Gong, Y. Zhou, H. Wu, D. Wu, Y. Huang, C.Q. Sun, Raman spectroscopy of alkali halide hydration: hydrogen bond relaxation and polarization. J. Raman Spectrosc. **47**(11), 1351–1359 (2016)

9. X. Zhang, Y. Zhou, Y. Gong, Y. Huang, C. Sun, Resolving H(Cl, Br, I) capabilities of transforming solution hydrogen-bond and surface-stress. Chem. Phys. Lett. **678**, 233–240 (2017)

10. Y. Zhou, D. Wu, Y. Gong, Z. Ma, Y. Huang, X. Zhang, C.Q. Sun, Base-hydration-resolved hydrogen-bond networking dynamics: Quantum point compression. J. Mol. Liq. **223**, 1277–1283 (2016)

11. J. Chen, C. Yao, X. Liu, X. Zhang, C.Q. Sun, Y. Huang, H_2O_2 and HO^- solvation dynamics: solute capabilities and solute-solvent molecular interactions. Chem. Select **2**(27), 8517–8523 (2017)

12. Y. Gong, Y. Xu, Y. Zhou, C. Li, X. Liu, L. Niu, Y. Huang, X. Zhang, C.Q. Sun, Hydrogen bond network relaxation resolved by alcohol hydration (methanol, ethanol, and glycerol). J. Raman Spectrosc. **48**(3), 393–398 (2017)

13. C. Ni, Y. Gong, X. Liu, C.Q. Sun, Z. Zhou, The anti-frozen attribute of sugar solutions. J. Mol. Liq. **247**, 337–344 (2017)

14. J. Chen, C. Yao, X. Zhang, C.Q. Sun, Y. Huang, Hydrogen bond and surface stress relaxation by aldehydic and formic acidic molecular solvation. J. Mol. Liq. **249**, 494–500 (2018)

Index

A

Alcohol, 12, 13, 55, 65, 192–196, 198, 199, 219, 285, 288, 291

Aldehyde, 12, 28, 55, 192, 200–203, 206, 207, 285, 288

Alkali cation, 11, 29, 58, 72, 106, 143, 156, 183, 217, 291

Anti-frozen, 65, 217, 218, 224

B

Bond contraction, 45, 50, 53, 56, 65, 96, 106, 108–110, 112, 114, 136, 153, 168–170, 264, 265, 274, 275, 278, 288, 293

Bond-order-deficiency, 3, 4, 6, 11, 103, 106, 108–111, 115, 117, 123, 290, 291, 293

Bond Order-Length-Strength (BOLS), 103, 106, 109, 112, 113

Bond switching, 68, 69, 72

C

Carboxylic acid, 131, 192, 193, 200, 202, 203, 214

Catalysis, 104, 294

Charge injection, 8, 10, 24, 27, 32, 37, 40, 45, 54, 74, 138, 151, 153, 154, 163, 180, 238, 286, 287, 289, 293–296

Collins notion, 135

Complex salt, 11, 29, 55, 75, 132, 156–160, 162, 163, 181, 183, 289

Compression, 5, 10–13, 24, 26, 31, 37, 40, 42, 45–49, 53, 54, 57, 61, 65, 74, 85, 90, 91, 103, 106, 108–113, 115–117, 123, 174, 179, 180, 198, 199, 203, 224, 235–243, 247–249, 253, 257–259, 261, 276–278, 288–290, 292, 293

Confinement, 5, 12, 45, 53, 237, 238, 240, 253, 259, 264, 266–268, 278, 292, 293

Conservation rule, 54, 289, 294, 295

Contact angle, 9–11, 13, 19, 30–32, 48, 49, 74, 106, 111, 140, 142, 148, 149, 159, 163, 172, 173, 176, 200, 203, 206, 208, 210, 213, 216, 237, 253, 258, 293

Coupling, 5, 8, 25, 39, 45, 46, 50, 53, 55, 105, 130, 166, 167, 273, 287, 294

D

Dangling bond, 4, 50, 51, 108, 112, 133, 146, 152, 153, 160, 198, 209, 224, 250–252, 254

Debye temperature, 31, 32, 41, 43, 65, 132, 179, 199, 219, 253, 257, 259, 291

Differential Phonon Spectrometrology (DPS), 9, 10, 13, 19, 20, 22, 27, 28, 32, 47, 48, 60, 72, 89–94, 96–99, 106, 108–110, 112, 123, 138–141, 143, 144, 146, 147, 150, 152–154, 159–162, 165, 166, 170, 171, 180, 194, 196, 198, 200, 202, 204, 205, 208, 209, 211, 212, 214–216, 223–225, 239, 250–253, 264–266, 278, 285, 286, 293, 296

Diffusivity, 8, 21, 29, 45, 129, 131–133, 135, 137, 175, 264, 267, 278, 285

Dipolar solute, 54, 60, 192

Dipole, 6, 8, 12, 27, 29, 31, 32, 37, 38, 40, 54, 58–60, 66–73, 75, 76, 86, 89, 92, 93, 96, 99, 130, 132, 142, 147, 148, 150, 152, 156, 161, 163, 171, 172, 180, 183, 194, 196, 202, 206, 208–210, 214, 216, 219, 224, 238, 265, 275, 286, 289–291, 294, 295

Printed in the United States
By Bookmasters